METHODS IN MOLECULAR BIOLOGY

Series Editor
John M. Walker
School of Life Sciences
University of Hertfordshire
Hatfield, Hertfordshire, AL10 9AB, UK

For further volumes:
http://www.springer.com/series/7651

Caspases, Paracaspases, and Metacaspases

Methods and Protocols

Edited by

Peter V. Bozhkov

Department of Plant Biology, Uppsala BioCenter, Swedish University of Agricultural Sciences and Linnean Center for Plant Biology, Uppsala, Sweden

Guy Salvesen

Sanford-Burnham Medical Research Institute, La Jolla, CA, USA

Humana Press

Editors
Peter V. Bozhkov
Department of Plant Biology
Uppsala BioCenter
Swedish University of Agricultural Sciences
and Linnean Center for Plant Biology
Uppsala, Sweden

Guy Salvesen
Sanford-Burnham Medical Research Institute
La Jolla, CA, USA

ISSN 1064-3745 ISSN 1940-6029 (electronic)
ISBN 978-1-4939-5405-6 ISBN 978-1-4939-0357-3 (eBook)
DOI 10.1007/978-1-4939-0357-3
Springer New York Heidelberg Dordrecht London

Printed on acid-free paper

Humana Press is a brand of Springer
Springer is part of Springer Science+Business Media (www.springer.com)

Preface

Among a plethora of known proteases, caspases are perhaps the ones that have attracted and continue to attract much more research than any other group of proteolytic enzymes. The reason for such an extraordinarily high interest to caspases is their pivotal regulatory role in cell death, cell differentiation, and inflammatory responses, with broad implications for human health and disease. However, caspases are just a tip of the iceberg, representing an apical and relatively small group of animal-specific enzymes within a huge superfamily of structurally related proteases found in all living organisms.

The discovery of caspase-related and apparently ancestral proteins called metacaspases and paracaspases in bacteria, protists, slime molds, fungi, and plants has initiated a "post-caspase" wave of research in studying the biochemistry and function of these proteins in the contexts of development, aging, stress response, pathogenicity, and disease resistance. This field of research moves very rapidly and has a motley pattern due to a wide evolutionary conservation and multifunctionality of para- and metacaspases, reflecting their diversity in molecular structure and enzymatic properties.

When planning this book, we pursued two opportunities. Firstly, as strange as it may seem, this is in fact the first collection of laboratory protocols to study caspases published in single cover. Secondly, we intended to break inter-kingdom barriers by including protocols for para- and metacaspases and in this way to support the rapid progress in these areas by providing common protocols that can be useful for distinct members of the caspase fold. Accordingly, the book consists of two parts. The first part presents methods to measure, detect, and inhibit activation and activity of a subset of or specific caspases in vitro and in several model systems and organisms, primarily in the context of programmed cell death. In addition, two chapters describe recently established protocols for high-throughput analysis of caspase substrate specificity and caspase substrates by employing chemistry and proteomics. The second part of the book provides experimental protocols for purification and in vitro and in vivo analysis of yeast, protozoan, and plant metacaspases, as well as of a human paracaspase MALT1.

Each technique in *Caspases, Paracaspases, Metacaspases Methods and Protocols* is described in an easy-to-follow manner with details so that the beginner can succeed with challenging techniques. The Notes section provides the researcher with valuable hints and troubleshooting advice. We wish to thank the authors for their valuable time in preparing these diligently written chapters.

Uppsala, Sweden
La Jolla, CA

Peter V. Bozhkov
Guy Salvesen

Contents

Contributors

ANDREAS BERGMANN • *Department of Cancer Biology, University of Massachusetts Medical School, Worcester, MA, USA*
DAVE BOUCHER • *Institute of Molecular Bioscience, University of Queensland, St. Lucia, QLD, Australia*
PETER V. BOZHKOV • *Department of Plant Biology, Uppsala BioCenter, Swedish University of Agricultural Sciences and Linnean Center for Plant Biology, Uppsala, Sweden*
BENJAMIN F. CRAVATT • *Department of Chemical Physiology, The Scripps Research Institute, La Jolla, CA, USA*
RASHID DARBANDI • *Department of Biochemistry, St. Jude Children's Research Hospital, Memphis, TN, USA*
JEAN-BERNARD DENAULT • *Department of Pharmacology, Faculty of Medicine and Health Sciences, Université de Sherbrooke, Sherbrooke, QC, Canada*
MELISSA M. DIX • *Department of Chemical Physiology, The Scripps Research Institute, La Jolla, CA, USA*
LORETTA DORSTYN • *Centre for Cancer Biology, SA Pathology, Adelaide, Australia; Division of Health Sciences, University of South Australia, Adelaide, Australia*
MARCIN DRĄG • *Division of Bioorganic Chemistry, Faculty of Chemistry, Wroclaw University of Technology, Wroclaw, Poland*
CATHERINE DUCLOS • *Department of Pharmacology, Faculty of Medicine and Health Sciences, Université de Sherbrooke, Sherbrooke, QC, Canada*
NICOLAS FASEL • *Department of Biochemistry, University of Lausanne, Lausanne, Switzerland*
CAITLIN E. FOGARTY • *Department of Cancer Biology, University of Massachusetts Medical School, Worcester, MA, USA*
KRIS GEVAERT • *Department of Medical Protein Research, VIB, Ghent, Belgium; Department of Biochemistry, Ghent University, Ghent, Belgium*
IVETH GONZALEZ • *Department of Biochemistry, University of Lausanne, Lausanne, Switzerland*
STEPHAN HAILFINGER • *Department of Biochemistry, University of Lausanne, Lausanne, Switzerland*
RANDY S. HAUN • *Central Arkansas Veterans Healthcare System, Little Rock, AR, USA; Department of Pharmaceutical Sciences, University of Arkansas for Medical Sciences, Little Rock, AR, USA*
CHRISTIAN HERZOG • *Department of Internal Medicine, University of Arkansas for Medical Sciences, Little Rock, AR, USA*
TOSHIHIKO HIBINO • *Shiseido Research Center, Tsuzuki-ku, Yokohama, Japan*
PAULINA KASPERKIEWICZ • *Division of Bioorganic Chemistry, Faculty of Chemistry, Wroclaw University of Technology, Wroclaw, Poland*
VARSHA KAUSHAL • *Biology Department, Hendrix College, Conway, AR, USA*

GUR P. KAUSHAL • *Central Arkansas Veterans Healthcare System, Little Rock, AR, USA; Department of Internal Medicine, University of Arkansas for Medical Sciences, Little Rock, AR, USA*
SHARAD KUMAR • *Centre for Cancer Biology, SA Pathology, Adelaide, Australia; Division of Health Sciences, University of South Australia, Adelaide, Australia*
ROBIN E.C. LEE • *Department of Cancer Biology, Dana Farber Cancer Institute, Boston, MA, USA; Center for Cancer Systems Biology, Dana Farber Cancer Institute, Boston, MA, USA; Department of Genetics, Harvard Medical School, Boston, MA, USA*
EUI SEUNG LEE • *Department of Molecular, Cellular, and Developmental Biology, University of Colorado, Boulder, CO, USA*
RICARDO MARTIN • *Department of Biochemistry, University of Lausanne, Lausanne, Switzerland*
FRANCIS MCCOY • *Department of Biochemistry, St. Jude Children's Research Hospital, Memphis, TN, USA*
KAREN MCLUSKEY • *Wellcome Trust Centre for Molecular Parasitology, Institute of Infection, Immunity and Inflammation, College of Medical, Veterinary and Life Sciences, University of Glasgow, Glasgow, UK*
LYNN A. MEGENEY • *Regenerative Medicine Program, Sprott Centre for Stem Cell Research, Ottawa Hospital Research Institute, The Ottawa Hospital, Ottawa, Ontario, Canada; Department of Cellular and Molecular Medicine, University of Ottawa, Ottawa, Ontario, Canada*
ELENA A. MININA • *Department of Plant Biology, Uppsala BioCenter, Swedish University of Agricultural Sciences and Linnean Center for Plant Biology, Uppsala, Sweden*
CATHERINE X. MOSS • *Wellcome Trust Centre for Molecular Parasitology, Institute of Infection, Immunity and Inflammation, College of Medical, Veterinary and Life Sciences, University of Glasgow, Glasgow, UK*
JEREMY C. MOTTRAM • *Wellcome Trust Centre for Molecular Parasitology, Institute of Infection, Immunity and Inflammation, College of Medical, Veterinary and Life Sciences, University of Glasgow, Glasgow, UK*
LETA K. NUTT • *Department of Biochemistry, St. Jude Children's Research Hospital, Memphis, TN, USA*
MAGNUS OLSSON • *Division of Toxicology, Institute of Environmental Medicine, Karolinska Institutet, Stockholm, Sweden*
CHRISTIANE PELZER • *Department of Biochemistry, University of Lausanne, Lausanne, Switzerland*
MARCIN PORĘBA • *Division of Bioorganic Chemistry, Faculty of Chemistry, Wroclaw University of Technology, Wroclaw, Poland*
GUY SALVESEN • *Sanford-Burnham Medical Research Institute, La Jolla, CA, USA*
AMIT SHRESTHA • *Regenerative Medicine Program, Sprott Centre for Stem Cell Research, Ottawa Hospital Research Institute, The Ottawa Hospital, Ottawa, Ontario, Canada; Department of Cellular and Molecular Medicine, University of Ottawa, Ottawa, Ontario, Canada*
GABRIEL M. SIMON • *Abide Therapeutics, La Jolla, CA, USA*
SIMON STAEL • *Department of Plant Systems Biology, VIB, Ghent, Belgium; Department of Plant Biotechnology and Bioinformatics, Ghent University, Ghent, Belgium*
ALEKSANDRA SZALEK • *Division of Bioorganic Chemistry, Faculty of Chemistry, Wroclaw University of Technology, Wroclaw, Poland*

MARGOT THOME • *Department of Biochemistry, University of Lausanne, Lausanne, Switzerland*

LIANA TSIATSIANI • *Biomolecular Mass Spectrometry and Proteomics, Bijvoet Center for Biomolecular Research and Utrecht Institute for Pharmaceutical Sciences, Utrecht University, Utrecht, The Netherlands; Netherlands Proteomics Center, Utrecht, the Netherlands*

FRANK VAN BREUSEGEM • *Department of Plant Systems Biology, VIB, Ghent, Belgium; Department of Plant Biotechnology and Bioinformatics, Ghent University, Ghent, Belgium*

PETRA VAN DAMME • *Department of Medical Protein Research, VIB, Ghent, Belgium; Department of Biochemistry, Ghent University, Ghent, Belgium*

DING XUE • *Department of Molecular, Cellular, and Developmental Biology, University of Colorado, Boulder, CO, USA*

MAMI YAMAMOTO-TANAKA • *Shiseido Research Center, Tsuzuki-ku, Yokohama, Japan; Department of Dermatology, Tokyo Medical University, Tokyo, Japan*

BORIS ZHIVOTOVSKY • *Division of Toxicology, Institute of Environmental Medicine, Karolinska Institutet, Stockholm, Sweden*

Part I

Caspases

Chapter 1

General In Vitro Caspase Assay Procedures

Dave Boucher, Catherine Duclos, and Jean-Bernard Denault

Abstract

One of the most valuable tools that have been developed for the study of apoptosis is the availability of recombinant active caspases. The determination of caspase substrate preference, the design of sensitive substrates and potent inhibitors, the resolution of caspase structures, the elucidation of their activation mechanisms, and the identification of their substrates were made possible by the availability of sufficient amounts of enzymatically pure caspases. The current chapter describes at length the expression, purification, and basic enzymatic characterization of apoptotic caspases.

Key words Caspase, Purification, Active-site titration, Enzymatic assays

1 Introduction

Since the identification of the first caspase in humans more than 20 years ago [1, 2], we have seen the unraveling of a new field of research that year after year still unveils fascinating new discoveries. By the same token, we have gained new understanding of many physiological and pathological processes, the most prominent being apoptotic cell death. This success is in part due to the availability of enzymatically pure recombinant caspase preparations. Moreover, the ever-growing recognition of the involvement of caspases in cellular processes [3] will require the use of recombinant caspases for years to come to understand the subtleties implied by this involvement. Earlier works using purified enzymes involved the characterization of the substrate preference of caspases [4, 5], which allowed the development of reliable peptidic substrates and potent inhibitors [6, 7], and the determination of many caspase structures in various molecular forms and complexes [8]. Through this work, important insight was also gained into the intricacies of caspase activation mechanisms.

In the past decade, the availability of recombinant caspases permitted the development of several proteomic approaches involving peptidases (degradomics) [9, 10], and along with several

Peter V. Bozhkov and Guy Salvesen (eds.), *Caspases, Paracaspases, and Metacaspases: Methods and Protocols*, Methods in Molecular Biology, vol. 1133, DOI 10.1007/978-1-4939-0357-3_1, © Springer Science+Business Media New York 2014

biochemical studies, these methods have populated a list of more than 1,400 caspase substrates [11, 12]. In most circumstances, the relevancy of these proteolytic events has not been determined, and yet again, the availability of caspases for in vitro assays will help to validate and study these substrates.

Over the years, we have developed an expertise in the expression, purification, and characterization of caspases. This chapter describes the basic protocols for caspase expression in *E. coli* as His-tagged proteins, their purification on immobilized metal affinity chromatography (IMAC) columns, and the in vitro characterization of their enzymatic activity. Caveats, pitfalls, and remedies for individual caspases are discussed, and a broader discussion of the production of specific molecular forms, and some protein engineering approaches for these enzymes are also presented. We propose a work flowchart allowing for the expression, purification, and characterization of recombinant caspases within a 5-day period (Fig. 1).

The physiological environment of caspases is the cytosol of a cell. In that respect, the osmolarity and reducing conditions found in both *E. coli* and mammalian cells are similar. Furthermore, there is no peptidase with similar functions or activity in *E. coli*, making this host ideal for expressing caspases. Finally, none of the posttranslational modifications of caspases that occur in mammalian cells (e.g., phosphorylation, ubiquitination, sumoylation)

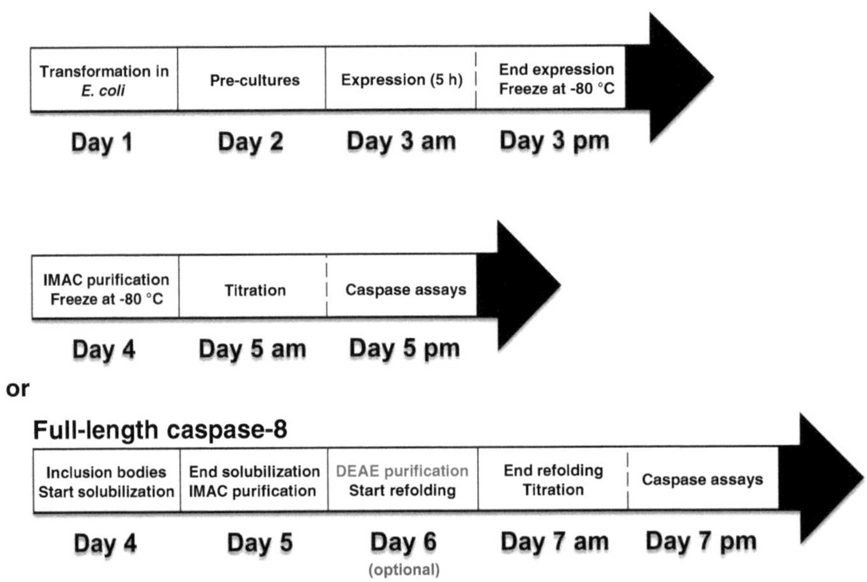

Fig. 1 Timeline of protocols. The basic expression protocol takes 3 days. Add an extra day if the time necessary to produce the protein is long (>12 h). Either way, purification is performed on the fourth day. The IMAC purification is quick (1 day), and the basic characterization also takes 1 day. Because full-length caspase-8 is insoluble when expressed in *E. coli*, the protocol is longer and involves denaturation of proteins, IMAC purification, an optional DEAE anion exchange chromatography, and a full day to refold the protein into an active enzyme. Purified caspase-8 is characterized immediately following purification

occur in *E. coli*, and caspases do not require any posttranslational modification to display full activity. Consequently, it is relatively easy to obtain enzymatically pure caspase preparations from *E. coli*.

2 Materials

2.1 Equipment

1. 15-mL bacterial culture tubes.
2. 1-L baffled culture flasks.
3. 250-mL baffled culture flask.
4. Benchtop microcentrifuge.
5. Benchtop centrifuge for 15/50-mL conical tubes.
6. 15-mL conical tubes.
7. 10,000 MWCO dialysis tube (Spectrum Laboratories or equivalent).
8. 0.45-μm 150-mL Durapore Stericup™ HV (Millipore) or equivalent.
9. Econo-Pac 0.7 × 5.0-cm column (Bio-Rad) or equivalent.
10. Floor centrifuge with 8 × 50 mL (Sorvall SW-34 or equivalent) and 6 × 250 mL (Sorvall SLA-1500 or equivalent) or higher volume capacity rotor.
11. 1.5-mL microfuge tubes.
12. Multichannel pipettors (8 channels, 200 and 10 μL).
13. 50/100-mL plastic beaker.
14. Repeating pipettor with various volume tips (20–200 μL).
15. Shaking incubator.
16. Spectrophotometer.
17. 10,000 MWCO spin concentrator (Millipore or equivalent).
18. Thermostatic fluorescence plate reader for 96-well plates.
19. Ultrasonic cell disruptor equipped with a large probe.
20. 96-well plates, preferentially black (*see* **Note 1**).

2.2 Reagents

1. 7-Amino-3-trifluoromethylcoumarin (Afc) 10 mM in dimethyl sulfoxide (DMSO; keep at −20 °C). *See* Subheading 3.2.1.1 for the preparation of the Afc standard solution.
2. Afc-based fluorogenic peptidic substrates, such as Ac-DEVD-Afc: 20 mM in DMSO (keep at −20 °C) (*see* **Note 2**).
3. Ampicillin solution: 100 mg/mL in water (filter-sterilized).
4. 2× executioner caspase buffer: 20 mM 1,4-piperazinediethanesulfonic acid (PIPES) at pH 7.2 (NaOH), 200 mM NaCl, 20 % w/v sucrose, 0.2 % w/v 3-[(3-cholamidopropyl) dimethylammonio]-1-propanesulfonate (CHAPS), 20 mM

DTT (freshly added), and 2 mM ethylenediaminetetraacetic acid (EDTA) (filter-sterilized).

5. Eukaryote lysis buffer: 50 mM HEPES at pH 7.4, 150 mM NaCl, 1 % NP-40 (*see* **Note 3**).
6. Chelating Sepharose Fast Flow resin (GE Healthcare Life Science).
7. Chloramphenicol solution: 34 mg/mL in ethanol (filter to remove insoluble material if any).
8. Competent BL21(DE3) *pLysS E. coli* (EMD Millipore, formerly Novagen).
9. 1 M dithiothreitol (DTT) in water (filter-sterilized).
10. Elution buffer: 50 mM Tris at pH 8.0, 0.1 M NaCl, and 0.2 M imidazole (filter-sterilized) (*see* **Note 4**).
11. Guanidine buffer: 50 mM Tris at pH 8.0, and 6 M guanidine hydrochloride.
12. 1.2× initiator caspase buffer: 60 mM 4-(2-hydroxyethyl)piperazine-1-ethanesulfonic acid (HEPES) at pH 7.4 (NaOH), 1.2 M sodium citrate, 60 mM NaCl, 0.012 % w/v CHAPS, and 12 mM DTT (freshly added) (filter-sterilized).
13. Isopropyl β-D-1-thiogalactopyranoside (IPTG; keep as powder at −20 °C).
14. Kanamycin solution: 20 mg/mL in water (filter-sterilized).
15. LB agar plates (1 L): 10 g tryptone, 5 g Bacto yeast extract, 5 g NaCl, 15 g agar (autoclave-sterilized); 100 μg/mL ampicillin or 20 μg/mL kanamycin, and 25 μg/mL chloramphenicol (*see* **Note 5** and Subheading 3.1.1.1 for antibiotic selection).
16. Bacterial lysis buffer: 50 mM Tris at pH 8.0 and 0.1 M NaCl (autoclave-sterilized).

    0.1 M $NiSO_4$ solution in water (filter-sterilized).
17. PBS: 10.2 mM Na_2HPO_4, 1.76 mM KH_2PO_4 at pH 7.4, 137 mM NaCl, and 2.7 mM KCl (autoclave-sterilized).
18. PBS-EGTA/EDTA: PBS, 1 mM EGTA (ethyleneglycoltetraacetic acid), and 1 mM EDTA.
19. Refolding buffer #1: 55 mM Tris at pH 8.0, 440 mM L-arginine, 400 mM NaCl, 10 mM DTT, 1 mM EGTA, and 0.88 mM KCl.
20. Refolding buffer #2: 50 mM HEPES at pH 8.0, 200 mM NaCl, 10 mM DTT, and 0.2 % Tween 20.
21. 3× SDS-PAGE gel loading buffer that is suitable for the SDS-PAGE gel system used.
22. Urea buffer: 50 mM Tris at pH 8.0, 8 M urea.

23. Washing buffer: 50 mM Tris at pH 8.0, and 0.5 M NaCl (autoclave-sterilized).
24. 2× TY media (1 L): 16 g tryptone, 10 g Bacto yeast extract, 5 g NaCl (autoclave-sterilized), 50 μg/mL ampicillin or 10 μg/mL kanamycin, and 25 μg/mL chloramphenicol (*see* **Note 5** and Subheading 3.1.1.1 for antibiotic selection).
25. Z-VAD-fmk solution: 100 μM in DMSO (keep at −20 °C in 10 μL aliquots).

3 Methods

3.1 Caspase Expression and Purification

There is no caspase in *E. coli*. However, because some *E. coli* proteins can be cleaved by caspases [13], it is not beneficial to the bacteria to express caspases. Therefore, caspases are best expressed using a system that leaks as little as possible, which is, in this case, the pET system (EMD Millipore, formerly Novagen) in the BL21(DE3) *E. coli* strain. This strain drives expression of the protein of interest via a T7 promoter. DE3 is a λ prophage carrying the T7 RNA polymerase gene and the lacIq repressor. An IPTG-inducible promoter drives the T7 RNA polymerase expression, which is repressed by lacIq. Furthermore, supplemental repression is obtained if the bacterium carries the p*LysS* plasmid, which encodes the T7 lysozyme, a T7 RNA polymerase inhibitor. Upon addition of IPTG to the growth medium, the lacIq repressor is neutralized, and the T7 RNA polymerase is expressed. The polymerase concentration then overcomes the T7 lysozyme inhibition and drives the T7 promoter that is found on the pET plasmid encoding the caspase. Although not absolutely necessary, the use of p*LysS* is recommended, as it will facilitate bacterial growth before protein expression induction and can prevent the selection of weakly expressing bacteria during the culture.

All full-length caspases must be expressed as C-terminally His-tagged proteins. This requirement arises because most caspases cleave themselves in the N-terminal domain, thus resulting in the loss of the catalytic domain if the purification tag is at the N-terminus. The N-termini of caspases contain regulatory domains that can be masked by the addition of a nearby tag. Furthermore, several groups have successfully fused fluorescent proteins at the C-termini of caspases [14–18]. However, initiator caspases expressed without the N-terminal domain can be purified as N-terminal His-tag fusion proteins.

Aside from a few exceptions (e.g., full-length caspase-8), caspases express as soluble proteins and can be purified from the soluble fraction of a bacterial lysate without the use of detergents.

Along with the regular protocol for purifying soluble active caspases (Subheading 3.1.1), protocols are provided for full-length caspase-8 fused at the C-terminus to YFP. These protocols describe expression and purification from inclusion bodies, followed by refolding the protein to recover its activity (Subheading 3.1.2).

3.1.1 General Protocol for Caspase Production in E. coli

The protocols are described for 1–2 L of bacterial culture, which is generally sufficient to produce enough caspase for most biochemical characterization. The protocols can be scaled up to accommodate larger expression volumes and protein yields. However, the centrifugation required to harvest bacteria may limit the total manageable volume when the zymogen forms of caspase-3 or caspase-7 are produced because of short expression time (<30 min). If this is the case, several smaller expression cultures should be planned unless there is access to a high-capacity rotor.

The procedure described below generates significant amounts of many caspases (Fig. 2). As shown in Fig. 2a, the absorbance profile of the imidazole-eluted fraction at 280 nm shows a typical broad peak indicative of His-tagged protein, in this case, the catalytic mutant of caspase-7 (35 kDa). If the yield is high, as is usually

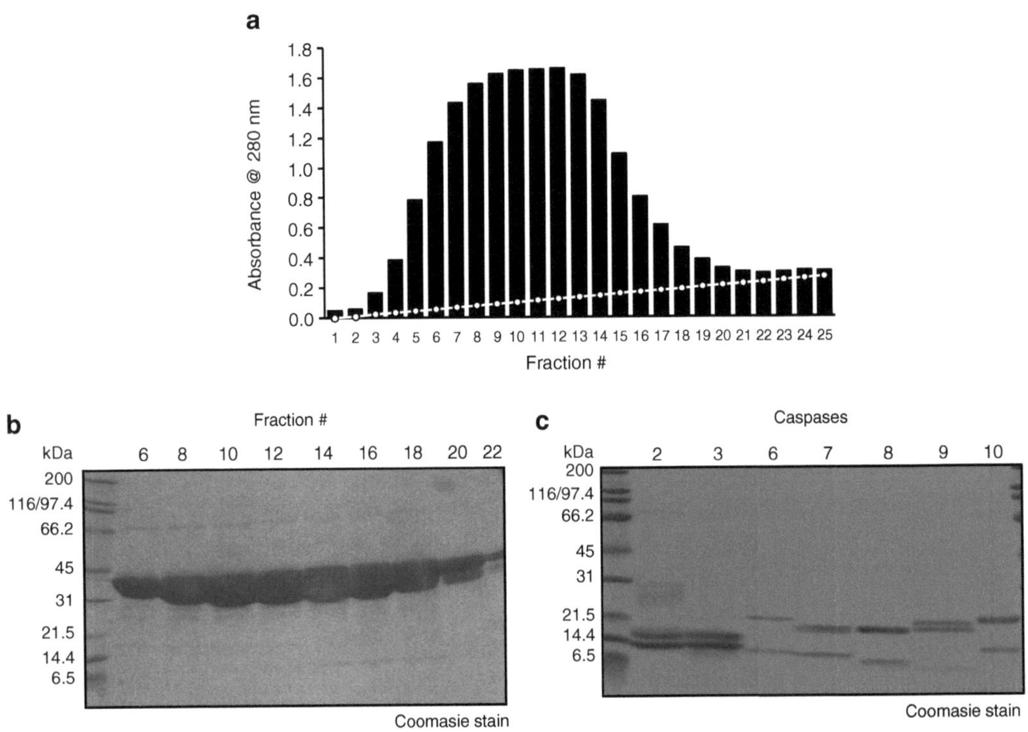

Fig. 2 Caspase purification. Caspases were expressed and purified according to the protocol described in Subheading 3.1.1. (**a**) Absorbance profile at 280 nm of each fraction for the expression of C-terminal His-tagged caspase-7 C285A mutant (catalytic cysteine mutated to alanine; Subheading 3.5 for details). The white trace represents the absorbance of the buffer alone. (**b**) Typical SDS-PAGE analysis of imidazole-eluted caspase-7 C285A (different purification than in (**a**). (**c**) SDS-PAGE analysis of various caspase preparations. Caspase-2 and caspase-9 lack the CARD; caspase-8 and caspase-10 lack DEDs

the case for the expression of inactive caspases, the purified protein is >95 % pure following IMAC chromatography (Fig. 2b). This procedure allows for the purification of all apoptotic caspase catalytic domains (Fig. 2c).

General Protocol for Caspase Expression in E. coli

1. Transform BL21(DE3)p*LysS* competent cells with the appropriate vector and spread the bacteria on LB agar plates with antibiotics. Incubate overnight at 37 °C. (*DAY 1*; *see* **Note 5**).
2. The following morning, inoculate 2 mL of 2× TY medium containing antibiotics with a small to medium colony of freshly transformed BL21(DE3)p*LysS*. Incubate in a 15 mL culture tube for ~8–10 h at 37 °C with vigorous shaking (250 rpm) (*DAY 2*).
3. In a 250 mL bacterial culture flask, dilute the primary culture 100-fold into fresh 2× TY medium containing antibiotics and incubate as in **step 1** for ~16 h (overnight). Prepare ~20 mL for each liter of final expression culture (*see* **Note 6**).
4. Set up 1 L baffled culture flasks (each containing 0.5 L of medium) by diluting the secondary culture 50-fold into 2× TY medium containing antibiotics. Incubate at 37 °C with vigorous shaking (250 rpm) until the optical density at 600 nm reaches between 0.5 and 0.7 (~2–4 h). Use sterile 2× TY medium as a spectrophotometer blank (*DAY 3*).
5. Decrease the temperature to 30 °C and induce expression by adding IPTG to a final concentration of 0.2 mM from a freshly made stock solution (48 mg/L of culture). Incubate at 30 °C with vigorous shaking (250 rpm) for ~5 h (*see* **Note 7**).
6. Once the expression period is over, transfer the culture to centrifuge bottles and collect the cells at 4 °C for 5 min at 3,900 ×*g*. Discard the supernatant.
7. Resuspend the cell pellet in 10–15 mL of bacterial lysis buffer per liter of original culture volume (**step 4**). Purify immediately or store at −80 °C for up to 6 months in 50 mL polypropylene disposable screw cap tubes or an equivalent (*see* **Note 8**).

General Caspase Purification Protocol

1. If frozen, thaw the bacterial suspension in tepid water. Do not let the suspension warm. Transfer the cell suspension into a 50/100-mL plastic beaker. Keep on ice (*DAY 4*).
2. Using an ultrasonic homogenizer (large probe), break cells for 2 min at 70 % power with a 50 % duty cycle (on for 0.5 s, then off for 0.5 s). Sonicate for 30–45 s/L of culture (*see* **Note 9**).
3. Transfer the lysate to centrifuge tubes and centrifuge at 4 °C for 30 min at 18,000 ×*g* (12,000 rpm in a Sorvall SM-34 rotor or equivalent). The soluble fraction contains the caspase.
4. During centrifugation, pour 0.5–5 mL of Chelating Sepharose resin into an empty chromatography column (1.0 cm or less in

diameter; e.g., Bio-Rad Econo-Pac 0.7×5.0-cm), and let the liquid drain. Sequentially rinse the resin with 5 bed volumes of Milli-Q water, 2 bed volumes of $NiSO_4$ solution, 5 bed volumes of Milli-Q water, and 5 bed volumes of bacterial lysis buffer. Let the column drain by gravity flow between each rinse. Do not let the resin dry. Keep the column at 4 °C. Following these steps, the column is ready to use (*see* **Note 10**).

5. Filter the lysate with a 0.45 μm Durapore Stericup HV filter unit. Rinse the filter with 5 mL of bacterial lysis buffer and pool with lysate (*see* **Note 11**).
6. Apply the lysate to the column and let drain by gravity flow.
7. Wash the resin five times with 10 mL of washing buffer. Let the liquid drain between each wash.
8. Re-equilibrate the column with 5 bed volumes of bacterial lysis buffer, and leave ~1 mL of buffer on top of the resin to prevent air bubbles from entering the resin.
9. Attach the gradient maker (valve closed) to the pump, the pump to the column flow adaptor, the adaptor to the column, and the column outlet to the fraction collector. Ensure that all tubing is full of bacterial lysis buffer.
10. Add 12.5 mL of bacterial lysis buffer in compartment 1 and 12.5 mL of elution buffer in compartment 2 of the gradient maker. Set the pump flow rate to 1 mL/min, open the gradient maker valve, and collect 1 mL fractions. Continuously stir compartment 1. Keep all fractions on ice (*see* **Note 12**).
11. Just before the end of the gradient, add 5 mL of elution buffer into compartment 2 of the gradient maker. This step will elute any remaining His-tagged protein.
12. Measure the absorbance of each fraction at 280 nm. Analyze 10 μL of every other fraction by SDS-PAGE (*see* **Note 13**).
13. Pool the purest and most concentrated fractions. Measure the absorbance of the pooled fractions at 280 nm, and estimate the initial caspase concentration using the Edelhoch relation (*see* **Note 14**).
14. Prepare 50–100 μL aliquots and freeze at −80 °C (*see* **Note 15**).
15. Perform an active site titration of the caspase preparation according to Subheading 3.2.2.

3.1.2 Protocol for Full-Length Caspase-8 Production in *E. coli*

Full-length caspase-8 does not express as a soluble protein in *E. coli*, it is exclusively found in inclusion bodies. Therefore, a purification strategy that employs denaturants to solubilize the caspase is required. The protein must then be refolded to recover the enzymatic activity. The following procedure allows for the production of small but enzymatically pure and active preparations of full-length caspase-8 (adapted from a protocol from Christina Pop, personal communication). The first purification step involves the preparation of crude inclusion bodies using centrifugation. Proteins are then solubilized using guanidine, and denatured caspase-8 is purified using two chromatography-based purification steps: (1) IMAC to recover all His-tagged proteins, and (2) optional ion exchange chromatography to remove caspase-8 fragments and concentrate the protein. Finally, refolding is performed by dialysis against an arginine buffer. Although the mechanisms of arginine-assisted refolding are not fully understood, it seems to reduce protein aggregation by interacting with amino acid side chains, increasing the free energy of protein–protein interactions, and increasing the stability and solubility of denatured proteins [19–22].

The procedure described below has been used to generate minute amounts of full-length caspase-8 fused at the C-terminus with yellow fluorescent protein (YFP) (Fig. 3). The recovery of YFP fluorescence was used as a mean to assess various refolding protocols, and the presence of YFP does not alter the enzymatic properties of the caspase. As shown in Fig. 3a, IMAC produces a

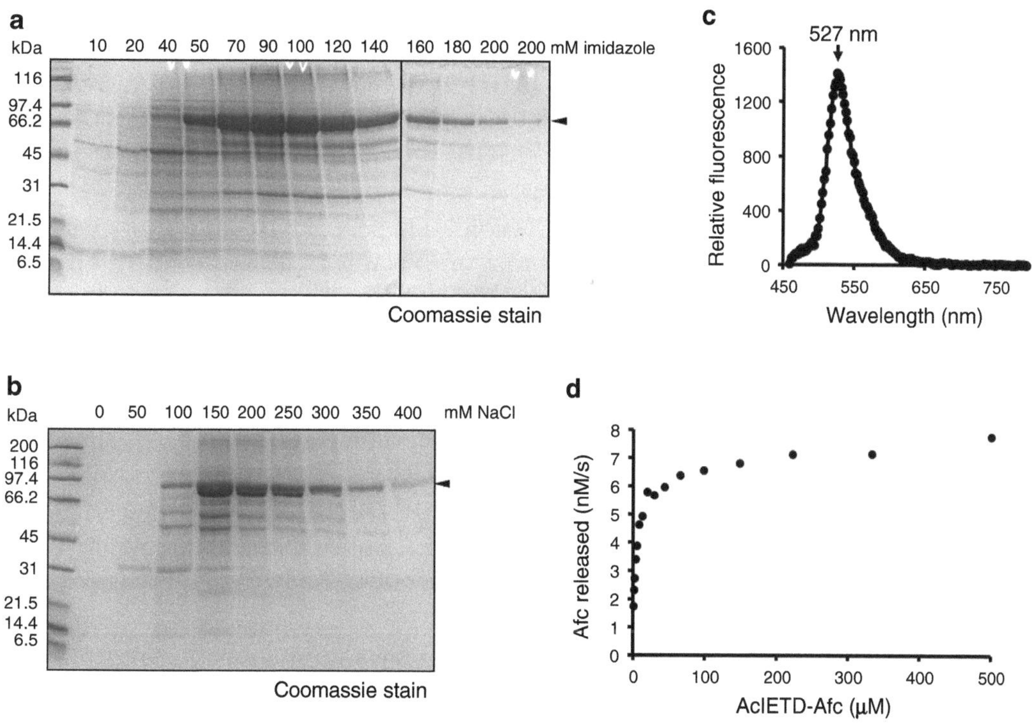

Fig. 3 Full-length caspase-8 production. (**a**) IMAC purification (Subheading 3.1.2) of denatured full-length caspase-8 C-terminally fused to YFP (~84 kDa; *arrowhead*). Proteins were eluted using a step gradient of imidazole (indicated above each lane). The procedure results is a protein preparation that is >80 % pure. (**b**) DEAE anion exchange chromatography of denatured full-length caspase-8 C-terminally fused to YFP (*arrowhead*). This results in a protein preparation that is >90 % pure. (**c**) Following refolding, the typical fluorescence spectrum of YFP is recovered showing a maximum emission of 527 nm. (**d**) The activity of caspase-8 is also recovered as demonstrated by the typical Michaelis–Menten substrate saturation curve. These data are consistent with a K_M and k_{cat} of 4.4 μM and 0.4 s^{-1}, respectively

series of fractions containing full-length caspase-8-YFP (~84 kDa) and main contaminants eluting prior to the pool of caspase. DEAE anion exchange chromatography (Fig. 3b), in addition to concentrating the caspase, allows for the removal of more impurities. Following refolding, the typical emission spectrum of YFP is recovered (Fig. 3c), along with enzymatic activity (Fig. 3d).

The following protocols are valid for bacterial cultures of 1–2 L and can be easily scaled up.

Full-Length Caspase-8 Expression Protocol

The expression protocol is based on the general procedure described in Subheading 3.1.1 with the modification that caspase-8 expression induction is performed using 0.4 mM IPTG at 37 °C for 5 h (**step 5**, Subheading 3.1.1.1) (*DAYS 1–3*).

Full-Length Caspase-8 Purification Protocol

1. If frozen, thaw the bacterial suspension in tepid water. Do not let the suspension warm. Transfer the cell suspension into a 50–100 mL plastic beaker. Keep on ice (*DAY 4*).
2. Using an ultrasonic homogenizer (large probe), break cells for 2 min at 70 % power with 50 % duty cycle (on for 0.5 s, then off for 0.5 s). Keep on ice.
3. Transfer the lysate to centrifuge tubes and centrifuge at 4 °C for 30 min at 18,000 × *g* (12,000 rpm in a Sorvall SM-34 rotor or equivalent). Discard the supernatant. The insoluble pellet contains the caspase.
4. Suspend the inclusion body pellet in 10–20 mL of guanidine buffer. Transfer to a small plastic beaker and stir overnight at room temperature to solubilize the proteins. *Keep everything at room temperature from this step forward* (*see* **Note 16**).
5. The next morning, centrifuge the solubilized proteins for 30 min at 18,000 × *g* (12,000 rpm in a Sorvall SM-34 rotor or equivalent) to eliminate remaining insoluble debris. The supernatant contains the solubilized caspase (*DAY 5*).
6. Prepare the Chelating Sepharose as described in Subheading 3.1.1.2, **step 4**, but equilibrate the column with 5 bed volumes of guanidine buffer. Allow 2 mL of resin to purify from 1 L of bacterial expression.
7. Resuspend the prepared resin with the supernatant from **step 5** in a 15–50 mL tube. Incubate with gentle shaking for ~2 h.
8. Recover the resin by centrifugation for 5 min at 800 × *g*.
9. Wash the resin twice with 10 bed volumes of urea buffer and recover the resin by centrifugation, as in **step 8**.
10. Transfer the resin into a 15 mL tube. Elute the caspase with a step gradient of imidazole in urea buffer (0–200 mM imidazole; 12 fractions; 1 bed volume per fraction) by suspending the resin in buffer then by centrifugation for 5 min at 800 × *g*.

11. Analyze a 10 μL aliquot from each fraction by SDS-PAGE. Pool the purest and most concentrated fractions. Proteins can be kept 3–4 days at room temperature before proceeding to refolding.

Optional: If higher protein concentration and purity are required, the following DEAE ion exchange purification procedure (**steps 12–16**) is recommended. This procedure will help to remove some cleaved caspase-8.

12. Pour 2 mL of DEAE Sephadex resin into an empty chromatography column (1.0 cm or less in diameter; e.g., Bio-Rad Econo-Pac 0.7 × 5.0-cm) and let the liquid drain by gravity flow. Rinse the resin twice with 5 bed volumes of Milli-Q water and then 5 bed volumes of urea buffer (*DAY 6*).
13. Apply the pooled fractions containing caspase-8 (**step 11**) to the column and let drain.
14. Wash the column with 5 bed volumes of urea buffer and let drain between each wash.
15. Elute the protein with a step gradient of NaCl in urea buffer (0–400 mM; 10 fractions; 2 mL/fraction).
16. Analyze a 10 μL aliquot from each fraction by SDS-PAGE. Pool the purest and most concentrated fractions. Proteins can be kept 3–4 days at room temperature before proceeding to refolding.

Refolding Full-Length Caspase-8

Note: Every step of this procedure is performed at room temperature.

1. Centrifuge the pooled fractions from **step 11** or **step 16** in Subheading 3.1.2.2 for 15 min at 18,000 × *g* to remove insoluble proteins and residual resin. Transfer the supernatant to a dialysis tube (10,000 MWCO; prepared according to manufacturer instructions) equilibrated in refolding buffer #1 (*DAY 6*).
2. Dialyze overnight at room temperature against 1 L of refolding buffer #1. Stir buffer at low speed.
3. The next day, dialyze for 5 h at room temperature against 1 L of refolding buffer #2. Stir buffer at low speed (*DAY 7*).
4. Recover the dialyzed protein with a pipet.
5. Centrifuge the protein solution for 30 min at 18,000 × *g* to remove any insoluble protein. Transfer the supernatant to a fresh tube.
6. Determine the final protein concentration using a protein assay that is compatible with refolding buffer #2 (e.g., Pierce BCA protein assay). Use fresh refolding buffer #2 as a blank.

Optional: Concentrate the caspase preparation using Millipore spin concentrators (10,000 MWCO).

7. Active site titrate the caspase-8 according to Subheading 3.2.2. Assume a refolding yield of 5–10 % (*see* **Note 17**).

3.2 Enzymatic Characterization of Caspases

Kinetic parameter determination is a very useful method to compare caspase activity based on their primary specificities. Therefore, it is highly recommended that every caspase preparation is characterized using a suitable substrate (Table 2). This criterion will ensure the quality of the caspase preparation and allow for the identification of problems that may have occurred during the purification process that may interfere with future experiments. Furthermore, it is essential to compare the enzymatic properties of caspase mutants to the wild-type enzyme. Too often, mutant caspases are used in cellular experiments, and conclusions are drawn without taking into account the possibility that kinetic properties of the mutated enzyme have changed (*see* **Note 18**).

Two fundamental parameters characterize an enzyme: the K_M and the k_{cat}. To be brief and by assuming that the enzyme follows or is well represented by the mechanism of Eq. 1, K_M (in molarity) compounds the dissociation rates, k_{-1} and k_2, of the enzyme–substrate complex and its association rate k_1 (Eq. 1): K_M equals $(k_{-1}+k_2)/k_1$.

$$E + S \underset{k_{-1}}{\overset{k_1}{\rightleftharpoons}} ES \xrightarrow{k_2} E + P \qquad (1)$$

Therefore, K_M represents more than the affinity of an enzyme for a substrate because it also takes into account the conversion of that substrate into products and not only the association and dissociation of the enzyme-substrate complex into free enzyme and free substrate. However, describing K_M as an affinity constant makes it more palatable. The parameter k_{cat} is the *catalytic constant*, also referred to the *turnover rate*. For enzymes such as caspases, for which a mechanism is well represented by Eq. 1, k_{cat} is similar to k_2. The k_{cat} represents the number of molecules per second converted from the enzyme–substrate complex into the free enzyme and the products.

Without going into the mathematics and theory, the rate of product formation dP/dt or v for Eq. 1 can be represented by the Michaelis–Menten equation that is often written as:

$$\frac{dP}{dt} = v = \frac{V_{max}[S]}{K_M + [S]} \qquad (2)$$

V_{max} is the rate of product formation when the enzyme is saturated, and equals $k_{cat}[E]$. Thus, V_{max} is the maximal rate of product generation by a specific amount of enzyme. V_{max} is used when the concentration of active enzyme is unknown, such as in an extract during purification procedures. One of the most revealing aspect of this equation is that when $K_M = [S]$, $v = 1/2\, V_{max}$. This is why K_M reflects the concentration of substrate at which the enzymatic reaction is at half its maximum rate. One can also realize that it becomes difficult to saturate an enzyme if K_M is relatively high, i.e., 10 K_M results in 0.91 V_{max}.

Finally, a useful term is k_{cat}/K_M (units of M^{-1} s^{-1}), which is the *catalytic specificity* of the enzyme. This term takes into account both K_M and k_{cat}. The catalytic specificity is a parameter that can be obtained even if both k_{cat} and K_M cannot be derived individually (*see* Subheading 3.3). The catalytic specificity is the best way to compare enzyme efficacy for various substrates and to compare enzymes in general.

To determine k_{cat}, one must determine the active enzyme concentration, which can be achieved by titration. Unfortunately, too many studies neglect this important step in caspase substrate characterization. Basing caspase concentration on protein quantity (e.g., μg/mL) or enzyme units (e.g., units/mL) should be discouraged because neither measurement reflects the active enzyme, but rather an amount of protein, active or not, or a quantity of enzymatic work on a standard substrate, respectively.

Many studies use peptidic substrates to characterize caspase activity. One of the most prevalent misconceptions is that many assume that the preferred recognition motif is specific for a given caspase (e.g., DEVD for caspase-3) [4, 5, 7]. This assumption is false if one understands how most peptidases, including caspases, work and how the preferred motifs were determined in the first place. Furthermore, this notion has been disproven experimentally [23]. Because caspases recognize their substrates' cleavage sites primarily via four subsites (S1 to S4; Schechter–Berger nomenclature [24]), tetrapeptides are used. Interestingly, caspase-2 has a substrate-binding pocket that extends to recognize at least five residues, and thus, pentapeptides are used for this enzyme [25]. To track hydrolysis, the peptide is linked at its C-terminus to a leaving group. Usually, this group is fluorescent, such as Afc (7-amino-4-trifluoromethylcoumarin) or Amc (7-amino-4-methylcoumarin), or is a colored compound such as *p*-nitroaniline (pNa). Cleavage of the peptide releases the fluorophore or the chromophore and enables the measurement of caspase activity with a spectrofluorometer (Afc and Amc) or a spectrophotometer (pNa) plate reader, respectively. It is important to note that several other substrates are available, but many of them have two peptides attached to the fluorophore [e.g., $(Z\text{-}DEVD)_2$-Rh110 or MR-$(DEVD)_2$]. Although many of these substrates are sensitive, their hydrolysis reactions do not conform to the mechanism depicted in Eq. 1, so the product resulting from the cleavage is still a substrate for the enzyme. Therefore, these dual-peptide substrates are not suitable for enzymatic characterization using the traditional Michaelis–Menten equation.

This section describes procedures to quantify the active site concentration in a caspase preparation and to determine the fundamental kinetic parameters of caspases. When all reagents are available and techniques are mastered, the full characterization (titration and kinetic parameter determination) can be accomplished in less than a day.

3.2.1 Substrate and Plate Reader Calibration

The determination of enzyme velocity requires that experimental values be represented in molar amount of product generated per unit of time (e.g., μM/s or nM/s). Therefore, the relationship between relative fluorescence units (RFU) given as readouts from the plate reader and the actual concentration of product must be determined. The conversion factor varies greatly between instruments and system settings for the same instrument. It also varies over time for the same instrument and settings. For instance, instruments that use a lamp for excitation will, over the lifetime of the lamp, display a decreased performance. Therefore, an accurate velocity measurement requires frequent determination of the relationship between RFU and the corresponding molar amount of product. To this end, one must first make a standard fluorophore solution. This solution is used to calibrate the spectrofluorometer.

In a similar fashion to the Afc standard, the substrate must also be calibrated. Indeed, the purity of commercially available fluorogenic substrates varies among providers and from lot to lot within the same manufacturer. A simple way to obtain accurate and reproducible values using commercial substrates is to calibrate them following the complete hydrolysis of a fixed amount of substrate. This process ensures that the concentration of "usable" substrate is determined.

To produce reliable and reproducible data, important technical considerations must be taken into account. First, it is critical to use the same brand and model of plates for calibration of the plate reader and enzymatic assays. This requirement arises because the plate used significantly influences the relationship between the fluorescence readout and the actual concentration of the fluorophore. It is also critical that the instrument settings used to calibrate the plate reader are exactly the same as the ones used for all enzymatic assays. Otherwise, the conversion of fluorescence readout into molar concentration of product will no longer stand. Consequently, it is recommended to generate several datasets of calibration for every group of settings that will be used. If available, softwares or instrument programs, templates or methods are useful to save the instrument parameters.

Standard Afc Solution

This procedure takes 1 h to be completed.

1. Weigh a small amount of Afc and prepare a 10 mM solution in DMSO.
2. Using the stock solution, prepare 1 mL of 50 μM Afc solution in water.
3. Measure the absorbance of the diluted Afc solution at 380 nm in a 1 cm path quartz cuvette. Determine the actual concentration of Afc using the molar extinction coefficient (ε) of Afc

($\varepsilon_{380\ nm} = 12{,}600\ M^{-1}\ cm^{-1}$). A 50 μM solution of pure Afc has an absorbance of 0.630 at 380 nm (*see* **Note 19**).

4. Correct the initial concentration of Afc accordingly. *This is the standard Afc solution*. Afc solution is stable for at least 1 year at −20 °C.

Plate Reader Calibration

This procedure takes 1 h to be completed.

1. Using the standard solution of fluorophore (**step 4**, Subheading 3.2.1.1), set up a series of 100 μL Afc samples in a microplate in 1× caspase buffer to cover a concentration range between 1 and 100 μM (15 samples). Plan more samples in the 1–10 μM range than in the higher range. Include a buffer-only sample (*see* **Note 20**).
2. Read the fluorescence at EXλ = 405 nm and EMλ − 510 nm for every setting of the instrument that will be used (e.g., various values of gain, wavelength bandwidth, flashes, integration time).
3. Plot the fluorescence of each sample against the Afc concentration and determine the slope of the straight portion of the trace by linear regression. The slope describes the relationship between RFU and Afc concentration.
4. Repeat the procedure regularly (*see* **Note 21**).

Fluorogenic Substrate Calibration

This procedure takes 2 h to be completed.

1. Based on the amount indicated on the product label, dissolve a vial of substrate in DMSO to obtain a final concentration of 30 mM.
2. Set up a 1 mL caspase reaction containing 50 μM Afc-based substrate in 1× executioner caspase buffer and add excess purified recombinant caspase-3 (100 nM) to completely hydrolyze the available substrate. Set up another sample without caspase. Incubate for 1 h at 37 °C in the dark (*see* **Note 22**).
3. Measure the absorbance of the reaction at 380 nm in a 1 cm path quartz cuvette. Use the caspase-free sample as a blank. Determine the actual concentration of "usable" substrate using the molar extinction coefficient (ε) of Afc ($\varepsilon_{380\ nm} = 12{,}600\ M^{-1}\ cm^{-1}$) (*see* **Notes 19** and **23**).
4. Correct the initial concentration of Afc-based substrate accordingly. Fluorogenic substrate solutions are stable for at least 1 year at −20 °C.

3.2.2 Caspase Active Site Titration

The goal of active site titration is to determine the molar amount of active sites in an enzyme preparation. Too often, enzymes (even caspases) are provided as unit amount, for example, restriction enzymes used in molecular cloning. Although this approach is satisfactory because enzymes are often used to perform a certain quantity of work (e.g., cleaving 10 μg of plasmid DNA) and because some of them are specific (e.g., restriction enzymes), peptidases are never truly specific, and the substrate used often changes from experiment to experiment. For instance, we use peptidases to screen peptide/inhibitor libraries, test various recombinant substrates, or search for new substrates by proteomics. In fact, knowing the molar amount of active sites permits the comparison of the proteolysis efficacy of several enzymes for a given substrate. Furthermore, the titer of an enzyme preparation does not vary with the inhibitor used, nor does it change based on the substrate employed in the titration procedure. Thus, enzyme titration is a cornerstone of the biochemical analysis of enzyme kinetics.

Active site titration of a caspase is a straightforward procedure once one understands the underlying principles. A titration reaction is a two-step procedure. First, an unknown concentration of caspase (estimated by protein quantification) is incubated with a known concentration of titrant. This titrant must be a covalent inhibitor that reacts stoichiometrically with its target. For caspases, the fluoromethyl ketone irreversible inhibitor Z-VAD-fmk is perfectly suited. Z-VAD-fmk forms a thioether adduct with the catalytic cysteine, thus irreversibly inhibiting the peptidase. Given enough time, every caspase active site will react with an inhibitor molecule in a 1:1 ratio until no more inhibitor is available. Therefore, if a series of Z-VAD-fmk samples, covering a range of concentrations that are below and above the effective caspase concentration, is incubated with an unknown amount of caspase, the lowest concentration of inhibitor that inactivates 100 % of the peptidase would be close to the active site concentration. The second step of the assay is evaluation of the fraction of uninhibited enzyme using a simple enzymatic assay. Plotting the reaction rates obtained against the Z-VAD-fmk concentration results in an initial downward straight line ($[I] < [E_0]$) that intercepts the *x*-axis at a value that is equal to the concentration of peptidase ($[I] = [E_0]$). The second portion of the graph is a flat line representing values of inhibitor concentration that are higher than that of the enzyme ($[I] > [E_0]$). Figure 4 shows the experimental data for the active site titration of wild-type caspase-7. Refer to the figure legend for an explanation. An example of the calculation done to obtain the enzyme titer is presented in Fig. 4d.

Because titration occurs in the first reaction (inhibitor plus caspase), the only enzymatic requirement for the second step of the assay (residual uninhibited caspase plus substrate) is that the hydrolysis rate of the reporter substrate remains constant over the measurement period. Additionally, because the goal of the second step is to report the fraction of uninhibited peptidase, it is not necessary to know the kinetic parameters of the enzyme for the substrate used. Furthermore, the concentration of substrate does not matter as long as no significant substrate depletion occurred.

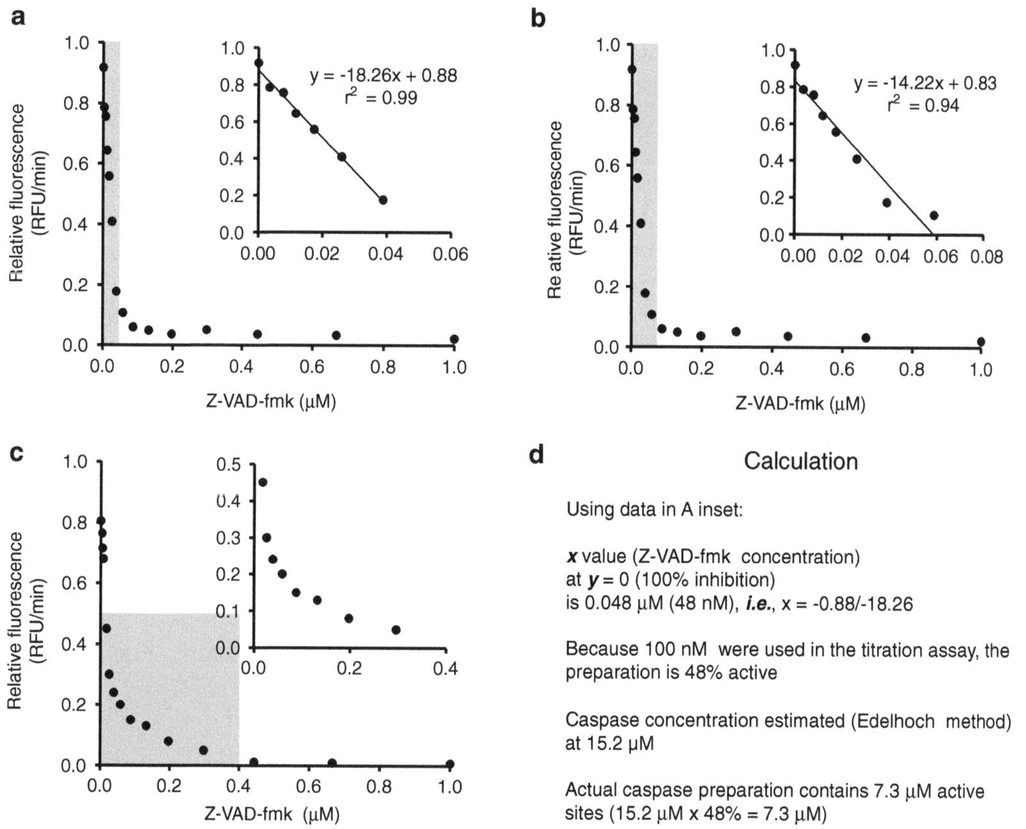

Fig. 4 Caspase titration. (**a**) Example of a good titration dataset of caspase-7 and good analysis of titration data. Data points used (*gray area*) to determine the titer are appropriately chosen resulting in an accurate titer determination. The data used follow the linear regression well. (**b**) The same titration dataset as in (**a**) but with poor analysis of titration data. Data points used (*gray area*) to determine the titer were poorly chosen, resulting in overestimation of the titer. The *inset* shows that the value at 0.06 μM Z-VAD-fmk clearly departed to the right of the linear regression set by all values, which suggests that inhibition was not complete in this sample. (**c**) Example of a titration experiment with incomplete inhibition leading to an inappropriate titration. The *inset* shows data describing a curve instead of a straight line. In this case, the titration should be repeated. (**d**) Calculation for caspase-7 titration using data in (**a**)

Executioner caspases (caspase-3, caspase-6, and caspase-7) are fully active and dimeric in a buffer that closely mimics the cytosol (e.g., executioner caspase buffer). On the contrary, initiator caspases are activated by dimerization on multimeric platforms such as the DISC (death-inducing signaling complex) for caspase-8 and caspase-10, the apoptosome for caspase-9, and the PIDDosome for caspase-2 [26]. These complexes recruit initiator caspases via homotypic interaction of death domain superfamily domains: caspase-8 and caspase-10 use tandem death effector domains (DEDs), and caspase-2 and caspase-9 employ a single caspase activation recruitment domain (CARD) [26]. Because initiator caspases are usually purified as a mixture of dimers (active) and monomers (inactive) from bacteria [27–30], the assay conditions must somehow force the dimerization of the caspase. To re-create this condition in vitro, a buffer that favors dimerization, such as one that contains kosmotropic salts (e.g., sodium citrate), is used [30, 31]. In addition to promoting dimerization, kosmotropic salts will also promote ordering/stabilization of crucial loops implicated in caspase activity.

Practically, to titrate a caspase preparation, a series of enzymatic reactions are set up in a micro-well plate with a serial dilution of Z-VAD-fmk. One reaction without the titrant is included. If the highest final Z-VAD-fmk concentration is 1 μM, a 2/3 serial dilution over 15 wells will cover values between 3.4 and 1,000 nM of caspase. For Z-VAD-fmk to inactivate as much caspase as possible in a reasonable time, relatively high concentrations of enzyme are required. These concentrations are necessary because, although Z-VAD-fmk is a pan-caspase inhibitor, it is not potent for all caspases. Most wild-type or truncated caspases will be inhibited within 30 min if 100 nM of enzyme is used in the appropriate assay buffer. After the inhibition reaction, aliquots of reactions are transferred to a new plate containing an appropriate substrate. The hydrolysis of that substrate is measured continuously.

This procedure takes 2 h to be completed.

Protocol

1. Thaw an aliquot of caspase on ice (this step takes approximately 30 min). Do not heat the sample.
2. In a 96-well plate, set up a 2/3 serial dilution of Z-VAD-fmk in 1× caspase buffer (50 μL/well, starting at 2 μM) over 15 wells. Set up one sample with buffer only (well 16) (*see* **Note 24**).
3. Based on the estimated caspase concentration, prepare 1 mL of a 200-nM solution of caspase in 1× caspase buffer. Add 50 μL of this solution to every well of the plate. The highest concentration of Z-VAD-fmk will be 1 μM, and the final concentration of caspase will be 100 nM.
4. Thoroughly mix using a microplate mixer or the mixing function of the plate reader. Seal the plate with paraffin film and incubate for 30 min at 37 °C.
5. Once the incubation is over, transfer an aliquot of the reaction to a new series of wells containing 1× caspase buffer (final volume of 80 μL). Thoroughly mix and incubate the plate for 5 min at 37 °C (*see* **Note 25**).
6. During this time, prepare 350 μL of a 500 μM solution of caspase substrate in 1× caspase buffer (*see* **Note 26**).
7. With a repeating pipettor, rapidly add 20 μL of the substrate solution to each well. Thoroughly mix and immediately read the fluorescence at EXλ = 405 nm and EMλ = 510 nm for 30 min at 37 °C. Ideally, take measurements every 5 s (*see* **Note 27**).

8. Extract the initial rate (straight portion of the fluorescence plotted against time) of every reaction and plot them against Z-VAD-fmk concentration (**step 3**). Utilize linear regression to obtain the *x*-axis intercept ($y=0$) using data starting from the lowest Z-VAD-fmk concentration up to the value at which ~10 % of enzyme activity remains. The intercept is equal to the concentration of caspase in the titration reaction (**step 3**) (*see* **Note 28**).
9. Because the estimated concentration of caspase in the titration reaction was set at 100 nM based on protein quantity, correct the value accordingly.

3.2.3 General Protocol to Determine K_M and k_{cat} of a Fluorogenic Substrate

Practically, a series of enzymatic reactions are set up in a micro-well plate with a serial dilution of the caspase substrate. The hydrolysis of that substrate is measured continuously in a plate reader. For most caspases, the hydrolysis rate will follow the typical Michaelis–Menten equation. The initial velocity of each sample is used to determine the kinetic parameters using nonlinear regression. The protocol describes how to determine the kinetic parameters for an Afc fluorogenic substrate. However, the same protocol can be used for other type of substrates, including chromogenic substrates. In the latter case, a spectrophotometer is used, and a standard curve of the free chromophore is used to convert the absorbance into molar amount of product generated.

See protocol in Subheading 3.2.2 and accompanying notes. This procedure takes 2 h to be completed.

Protocol

1. Thaw an aliquot of active site-titrated caspase on ice (this step takes approximately 30 min). Do not heat the sample (*see* **Note 29**).
2. In a 96-well plate, set up a 3/4 serial dilution of Afc fluorogenic substrate in 1× caspase buffer (50 μL/well, starting at 300 μM or higher if necessary) over 16 wells (*see* **Notes 30** and **31**).
3. *For initiator caspases only*: prepare 100 μL of 1 μM caspase solution in 1× initiator caspase buffer. Incubate at 37 °C for 30 min. This solution is used to prepare the diluted caspase solution in **step 4**.
4. Prepare 1 mL of a twofold concentrated caspase solution (i.e., 2–40 nM) in 1× caspase buffer. With a repeating pipettor, rapidly add 50 μL of this solution to every well of the plate. Thoroughly mix and immediately read the fluorescence at EXλ = 405 nm and EMλ = 510 nm for 30 min at 37 °C. Ideally, take measurements every 5 s.
5. Extract the initial rate of every reaction, convert the rates to μM/min or nM/s using the RFU to μM/nM relationship (Subheading 3.2.1.2) for the specific instrument settings used, and plot rates against substrate concentration. Use nonlinear regression and Eq. 2 to extract V_{max} and K_M. Calculate k_{cat} using $k_{cat} = V_{max}/[E]$ (*see* **Note 32**).

3.3 Studying Protein Caspase Substrates

Many in vitro methods enable the assessment of specificity. However, one must be careful in using the primary specificity of a caspase to assess its preference on natural substrates. The recent report that caspase determinants located outside the substrate-binding pocket can overcome a poor cleavage site motif demonstrates the importance of studying protein substrates [32]. Practically, the determination of kinetic parameters for a protein, as described in the previous subheading, is difficult. First, it is problematic to set up assays with protein substrate concentrations above K_M (e.g., >10 μM). Second, not all protein substrates can be produced as recombinant proteins or purified to homogeneity from tissues or cells.

In pseudo-first order conditions ($[S] \ll K_M$), Eq. 2 can be simplified to $v \approx V_{max}[S]/K_M$. The proportion of substrate used by an enzyme is described by Eq. 3, which can be rearranged to Eq. 4 [25]:

$$p = 1 - e^{-k[E]^t} \tag{3}$$

$$k = -\ln(1 - p) / [E]t \tag{4}$$

where p is the proportion of substrate cleaved (varies between 0 and 1) by the enzyme $[E]$ at time t in seconds. The parameter k approximates k_{cat}/K_M. However, unless pseudo-first order conditions are met, it is better to refer to k as $k_{cat,app}/K_{M,app}$. It is noted that the substrate does not appear in Eqs. 3 or 4. Thus, it is not necessary to know the protein substrate concentration to estimate k_{cat}/K_M in pseudo-first order conditions.

Another interesting aspect of pseudo-first order kinetics is that it is possible to use a mixture of proteins, such as a cell lysate, as a source of substrate. Importantly, cell lysate contains other substrates. However, the competition for the enzyme will be negligible as long as those other substrates, as a whole, also meet pseudo-first order conditions. This statement is valid because the rate of hydrolysis of a substrate in the presence of a competitor substrate S', which has a Michaelis–Menten's constant K_M', is:

$$\frac{dP}{dt} = v = \frac{V_{max}[S]}{K_M\left(1 + \frac{[S']}{K_M'}\right) + [S]} \tag{5}$$

If S' is much smaller than K_M' as in a sufficiently diluted lysate, the denominator approaches $K_M + [S]$, and the enzyme behaves as if no other substrate were present. Thus, this condition allows the estimation of kinetic parameters from a cellular lysate using Eq. 3. Figure 5 shows the analysis of poly(ADP ribose) polymerase 1 (PARP-1) cleavage in a cell lysate by caspase-3 and caspase-7.

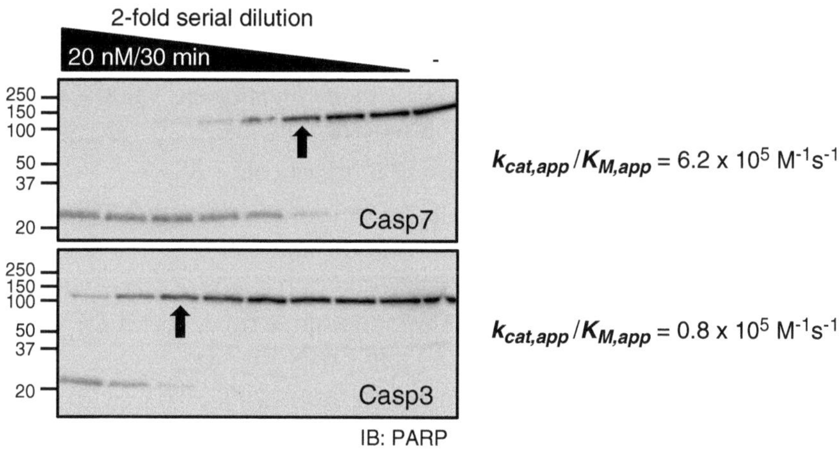

Fig. 5 PARP-1 cleavage by two executioner caspases. MCF-7 cell extracts were incubated for 30 min with twofold serial dilution of the indicated recombinant caspase in executioner caspase buffer starting at 20 nM. Apparent k_{cat}/K_M values were estimated as described in Subheading 3.3.2. Samples were analyzed by immunoblotting using an antibody recognizing the N-terminus of PARP. *Arrows* mark the point at which 50 % of PARP-1 is cleaved. These results were originally published in the *Proceedings of the National Academy of Science of the USA* [61] © National Academy of Sciences

Here, a simple protocol is described to characterize the cleavage of a natural substrate by a caspase in vitro. The procedure uses a cell lysate as a source of substrate, but the same protocol is suitable for a recombinant protein. For recombinant substrates, it is useful to label the protein with fluorescein using an amine or a sulfhydryl reacting reagent such as *N*-hydroxysuccinimide ester (NHS) or *N*-ethylmaleimide (NEM)-fluorescein, respectively. This step will enable a more sensitive measurement of cleavage rates and render antibody use unnecessary. However, it is essential to verify that the labeling strategy used does not affect caspase cleavage. Such an approach has been used to study the cleavage of the Hsp90 co-chaperone p23 by caspase-7 [32].

3.3.1 Preparing Cytosolic Extracts from Mammalian Cells

Lysates made from cell lines deficient in specific caspases are very useful. For example, breast cancer carcinoma MCF-7 cells, which are deficient in caspase-3 and caspase-10a [33, 34], limit experimental bias when studying caspase-7 and caspase-8, respectively. Prepare large quantities of cell extracts (5–10 plates) and freeze the lysate in small aliquots. Doing so will make it easier to compare substrate cleavage by various caspases over several weeks or a few months. This protocol is described for cells grown as a monolayer in one 15 cm tissue culture dish but can easily be scaled up.

The procedure takes 1 h to be completed.

1. Rinse cells twice with 10 mL of cold PBS. Keep the tissue culture dish on ice.
2. Detach the cells using 10 mL of cold PBS-EGTA/EDTA. Incubate for 5 min on ice.

3. Using a pipettor, gently resuspend the cells and transfer them to a 15 mL conical tube.
4. Recover the cells by centrifugation at 1,000 × *g* for 5 min at 4 °C. Discard the supernatant.
5. Wash the cells using 5 mL of cold PBS.
6. Recover the cells by centrifugation at 1,000 × *g* for 5 min at 4 °C. Discard the supernatant.
7. Resuspend the cell pellet in 0.5 mL of eukaryotic lysis buffer, transfer to a 1.5 mL microfuge tube, vortex for 10 s and incubate on ice for 30 min (*see* **Note 33**).
8. Centrifuge the lysate at 7,000 × *g* for 10 min at 4 °C. Transfer the supernatant to a new microfuge tube.
9. Determine the protein concentration using a protein assay that is compatible with the eukaryotic lysis buffer (e.g., Pierce BCA protein assay).
10. Dispense into 30 μL aliquots and freeze at −80 °C.

3.3.2 Determining the Kinetic Parameter $k_{cat,app}/K_{M,app}$ for a Protein Substrate

1. Thaw an aliquot of active site titrated caspase and lysate (this step takes approximately 30 min) on ice. Do not heat the samples.
2. Set up 8 samples of 25 μL of a 2/3 serial dilution of the caspase in 1× caspase buffer, starting with a caspase concentration of 100 nM. Include one sample with buffer only (sample 9). Incubate for 5 min at 37 °C (*see* **Note 34**).
3. Prepare 0.25 mL of a 2 mg/mL lysate solution (or 200 nM of purified protein substrate) in 1× caspase buffer and incubate at 37 °C for 5 min (*see* **Note 35**).
4. With a repeating pipettor, add 0.25 μL of diluted lysate to each of the samples, mix, and incubate for 30 min at 37 °C.
5. Stop the reaction by adding 0.5 volume of 3× gel loading buffer.
6. Analyze the samples by immunoblotting using an antibody directed against the protein of interest.
7. Using imaging software, determine the concentration of caspase at which ~50 % of the substrate is cleaved. Use this value to determine *k* using Eq. 4 (*see* **Note 36**).
8. Repeat the experiment and adjust caspase concentration, lysate concentration, time, or a combination thereof to set the ~50 % cleavage sample between samples 4–6 (*see* **Note 37**).

3.4 Particularities of Caspases

Caspase-2

Caspase-2 has an extended substrate-binding pocket that recognizes amino acids in position P5 (Schechter-Berger nomenclature [24, 25]). Consequently, AcVDVAD-Afc is used to characterize the activity of this caspase. However, because of that extended substrate-binding pocket, active site titration requires conditions that force inhibition by Z-VAD-fmk. Caspase-2 can be successfully titrated using 2 μM of estimated concentration of the peptidase and serial dilution of Z-VAD-fmk started at 20 μM.

Caspase-3

Caspase-3 is the easiest executioner caspase to produce and purify. This enzyme is highly active and titrates well using the Z-VAD-fmk inhibitor. Caspase-3 is the caspase with the highest intrinsic activity and displays the highest activity on AcDEVD-Afc and a wide variety of other sequences [4, 5, 23]. Consequently, one must be careful when studying caspase activity in cells and in vitro because caspase-3 can often overpower other caspases, even on their best substrates. To express the zymogen form of caspase-3, a short time of expression (25–30 min) is used to prevent auto-proteolysis, and purification of the protein must be performed immediately following expression (without freezing), using a minimal amount of Chelating Sepharose resin (0.5–1.0 mL). Because yields are low, it is recommended to use at least 4 L of culture.

Caspase-6

Better yields are obtained if caspase-6 is expressed slowly with low IPTG concentration (0.02–0.05 mM) and long expression time (16–20 h). The relatively long expression time also ensures removal of the N-terminal peptide and efficient cleavage of the linker. Caspase-6's activity is not stable over time. Consequently, caspase-6 must be used quickly after thawing. For example, for titration, the inhibition step is performed for 15 min instead of 30 min to enable the measurement of a proper level of remaining activity.

Caspase-7

Eight hours of expression time for caspase-7 is enough to obtain a fully processed enzyme (N-terminal peptide removed and linker cleaved). Longer expression time results in unwanted cleavage at Asp192 (Asp291 according to caspase-1 nomenclature) that inactivates the enzyme [35]. It is also important to pay particular attention to the pH of the buffers that are used for the purification of some specific forms of caspase-7. Indeed, the zymogen form has a p*I* of 5.7 and can be efficiently purified with buffers at pH 8.0. However, for the mature form or for any forms that do not carry the N-terminal peptide (residues 1–23), the p*I* is 8.3. Thus, it is suggested to use purification buffers at pH 7.2 to limit protein precipitation. See caspase-3 remarks for the expression of the zymogen form of caspase-7.

Caspase-8

The truncated form of caspase-8 expresses well as a soluble enzyme in *E. coli* and is easily purified using the procedure described in Subheading 3.1.1. The absence of its DEDs does not affect caspase-8 substrate specificity on small peptidic substrates [36]. However, because full-length caspase-8 aggregates, this form is insoluble, and the alternative protocol described in Subheading 3.1.2 must be used. Alternatively, F122Y and L123S mutations in the second DED of caspase-8 may be used to prevent aggregation and render the protein soluble [36], but it is important to keep in mind that it is no longer the wild-type enzyme.

Caspase-9
Caspase-9 is the easiest initiator caspase to produce, either as a truncated form or as a full-length protein. Complete processing occurs within 5 h of expression. Indeed, it is impossible to produce uncleaved caspase-9 without the use of cleavage-site mutants [37]. Full-length and CARD-less caspase-9 have the same substrate preference [38], at least on small peptidic substrates, and for reasons that are still poorly understood, the complete enzyme is less active than the truncated caspase-9 [38].

Caspase-10
Caspase-10 expresses similarly to caspase-8. However, the large subunit and small subunit of the catalytic domain bind each other with much less strength so that some of the large subunit is lost during purification. Consequently, care must be taken to keep fractions that, once pooled, result in a 1:1 subunit ratio. It is assumed that caspase-10, similarly to caspase-8, will be essentially insoluble as a full-length form.

3.5 Useful Molecular Forms of Caspases

Over the years, many variants of caspases were produced so that their mechanisms could be studied. The most prominent classes of those variants are either inactive forms or forms that carry cleavage-site mutations. For example, catalytic mutants of caspase-3 and caspase-7 can be used as substrates in cleavage assays by initiator caspases. Catalytic mutants of caspases are easily expressed (much easier than the corresponding zymogen forms), with high yields using the protocol described in Subheading 3.1.1. The use of catalytic mutants of caspases can also be appropriate as a negative control. Inactive caspases are usually generated by mutating the catalytic cysteine into an alanine residue [C285A mutation, caspase-1 structural nomenclature [8]], but some researchers have replaced the catalytic cysteine with a serine residue [39]. Mutation of histidine 237 to an alanine (H237A) has the same effect as the cysteine mutation. This is because those two amino acids, Cys285 and His237, form the caspase catalytic dyad of the peptidase, and mutation of either one of them is sufficient to completely abrogate enzymatic activity [8, 40].

Caspases participate in a proteolytic cascade and often cleave themselves during apoptosis. Those cleavage events may activate the caspases, regulate their association with activation platforms, alter their substrate specificity or provide a way to regulate their interactions with endogenous inhibitors and molecular partners. Mutation of the aspartate residue at the cleavage site abrogates proteolysis. These mutations are quite useful to study the activation of

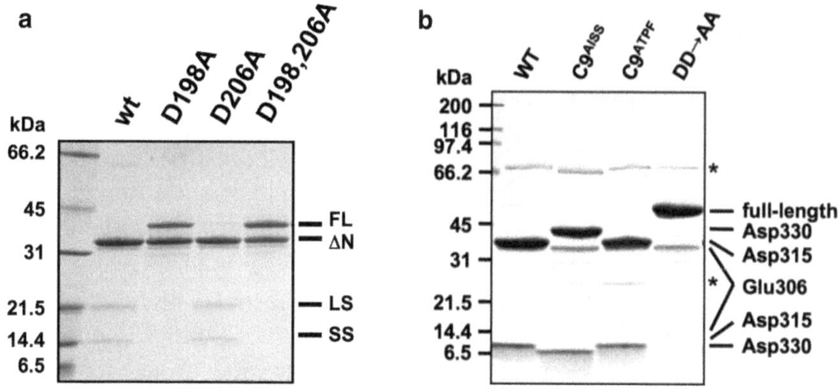

Fig. 6 Various molecular forms of caspases. (**a**) Zymogen (inactive) form of caspase-7 cleavage-site mutants. The zymogen forms of wild-type (wt) caspase-7 or cleavage-site mutants were expressed for 30 min and purified using the protocol described in Subheading 3.1.1. *FL* full-length, Δ*N* no N-terminal peptide (residues 1–23), *LS* large subunit, *SS* small subunit. These results were originally published in *Molecular Cell* [35] © Elsevier B.V. B) Cleavage site mutants of full-length caspase-9. Active wild-type caspase-9 (WT) or caspase-9 proteins cleaved at site 1 ($C9^{AISS}$) or site 2 ($C9^{ATPF}$) in the linker or a double cleavage-site mutant (DD → AA) were expressed as described in Subheading 3.1.1. Gels were stained using Coomassie Blue. These results were originally published in *Biochemical Journal* [60] © The Biochemical Society

caspases and to understand the impact of specific cleavage events on caspase activity and specificity. Indeed, although it was known for years that the initiator caspase-9 does not require cleavage to display full enzymatic activity [37], other initiators are more stable when cleaved [41]. Cleavage may also increase the activity of initiator caspases without affecting their substrate specificity in vitro [29]. Examples of caspase-7 and caspase-9 cleavage-site mutants purified as described in this chapter are presented in Fig. 6.

Expression of wild-type zymogen forms of executioner caspase-3 and caspase-7 can be useful to study activation of these peptidases by initiator caspases [35, 38, 42, 43]. Because of the intrinsic activity of initiator caspases and their propensity to dimerize and become active, it is impossible to prepare uncleaved zymogen forms of these caspases without resorting to cleavage-site mutants (Fig. 6a).

Several other caspase mutants that result in the production of particular molecular forms have been described in the literature. These include mutations that prevent dimerization of caspase-8 [30], force caspase-9 dimerization [44], allow the activity of the zymogen form of executioner caspases [45], and produce caspase chimeras [31, 32]. However, it must be stressed that although these mutations were made to affect specific properties of caspases, they may also alter the kinetic properties of the enzyme in vitro and in cells.

4 Notes

1. Black plates are designed to reduce well-to-well cross talk and background for fluorescence assays, whereas white plates are designed to reduce well-to-well cross talk and background for luminescence assays. White plates show higher fluorescence background and amplify the fluorescence signal. Thus, more accurate data can be obtained using black plates.
2. Stock solutions of peptidic substrates are prepared at 20 mM to limit the final concentration of DMSO in enzymatic assays. DMSO above 3 % can alter the catalytic activity of caspases. *See* Subheading 3.2.1.1 for the preparation of Afc substrate solutions.
3. Protease inhibitors can be added fresh to the lysis solution. However, it is important that none of them affects the caspase activity; thus, all-in-one tablets are not recommended. General non-caspase protease inhibitors one can use are: 1,10-orthophenanthroline (1 mM), 1 mM EDTA (already in the executioner caspase buffer), E-64 (10 μM), leupeptin (10 μM), 3,4-dichloroisocoumarin (10 μM), and MG-132 (1 μM).
4. The purity of the imidazole is very important. It must have a low absorbance at 280 nm. Imidazole meeting ACS specifications is not suitable and will produce significant background absorbance. The reagent needs to have 0.005 % β-NADH equivalent as a fluorescence blank (e.g., Sigma cat. no. I-0250). Some lower-grade imidazole alters the caspase activity.
5. The pET vectors are either ampicillin- (pET-15b/23b(+)) or kanamycin- (pET-28b(+)) resistant, whereas the p*LysS* plasmid confers resistance to chloramphenicol. Always use a freshly transformed colony (1–2 days) to initiate the culture. Do not keep cDNA in BL21 strains because of potential stability issues.
6. Sufficient yields of most caspases are obtained from BL21(DE3) p*LysS* bacteria grown in rich medium such as 2× TY. However, higher yields can be obtained using a richer medium (supplemented with 5 % glycerol or 2 % glucose) or a buffered medium (Terrific Broth/TB medium). Those media generally result in higher bacterial density, thus increasing yields. However, 2× TY is much simpler to prepare and can be fully sterilized by autoclaving. It is recommended to use baffled flasks instead of regular flasks because the former provide better aeration and can be filled up to half the volume. Pre-heat the bacterial medium at 37 °C before starting your expression culture. This step ensures that the optimal bacterial density is reached more quickly.
7. Not all caspases express well under these conditions. For instance, more incubation time must be allowed for caspase-6

and caspase-7 to self-process fully as compared to caspase-3. Additionally, better yields are obtained using a slower expression protocol (less IPTG) for caspase-2 and caspase-6. Refer to Table 1 for optimal temperature, IPTG concentration and

Table 1
Caspase expression construct

Caspases	Residues	Expression vector	Histidine tag	Expression time (h) IPTG (mM) temp. (°C)	UniProt reference
Caspase-2 truncated (no CARD)	159–452	pET-23b(+)	C-terminal LEHHHHHH	18 h 0.02 mM 16 °C	P42575
Caspase-3 full-length	1–277	pET-23b(+)	C-terminal LEHHHHHH	5 h 0.2 mM 30 °C	P42574
Caspase-3 zymogen	1–277	pET-23b(+)	C-terminal LEHHHHHH	0.3–0.5 h 0.2 mM 30 °C	P42574
Caspase-6 full-length	1–293	pET-23b(+)	C-terminal LEHHHHHH	18 h 0.02 mM 30 °C	P55212
Caspase-7 full-length	1–303	pET-23b(+) pET-28b(+)	C-terminal LEHHHHHH	8 h 0.2 mM 30 °C	P97864
Caspase-7 zymogen	1–303	pET-23b(+)	C-terminal LEHHHHHH	0.5 h 0.2 mM 30 °C	P97864
Caspase-8 full-length (-YFP/CFP)	1–479	pET28b(+)	C-terminal HHHHHH	5 h 0.4 mM 37 °C	Q14790
Caspase-8 truncated (no DED)	216–479	pET-15b	N-terminal HHHHHH	6 h 0.2 mM 30 °C	Q14790
Caspase-9 full-length	1–416	pET-15b	C-terminal LEHHHHHH	5 h 0.2 mM 30 °C	P55211
Caspase-9 truncated (no CARD)	139–419	pET-15b	C-terminal LEHHHHHH	5 h 0.2 mM 30 °C	P55211
Caspase-10 truncated (no DED)	202–521	pET-15b	N-terminal HHHHHH	5 h[a] 0.2 mM 30 °C	Q92851

[a]The expression time is based on full-length caspase-8, not on experimental data

expression time for various caspases. Be consistent with the expression time. A longer or a shorter expression period can influence the activation and activity of the caspase and can consequently affect results.

8. Even if proteins are purified immediately, a freeze–thaw cycle will assist bacterial lysis.
9. The efficacy of lysis is somewhat difficult to assess. The lysate will initially turn gooey as the bacterial DNA is released. Sonication will break the DNA over time. The lysate will turn from a thick, whole milk-like to a skim milk-like appearance. For very large expressions, lysozyme can be added to a final concentration of 1 mg/mL. Note that lysozyme is inhibited by indole derivatives such as imidazole [46]. Therefore, do not use a lysis buffer that includes imidazole, which is often used to prevent binding of unwanted proteins on the IMAC resin.
10. The amount of resin varies based on the anticipated yield. As a rule of thumb, 1 mL of resin will bind ~5 mg of caspase. It is recommended to use just enough resin to bind the caspase and not much more. However, if yield is low (<100 μg) and too little purification resin is used, the *E. coli* SlyD protein, a histidine-rich peptidyl prolyl *cis-trans* isomerase (UniProt identification number P0A9K9) will compete with the His-tagged protein for the resin. Consequently, do not use too little resin (<0.5 mL). If pre-charged resin is used, omit **step 4**.
11. Lysate filtration helps to obtain a purer caspase preparation by removing large floating debris. Durapore HV membranes have a low protein-binding membrane, thus permitting more lysate to be filtered before clogging.
12. If no gradient maker is available, step elution can be used (0 %, 10 %, 20 %, etc. ratio of elution buffer/bacterial lysis buffer). However, gradient elution results in a preparation with a higher degree of purity. A fast protein liquid chromatography (FPLC) system can also be used. However, each run usually takes more time than manual purification. Furthermore, the utilization of an FPLC system does not allow for tailoring of the amount of resin used and requires more buffer, which increases cost; for example, high-grade imidazole is relatively expensive.
13. To adequately visualize caspase subunits, the SDS-PAGE gel system used must be carefully chosen. This is especially true for caspases because the small subunit of caspases has an apparent mobility of ~10 kDa. We recommend the use of gradient acrylamide gels using the ammediol buffer system, which is particularly suited for the resolution of small proteins [47]. SDS-PAGE analysis of the active caspase preparation will show two proteins: a large subunit (~20 kDa plus N-terminal domain) and a

small subunit (~10–12 kDa). The generation of those subunits is the result of the auto-catalytic activation of caspases during overexpression in *E. coli* but does necessarily reflect the normal activation mechanism of caspases in eukaryotic cells. With some caspase mutants, this process may not be complete. Usually, increasing the expression time will resolve this problem.

14. The Edelhoch relationship uses the observed absorbance of the tryptophan, tyrosine and reduced/oxidized cysteine residue content of a protein as a basis for calculating the extinction coefficient [48]. The ProtParam algorithm provides the estimated extinction coefficient and is available online at http://www.expasy.org. The value obtained is used to calculate the initial enzyme concentration for active site titration.
15. Do not refreeze caspase aliquots. Like many enzymes, caspases do not appreciate freeze–thaw cycles. The caspase preparation will be stable for months at −80 °C. However, caspase-6 has a shorter shelf life and should be active site-titrated every time. We highly suggest performing active site titration every month or prior to use.
16. From this step and until further notice, it is important to proceed at room temperature to avoid guanidine and urea crystallization and full-length caspase-8 precipitation.
17. Caspase-8 prepared using the above protocol is unstable and must be analyzed immediately. The more quickly this analysis is performed, the higher the active site titer of the preparation will be; this will also be reflected in the kinetic parameter, k_{cat}.
18. Because this text is oversimplified, it is strongly recommended that readers consult enzymology textbooks for a more thorough discussion of enzyme kinetics.
19. As an alternative to Afc fluorophores, 7-amino-4-methylcoumarin (Amc; $\varepsilon_{354\,nm} = 17{,}800\ M^{-1}\,cm^{-1}$; EXλ = 380 nm and EMλ = 460 nm)-based fluorogenic substrates can be used. Amc fluorophore has a smaller Stokes' shift (separation between excitation and emission spectra) upon cleavage than Afc. Chromogenic substrates based on *p*-nitroaniline (pNA; $\varepsilon_{410\ nm} = 8{,}800\ M^{-1}\ cm^{-1}$) could be used, but because their dynamic range of absorbance is much smaller than that of most fluorophores, pNA-based substrates are less useful. Prepare Amc and pNA standard solutions in DMSO.
20. The buffer used will alter the emission and excitation intensities, a phenomenon called quenching. Therefore, the plate reader must be calibrated for every buffer (e.g., executioner vs. initiator caspase buffer). It is also important that the volume of the sample is exactly the same as the one used in enzymatic assays.

21. LED-based systems are more stable and should be calibrated every 6 months, whereas lamp-based instruments should be calibrated every other month. Recalibrate every time a performance variation is suspected.
22. After this period, all available Afc substrate should be cleaved. Although caspase-3 may not be the best enzyme to hydrolyze a specific substrate, it is the most reliable caspase available, and excess enzyme will compensate for a less optimal substrate. Other caspases can be substituted, but the reaction must be carefully monitored to ensure complete substrate hydrolysis (monitored in **step 3**).
23. It is important that the reaction is complete when the absorbance is measured. Therefore, if the absorbance is not stable after 1 h, it is recommended to incubate the reaction for an additional 30 min and read the absorbance again. This procedure works because the executioner caspase buffer does not significantly affect the absorbance of Afc. If the buffer is modified, one must calibrate the substrate by comparing the 10 μM hydrolyzed substrate sample (**step 2**) to dilutions of Afc standard (Subheading 3.2.1.1) in the same buffer.
24. Several techniques can be used to generate a serial dilution of a reagent. However, although it may not be the most accurate way of doing so, the following approach works fine. Fifty microliters of 1× caspase buffer is added to wells #2–16, 150 μL of 2 μM Z-VAD-fmk is added to well #1, and 100 μL is transferred from well to well (wells #1–15) by repeatedly pipetting up and down to mix. Use either executioner or initiator caspase buffer throughout this procedure.
25. The final concentration of caspase that is used to determine the uninhibited enzyme varies based on the intrinsic activity of the enzyme: 1–2 nM caspase-3; 2–5 nM caspase-2, caspase-6, and caspase-7; and 10–20 nM caspase-8, caspase-9, and caspase-10. It is important to note that mutant caspases may display enzymatic activity different from that of the wild-type enzyme. Thus, the volume of enzyme transferred must be tailored based on the caspase.
26. We used a different substrate for each caspase (Table 2).
27. Continuous reading is preferred to end point kinetics. Continuous reading allows the accumulation of several data points at early times in the assay that will provide a better estimation of the initial rate. Continuous reading also allows for the identification of potential substrate depletion, initial issues with temperature reaction equilibration, and unusual behavior such as sample precipitation.
28. Do not include values that show >90 % inhibition for the calculation of the titer, as these values may cause an overestimation of caspase concentration. It is important that the data

Table 2
Caspase internal cleavage site and specificity

Molecular form expressed	Internal cleavage site ($P_4P_3P_2P_1$)[a]	Substrate preference ($P_4P_3P_2P_1$)	Synthetic substrate used ($P_4P_3P_2P_1$)
Caspase-2	DQQD (linker)	(V)DEHD	VDVAD[b]
Caspase-3	ESMD (N-terminal domain) IETD (linker)	DEVD	DEVD
Caspase-6	DVVD (linker) TEVD (linker) TETD (N-terminal domain)	VEHD	VEID DEVD
Caspase-7	DSVD (N-terminal domain) IQAD (linker) NDTD (linker)	DEVD	DEVD
Caspase-8	VETD (linker) LEMD (linker) REQD (N-terminal domain)	(I/L)ETD	IETD VEID DEVD
Caspase-9	PEPD (linker) DQLD (linker)	(I/L)EHD	LEHD IETD
Caspase-10	IEAD (linker) SQTD (N-terminal domain)	LEHD	LEHD DEVD

[a]Internal cleavage sites occurring between the N-terminal domain and the catalytic site or within the linker that separates the large and small subunit of the catalytic domain are indicated. All those cleavages can be observed during overexpression in *E. coli*, but not always during apoptosis

[b]The most complete studies used a tetrapeptide library to establish caspase-2 peptidic substrate preference [4]. However, others have shown that caspase-2 has an extended substrate-binding pocket, which explains why this caspase is more active on pentapeptides [25]

used for linear regression result in a good correlation factor ($r^2 > 0.98$). Most caspase preparations will have a titer that is slightly below (75–95 %) the protein concentration estimated using the Edelhoch method. A titer that is higher often reveals inefficient inhibition by Z-VAD-fmk. In this case, inhibition can be prolonged for 1 h, and/or caspase and Z-VAD-fmk concentrations can be proportionally increased. Nevertheless, Z-VAD-fmk may not be potent enough to react effectively with some caspase mutants. In this case, we recommend titration using the baculoviral protein p35. This pan-caspase inhibitor is much more potent at inhibiting caspases [49, 50]. p35 is expressed as a C-terminal His-tagged protein and purified using the same procedure as the one described for caspase-3 (Subheading 3.1.1). However, because it is a protein, p35 must be titrated before using it as a titrant. A titration reaction is set up with a wild-type caspase-3 preparation that has already been titrated using Z-VAD-fmk. In this case, the *x*-axis intercept ($y = 0$) corresponds to the concentration of p35 that is required to completely inhibit caspase-3. For example, if the titration was performed using 100 nM caspase-3, an intercept at 500 nM means that only 20 % of the p35 preparation is active, which is a reasonable value. Thus, the value of active p35 is 20 % of the concentration that has been determined using the Edelhoch method.

29. Because enzyme addition to the reaction takes some time, more reliable data are obtained if no more than two series of assays are performed simultaneously. However, one can determine the kinetic parameters for multiple substrates simultaneously if an automatic injector is used to add the enzyme and reading can be programmed to alternate injection and reading.

30. To set up the 3/4 serial dilution: add 50 μL of 1× caspase buffer to wells #2–16, add 200 μL of 600 μM substrate to well #1, and transfer 150 μL from well to well (wells #1–16) by repeatedly pipetting up and down to mix at every transfer.

31. Make sure that the final DMSO concentration is not >3 %. DMSO will negatively affect caspase activity at high substrate concentration and consequently artificially alter K_M and k_{cat} values.

32. It is important that the data used for linear regression result in a very good correlation factor (r^2>0.98). Otherwise, this may indicate that mistakes have been made during the experiments or that the enzyme does not follow the classical Michaelis–Menten mechanism, for example through enzyme cooperativity, product inhibition, inhibition by DMSO. The latter often occurs for the 300 μM substrate sample. If this is the case, this data point can be omitted. Care must be taken to select the time interval used for parameter determination. Do not take all measurements, but only an early interval during which the data show constant rates (2–10 min). This instruction is given because later time point values underestimate caspase activity due to substrate depletion or enzyme fatigue (e.g., caspase-6).

    Make sure that the enzyme concentration, rates, and RFU to product concentration conversion number are in the same molar unit. For example, if $[E]$ is in nanomolar, convert rates to nM/s and RFU numbers to nM. Remember that k_{cat} is a number of molecules per second and not per minute.

    If K_M is high (>100 μM), the above procedure will result in less accurate K_M and k_{cat} values. This phenomenon arises because not enough assays are above K_M. Ideally, half the data points should be above K_M, and the best estimates are obtained with data that populate the inflection point of the Michaelis–Menten curve, between 0.1 and 10 times K_M. If K_M is too high or too low, it is recommended to repeat the experiment using a different range of substrate concentrations. Accordingly, it may be necessary to prepare substrate stocks that are more concentrated (e.g., 50 mM instead of 20 mM).

If K_M is too high (>200 μM), it is better to determine the k_{cat}/K_M value using pseudo-first order conditions. If $[S] \ll K_M$, Eq. 2 can be simplified to $v \approx V_{max}[S]/K_M$, which can be rearranged to $V_{max}/K_M \approx v/[S]$. If pseudo-first order conditions are met, a plot of rates against substrate concentrations gives a straight line with a slope equivalent to V_{max}/K_M, and $k_{cat}/K_M = V_{max}/K_M[E]$. As explained in the introduction of this section, k_{cat}/K_M is a useful value to compare enzymes. Because the obtained value is an approximation, it is more appropriate to refer to an apparent value, such as $k_{cat,app}/K_{M,app}$. If one cannot assume that pseudo-first order conditions are met, the value can be referred to as a rate constant.

33. An alternative protocol is to employ the mRIPA lysis buffer (50 mM Tris at pH 7.4, 100 mM NaCl, 1 % NP-40, 0.5 % deoxycholic acid, 0.1 % sodium dodecylsulfate (SDS), and 1 mM EDTA). This solution extracts more proteins, but not only cytosolic proteins.
34. Use initiator caspase buffer if working with initiator caspases, and incubate a 1 μM initiator caspase solution for 30 min at 37 °C before setting up the serial dilution. To prepare the 2/3 serial dilution, 25 μL of 1× caspase buffer is added to tubes #2–8, 75 μL of 200 nM of caspase is added to tube #1, and 50 μL is transferred from tube to tube (tubes #1–8) by repeatedly pipetting up and down to mix.
35. Initiator caspase buffer may cause the precipitation of lysate protein or recombinant proteins. It is recommended to test the solubility of the substrate in buffer before performing the assay. If precipitation occurs, centrifuge the tube and test the supernatant for the protein of interest. Diluting the substrate/lysate will reduce precipitation and help meet the pseudo-first order condition.
36. A more accurate estimation of k is obtained by calculating the value for all samples and averaging them. The values obtained should be roughly similar for a pseudo-first order condition. Alternatively, plot $\ln(1-p)/t$ against $[E]$ for all samples. The slope will be $-k$.
37. Because the efficacy of cleavage may vary greatly, it is likely that results will show full cleavage of the protein of interest or barely any cleavage. This situation is why it is suggested to start with a caspase concentration that is relatively high, yet physiological. The experiment is then repeated using a different and narrower range of caspase concentrations and a different reaction time to more accurately determine k_{cat}/K_M. Whenever feasible, do not use concentrations of caspase that are too high (>100 nM), and lower the amount of substrates in the assay. The lower the substrate concentration is, the more likely the reaction will occur in a pseudo-first order condition. If samples

are too diluted for analysis, set up larger sample volumes and precipitate proteins with 10 % trichloroacetic acid (TCA) for 2 h at −20 °C. The proteins are then recovered by centrifugation for 30 min at 18,000 × g at 4 °C and washed twice with acetone before solubilizing the proteins in 1× SDS-PAGE loading buffer.

5 Discussion

The protocols described in this chapter permit the expression, purification, and basic characterization of apoptotic caspases. The techniques and instrumentation that are necessary to perform these procedures are available to most laboratories. Therefore, it is relatively easy to biochemically assess most caspase mutants to ensure that enzymatic properties are not significantly altered prior to performing cellular studies.

Caspases are versatile enough to allow the production of various molecular forms. Indeed, by using cleavage site mutants, varying expression conditions, and using biochemical tricks (e.g., kosmotropic salts), all known molecular forms of caspases that occur in cells can be produced. It has been suggested recently that cleavage of the initiator caspase-8 regulates proteolysis of key substrates, but not during necroptosis [51]. Thus, the ability to produce active but uncleaved full-length caspase-8 will be a valuable tool to characterize these two molecular forms of initiator caspase. Further protein engineering has allowed the "on-demand" activation of initiator caspases in cells using chemically induced dimerization (CID) domains [36, 51–53].

The linker that separates the large and small caspase subunits is very flexible and is thus amenable to modification. Methods to activate executioner caspases that bypass initiator caspases have been developed in which an exogenous peptidase is used to cleave the caspase. The tobacco etch virus (TEV) protease is a good candidate for this type of experiment, as this peptidase is highly specific and has no substrates in human cells. The approach uses exogenously expressed TEV protease to cleave an engineered cleavage site in the linker of the catalytic domain [52, 54]. However, experimenters must be careful with the location used to introduce the cleavage site because the linker has specific determinants that are involved in stabilization of the active caspase form [38, 55]. This approach has also been used to study the specificity of executioner caspases in cells [54]. Other peptidases, such as thrombin, have been proposed as exogenous peptidases to cleave caspases [41], but their cellular utilization must be carefully assessed.

If the remnants of an engineered cleavage site are potentially detrimental, the large and the small caspase subunits can be expressed separately on the same vector or by co-expression.

In *E. coli*, a fraction of the protein subunits will reassemble into a fully cleaved dimer. This approach can be particularly useful to obtain fully processed but inactive or poorly active caspase mutants.

More elaborate expression schemes have been used to generate caspase dimers with different catalytic domains [35]. These heterodimeric caspases were produced by co-expressing two different caspase-7 molecules, each carrying a different C-terminal purification tag, for example, a His-tag on one caspase-7 molecule and a FLAG-tag on the other. Briefly, after IMAC, the purified caspase was purified on anti-FLAG-coupled IgG beads and eluted with free FLAG peptide. The resulting product is a true hybrid dimer of caspase-7. This method allows for the detailed characterization of an executioner caspase [35].

Recently, an engineered caspase-7 that is sensitive to redox conditions has been described [56]. In oxidizing conditions, two substrate-binding loops remain linked by a disulfide bound between two designed cysteine residues, inactivating the caspase. In reducing condition, the disulfide bond is broken, and the caspase becomes fully active.

Another recent study reports the presence of additional determinants on caspases involved in substrate cleavage [32]. Those determinants, called exosites, are located outside the substrate-binding pocket and add an additional layer of specificity to the enzyme. Currently, only one exosite has been characterized, but much evidence in the literature suggests the presence of other exosites. Allosteric sites also exist in caspases [57–59]. The identification of exosites and allosteric sites brings opportunities to develop more specific caspase inhibitors.

In conclusion, the expression and purification of caspases have been useful since the time these enzymes were discovered more than 20 years ago and will remain a valuable and necessary tool to study the many processes in which caspases are involved.

References

1. Cerretti DP et al (1992) Molecular cloning of the interleukin-1β converting enzyme. Science 256:97–100
2. Thornberry NA et al (1992) A novel heterodimeric cysteine protease is required for interleukin-1beta processing in monocytes. Nature 356:768–774
3. Yi CH, Yuan J (2009) The Jekyll and Hyde functions of caspases. Dev Cell 16:21–34
4. Thornberry NA et al (1997) A combinatorial approach defines specificities of members of the caspase family and granzyme B. Functional relationships established for key mediators of apoptosis. J Biol Chem 272: 17907–17911
5. Stennicke HR, Renatus M, Meldal M, Salvesen GS (2000) Internally quenched fluorescent peptide substrates disclose the subsite preferences of human caspases 1, 3, 6, 7 and 8. Biochem J 350:563–568
6. Lavrik IN, Golks A, Krammer PH (2005) Caspases: pharmacological manipulation of cell death. J Clin Invest 115:2665–2672
7. Garcia-Calvo M et al (1998) Inhibition of human caspases by peptide-based and macromolecular inhibitors. J Biol Chem 273:32608–32613
8. Fuentes-Prior P, Salvesen GS (2004) The protein structures that shape caspase activity, specificity, activation and inhibition. Biochem J 384: 201–232

9. Agard NJ, Wells JA (2009) Methods for the proteomic identification of protease substrates. Curr Opin Chem Biol 13:503–509
10. Demon D et al (2009) Caspase substrates: easily caught in deep waters? Trends Biotechnol 27:680–688
11. Luthi AU, Martin SJ (2007) The CASBAH: a searchable database of caspase substrates. Cell Death Differ 14:641–650
12. Igarashi Y et al (2007) CutDB: a proteolytic event database. Nucleic Acids Res 35: D546–D549
13. Timmer JC et al (2009) Structural and kinetic determinants of protease substrates. Nat Struct Mol Biol 16:1101–1108
14. Muppidi JR et al (2006) Homotypic FADD interactions through a conserved RXDLL motif are required for death receptor-induced apoptosis. Cell Death Differ 13:1641–1650
15. Bouchier-Hayes L et al (2009) Characterization of cytoplasmic caspase-2 activation by induced proximity. Mol cell 35:830–840
16. Vegran F, Boidot R, Solary E, Lizard-Nacol S (2011) A short caspase-3 isoform inhibits chemotherapy-induced apoptosis by blocking apoptosome assembly. PLoS One 6:e29058
17. Kamada S, Kikkawa U, Tsujimoto Y, Hunter T (2005) Nuclear translocation of caspase-3 is dependent on its proteolytic activation and recognition of a substrate-like protein(s). J Biol Chem 280:857–860
18. Beaudouin J, Liesche C, Aschenbrenner S, Horner M, Eils R (2013) Caspase-8 cleaves its substrates from the plasma membrane upon CD95-induced apoptosis. Cell Death Differ 20:599–610
19. Arakawa T et al (2007) Suppression of protein interactions by arginine: a proposed mechanism of the arginine effects. Biophys Chem 127:1–8
20. Baynes BM, Wang DI, Trout BL (2005) Role of arginine in the stabilization of proteins against aggregation. Biochemistry 44:4919–4925
21. Arakawa T, Tsumoto K (2003) The effects of arginine on refolding of aggregated proteins: not facilitate refolding, but suppress aggregation. Biochem Biophys Res Commun 304: 148–152
22. Reddy KR, Lilie H, Rudolph R, Lange C (2005) L-Arginine increases the solubility of unfolded species of hen egg white lysozyme. Protein Sci 14:929–935
23. McStay GP, Salvesen GS, Green DR (2008) Overlapping cleavage motif selectivity of caspases: implications for analysis of apoptotic pathways. Cell Death Differ 15:322–331
24. Schecter I, Berger M (1967) On the size of the active site in proteases. Biochem Biophys Res Commun 27:157–162
25. Talanian RV et al (1997) Substrate specificities of caspase family proteases. J Biol Chem 272: 9677–9682
26. Mace PD, Riedl SJ (2010) Molecular cell death platforms and assemblies. Curr Opin Cell Biol 22:828–836
27. Renatus M, Stennicke HR, Scott FL, Liddington RC, Salvesen GS (2001) Dimer formation drives the activation of the cell death protease caspase 9. Proc Natl Acad Sci U S A 98:14250–14255
28. Baliga BC, Read SH, Kumar S (2004) The biochemical mechanism of caspase-2 activation. Cell Death Differ 11:1234–1241
29. Wachmann K et al (2010) Activation and specificity of human caspase-10. Biochemistry 49:8307–8315
30. Boatright KM et al (2003) A unified model for apical caspase activation. Mol Cell 11:529–541
31. Pop C, Timmer J, Sperandio S, Salvesen GS (2006) The apoptosome activates caspase-9 by dimerization. Mol Cell 22:269–275
32. Boucher D, Blais V, Denault JB (2012) Caspase-7 uses an exosite to promote poly(ADP ribose) polymerase 1 proteolysis. Proc Natl Acad Sci U S A 109:5669–5674
33. Janicke RU, Sprengart ML, Wati MR, Porter AG (1998) Caspase-3 is required for DNA fragmentation and morphological changes associated with apoptosis. J Biol Chem 273: 9357–9360
34. Soule HD, Vazguez J, Long A, Albert S, Brennan M (1973) A human cell line from a pleural effusion derived from a breast carcinoma. J Natl Cancer Inst 51:1409–1416
35. Denault JB et al (2006) Engineered hybrid dimers: tracking the activation pathway of caspase-7. Mol Cell 23:523–533
36. Pop C et al (2011) FLIPL induces caspase 8 activity in the absence of interdomain caspase 8 cleavage and alters substrate specificity. Biochem J 433:447–457
37. Stennicke HR et al (1999) Caspase-9 can be activated without proteolytic processing. J Biol Chem 274:8359–8362
38. Boucher D, Blais V, Drag M, Denault JB (2011) Molecular determinants involved in activation of caspase 7. Biosci Rep 31:283–294
39. Araya R, Takahashi R, Nomura Y (2002) Yeast two-hybrid screening using constitutive-active caspase-7 as bait in the identification of PA28gamma as an effector caspase substrate. Cell Death Differ 9:322–328

40. Stennicke HR, Salvesen GS (1999) Catalytic properties of the caspases. Cell Death Differ 6:1054–1059
41. Pop C, Fitzgerald P, Green DR, Salvesen GS (2007) Role of proteolysis in caspase-8 activation and stabilization. Biochemistry 46:4398–4407
42. Zhou Q, Salvesen GS (1997) Activation of pro-caspase-7 by serine proteases includes a non-canonical specificity. Biochem J 324:361–364
43. Stennicke HR et al (1998) Pro-caspase-3 is a major physiologic target of caspase-8. J Biol Chem 273:27084–27090
44. Chao Y et al (2005) Engineering a dimeric caspase-9: a re-evaluation of the induced proximity model for caspase activation. PLoS Biol 3:e183
45. Walters J et al (2009) A constitutively active and uninhibitable caspase-3 zymogen efficiently induces apoptosis. Biochem J 424:335–345
46. Swan ID (1972) The inhibition of hen egg-white lysozyme by imidazole and indole derivatives. J Mol Biol 65:59–62
47. Bury A (1981) Analysis of protein and peptide mixtures: evaluation of three sodium dodecyl sulphate-polyacrylamide gel electrophoresis buffer systems. J Chromatogr 213:491–500
48. Edelhoch H (1967) Spectroscopic determination of tryptophan and tyrosine in proteins. Biochemistry 6:1948–1954
49. Zhou Q, Salvesen GS (2000) Viral caspase inhibitors CrmA and p35. Methods Enzymol 322:143–154
50. Zhou Q et al (1998) Interaction of the baculovirus anti-apoptotic protein p35 with caspases: specificity, kinetics, and characterization of the caspase/p35 complex. Biochemistry 37:10757–10765
51. Oberst A et al (2011) Catalytic activity of the caspase-8-FLIP(L) complex inhibits RIPK3-dependent necrosis. Nature 471:363–367
52. Oberst A et al (2010) Inducible dimerization and inducible cleavage reveal a requirement for both processes in caspase-8 activation. J Biol Chem 285:16632–16642
53. van Raam BJ, Ehrnhoefer DE, Hayden MR, Salvesen GS (2013) Intrinsic cleavage of receptor-interacting protein kinase-1 by caspase-6. Cell Death Differ 20:86–96
54. Gray DC, Mahrus S, Wells JA (2010) Activation of specific apoptotic caspases with an engineered small-molecule-activated protease. Cell 142:637–646
55. Witkowski WA, Hardy JA (2009) L2′ loop is critical for caspase-7 active site formation. Protein Sci 18:1459–1468
56. Witkowski WA, Hardy JA (2011) A designed redox-controlled caspase. Proc Natl Acad Sci U S A 20(8):1421–1431
57. Hardy JA, Lam J, Nguyen JT, O'Brien T, Wells JA (2004) Discovery of an allosteric site in the caspases. Proc Natl Acad Sci U S A 101:12461–12466
58. Scheer JM, Romanowski MJ, Wells JA (2006) A common allosteric site and mechanism in caspases. Proc Natl Acad Sci U S A 103:7595–7600
59. Velazquez-Delgado EM, Hardy JA (2012) Zinc-mediated allosteric inhibition of caspase-6. J Biol Chem 287:36000–36011
60. Denault JB, Eckelman BP, Shin H, Pop C, Salvesen GS (2007) Caspase 3 attenuates XIAP (X-linked inhibitor of apoptosis protein)-mediated inhibition of caspase 9. Biochem J 405:11–19
61. Boucher D, Blais V, Denault JB (2012) Caspase-7 uses an exosite to promote poly(ADP ribose) polymerase 1 proteolysis. Proc Natl Acad Sci U S A 109(15):5669–5674

Chapter 2

Positional Scanning Substrate Combinatorial Library (PS-SCL) Approach to Define Caspase Substrate Specificity

Marcin Poręba, Aleksandra Szalek, Paulina Kasperkiewicz, and Marcin Drąg

Abstract

Positional scanning substrate combinatorial library (PS-SCL) is a powerful tool for studying substrate specificity of proteolytic enzymes. Here, we describe the protocol for analyzing S4-S2 pockets preferences of caspases using PS-SCL. Additionally, we describe procedures for the identification of optimal substrates sequence after PS-SCL, solid phase synthesis, and purification of selected fluorogenic substrates, as well as their kinetic analysis.

Key words Substrate specificity, Caspase, Fluorogenic substrate, Combinatorial library, Cysteine protease

1 Introduction

The accurate knowledge of substrate specificity of an enzyme provides important information about its functions and aids in the understanding of molecular pathways [1]. Moreover, this knowledge is invaluable in the development of potent and specific inhibitors and activity-based probes. One of the best understood family of proteases are caspases, which are involved in apoptosis regulation and inflammatory response [2–4]. Most caspases cleave substrates after aspartic acid residue in the P1 position, and amino acids after aspartate (P2, P3, P4…) what collectively determines their substrate specificity [5].

At present, there are several strategies to identify the substrate specificity of proteolytic enzymes [6]. Positional scanning substrate combinatorial library (PS-SCL) is one of the most reliable and powerful tools in determining substrate-enzyme interaction in binding pockets around protease active site. In this method, libraries of substrate mixtures with conjugated fluorescent molecule are synthesized. Fluorescent molecules or fluorophores are fixed in

Peter V. Bozhkov and Guy Salvesen (eds.), *Caspases, Paracaspases, and Metacaspases: Methods and Protocols*, Methods in Molecular Biology, vol. 1133, DOI 10.1007/978-1-4939-0357-3_2, © Springer Science+Business Media New York 2014

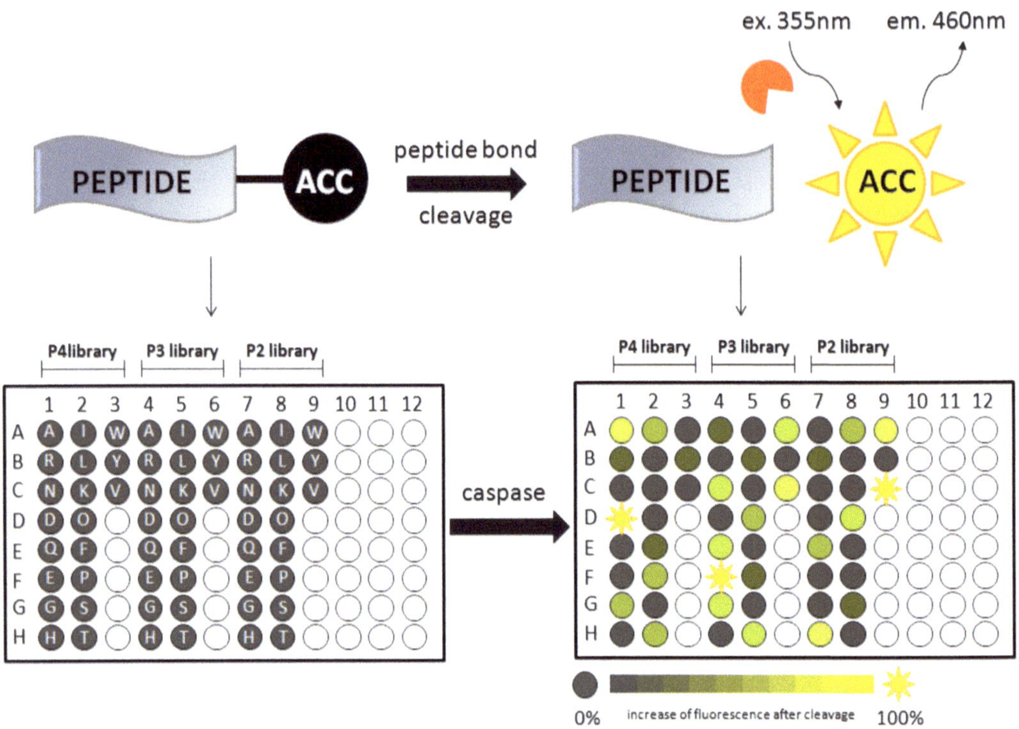

Fig. 1 General outline for caspase substrate specificity analysis using SCL screening with ACC as fluorescent reporter group

P1′ position. In intact substrates, fluorescence emission is either absent or very weak. Following protease hydrolysis, fluorophore is released and emits fluorescence when excited by an appropriate wavelength. Fluorescence can be quantified thereby providing the data on reaction kinetics. This method enables fast and reliable analysis of protease specificity. Noteworthy, apart from fluorophores, chromophores and luminophores can be also used as reporter groups. However, the sensitivity of each group differs, with luminophores being the most sensitive and chromophores possessing the lowest sensitivity. Fluorophores are the best choice of reporter group, since they are quite easy to synthesize and yield strong fluorescence in biological tests [6] (Fig. 1).

PS-SCL are composed of mixtures of peptidic substrates, which are divided into sublibraries. In each sublibrary, one or more positions are fixed with a defined amino acid, whereas the remaining positions contain equimolar concentrations of amino acids. It allows to establish the effect of the fixed amino acids independently. The same degree of substitution for each amino acid is necessary to obtain reliable results [6]. A procedure for the synthesis of equimolar mixture of natural amino acids on solid support was developed in 1994 by Ostresh et al. [7]. First, the level of substitution of the particular amino acids was established, which was followed by the identification of the exact composition of isokinetic mixture determining equimolar levels of substitution. The following Fmoc-amino acid composition was established (the numbers in parentheses are percentages of molar fraction): Fmoc-Ala-OH (3.4), Fmoc-Arg(Pbf)-OH (6.5), Fmoc-Asn(Trt)-OH (5.3), Fmoc-Asp(O-*t*-Bu)-OH (3.5), Fmoc-Glu(O-*t*-Bu)-OH (3.6), Fmoc-Gln(Trt)-OH (5.3), Fmoc-Gly-OH (2.9), Fmoc-His(Boc)-OH (3.5), Fmoc-Ile-OH (17.4), Fmoc-Leu-OH (4.9), Fmoc-Lys(Boc)-OH (6.2), Fmoc-Nle-OH (3.8), Fmoc-Phe-OH (2.5), Fmoc-Pro-OH (4.3), Fmoc-Ser(O-*t*-Bu)-OH (2.8), Fmoc-Thr(O-*t*-Bu)-OH (4.8), Fmoc-Trp(Boc)-OH (3.8), Fmoc-Tyr(O-*t*-Bu)-OH (4.1), and Fmoc-Val-OH (11.3). Nowadays, Ostresh procedure is commonly used during synthesis of SCL. Utilizing combinatorial chemistry techniques, the libraries can be synthesized quickly and efficiently.

The first broad study of caspase substrate specificity using PS-SCL was carried out in 1997 by Rano et al. [8] for interleukin-1β converting enzyme (ICE, caspase-1). In this study, three separate combinatorial sublibraries of tetrapeptides conjugated with 7-amino-4-methyl-coumarin (AMC) were synthesized. Each sublibrary contained 8,000 compounds. The P1 position was occupied by aspartic acid residue, what was in line with previous findings of a strong requirement for Asp in that position [9–11]. In each sublibrary, one position was fixed with a defined amino acid and the other contained equimolar concentrations of natural amino acids (cysteine was omitted and methionine was replaced by norleucine in order to avoid their oxidation), according to Ostresh et al. procedure [7]. This study has revealed that the WEHD sequence is the most favorable tetrapeptide recognition motif for caspase-1 [8]. The result was surprising as it was inconsistent with formerly found optimal sequence YVAD [12] and differed from caspase 1 cleavage site YVHD present in its natural substrate pro-IL-1β. Further investigations confirmed that WEHD is an optimal caspase 1 substrate and proved that PS-SCL is a trustworthy method [8].

In the same year, Thornberry group used an identical library to examine the substrate specificity of almost all members of caspase family [13]. It was a seminal study, which has resulted in a division of caspases into three main groups based on their substrate specificity. The P4 position played an important role in this classification. The first group constituted the inflammatory caspases (caspases 1, 4, and 5), which prefer the (W/L)EHD tetrapeptide sequences. Group II consisted of caspases 2, 3, and 7, all favoring DEXD (X indicates that several amino acids are tolerated in this position), and group III was composed of caspases 6, 8, and 9 showing specificity for (L/V)EXD. Soon after, Garcia-Calvo et al. [14] examined the substrate specificity of caspase 10 using the same method, and the enzyme was classified to the third group based on the optimal recognition sequence LEXD. Further studies conducted by Wachmann et al. [15] with a use of PS-SCL confirmed this finding and established LEHD as the best substrate for caspase-10. Mikolajczyk et al. [16] studied optimal substrates for caspase 14 employing the same type of library as Garcia-Calvo, but with AFC as a reporter group. It was found that the substrate specificity of caspase 14 matches that of group I.

All caspases show strong preference for aspartic acid and glutamic acid residues in the P1 and P3 positions, respectively. The library constructed by Thornberry et al. [13] can be used only for proteases with strong preferences for aspartic acid in the P1 position of their substrates. Later, some modifications to this method were published [17, 18]; however, it is still not easy to examine proteases with unknown P1 specificity using such a library.

In 2000, new strategy to synthesize fluorogenic libraries, which employed bifunctional fluorogenic group 7-amino-4-carbamoylmethylcoumarin (ACC) was reported [19]. In this method, fluorophore ACC is bound to the solid support and the amino acids are being directly attached one by one [20], significantly simplifying the synthesis of fluorogenic libraries. The method enables incorporating any amino acid in any position, and consequently, complete diversification of peptide libraries is possible. Moreover, the assay sensitivity is enhanced as ACC has higher fluorescent yield than AMC [19]. When caspase 3 substrate specificity was investigated using this approach [21], the results were identical to those obtained by Thornberry et al. [13] demonstrating that the type of fluorophore does not affect substrate specificity. Various proteases were examined with ACC-based libraries, including papain, bromelain, human cathepsins [22], kallikreins [23], human paracaspase MALT1 [24], and DUBs (deubiquitinating enzymes) [25].

In this chapter, we describe how to profile substrate specificity of caspases using PS-SCL. In addition, we present step by step the synthesis of individual tetrapeptide substrates with ACC fluorescent tag. Finally, we describe kinetic analysis of caspase fluorogenic substrates.

2 Materials

2.1 Caspase Profiling by SCL

1. Combinatorial library of tetrapeptides conjugated with fluorescent tag (AMC or ACC) and composed of the sublibraries (*see* Table 1):
   - P4 sublibrary: Ac-Aaa-Mix-Mix-Asp-ACC (or AMC).
   - P3 sublibrary: Ac-Mix-Aaa-Mix-Asp-ACC (or AMC).
   - P2 sublibrary: Ac-Mix-Mix-Aaa-Asp-ACC (or AMC).

   The sublibraries are dissolved in peptide grade DMSO to the final concentration of 5 mM (*see* **Notes 1, 2**).

Table 1
Complete tetrapeptide-ACC libraries for studying P4-P2 substrate specificity of caspases

Library	Structure	Number of sublibraries	Compounds
P4	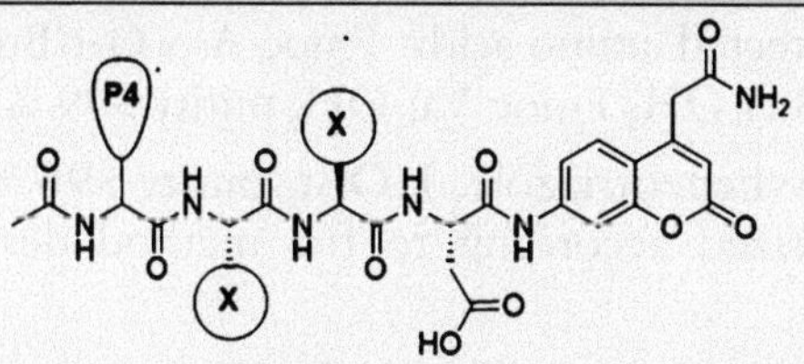	19	361
P3	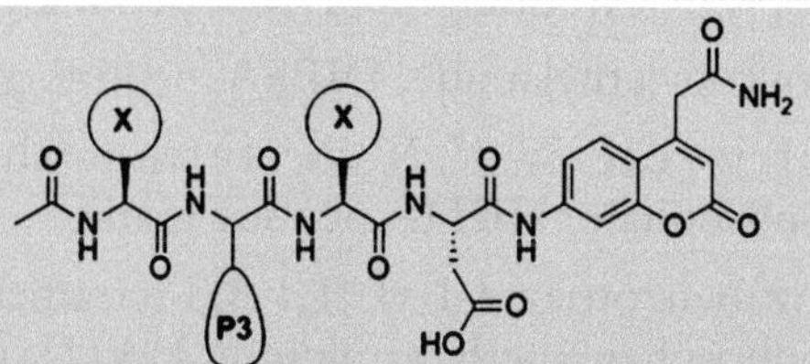	19	361
P2	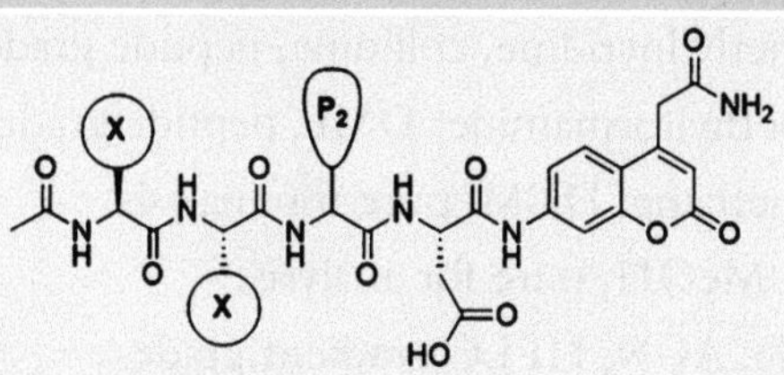	19	361

(X) Equimolar mixture of natural amino acids (cysteine was omitted and methionine was replaced by norleucine)

(Px) Fixed natural amino acid residue

2. Human caspases expressed according to standard procedures described in literature [8, 13, 26, 27] or purchased from commercial sources (*see* **Notes 3, 4**).
3. Standard caspase reaction buffer: 20 mM pipes, 100 mM NaCl, 10 % (w/v) sucrose, 10 mM DTT, 1 mM EDTA, 0.1 % (w/v) CHAPS, pH 7.2. Sometimes, caspase buffer can be supplemented with 0.7–0.75 M sodium citrate (kosmotropic salt) to increase caspase activity [28] (*see* **Notes 5, 6**).
4. Plate reader to monitor fluorescence upon substrate hydrolysis (e.g., Molecular Devices SpectraMax Gemini).
5. Round-bottom 96-well plates suitable for plate reader (e.g., Corning® 96-well plates, opaque bottom).
6. A set of single channel pipettes with different capacity and one multichannel pipette for delivering 8 × 100 μl.
7. 15 ml Falcon tubes for caspase incubation.
8. Reagent reservoir (50 ml size) for 8 channel pipettes.
9. 37°C incubator.
10. Vortex mixer.

2.2 Synthesis of ACC-Conjugated Tetrapeptide Substrates

1. Rink amide RA resin, particle size 200–300 mesh, loading 0.48 mmol/g (*see* **Note** 7).
2. 7-Fmoc-aminocoumarin-4-acetic acid, purity >98 %, commercial or synthesized according to the method described by Maly et al. [20].
3. Fmoc-protected amino acids: Fmoc-Asp(O-*t*-Bu)-OH, Fmoc-Glu(O-*t*-Bu)-OH, Fmoc-Val-OH, purity >99 %.
4. *N*-Hydroxybenzotriazole, HOBt, purity >98 %, commercial or synthesized according to the method described by Fu et al. [29].
5. *N,N′*-Diisopropylcarbodiimide, DICI, peptide grade.
6. *N,N′*-Diisopropylethylamine, DIPEA, peptide grade.
7. *O*-Benzotriazole-*N,N,N′,N′*-tetramethyluronium-hexafluorophosphate, HBTU, peptide grade.
8. 2-(1H-7-Azabenzotriazol-1-yl)-1,1,3,3-tetramethyl uranium hexafluorophosphate methanaminium, HATU, peptide grade.
9. 2,4,6-Trimethylpyridine, collidine, peptide grade.
10. *N,N′*-Dimethylformamide, DMF, peptide grade.
11. Dichloromethane, DCM, pure for analysis.
12. Methanol, MeOH, pure for analysis.
13. Acetonitrile, ACN, HPLC gradient grade.
14. Diethyl ether, Et_2O, pure for analysis.
15. Piperidine, PIP, purity >99 %.
16. Cleavage mixture: 1.9 ml trifluoroacetic acid (TFA, purity 99 %), 50 μl triisopropylsilane (TIPS, purity 99 %), and 50 μl H_2O.
17. Distilled H_2O.
18. Acetic acid, AcOH, purity >98 %.
19. Phosphorus pentoxide, P_2O_5, purity 98 %.
20. Vacuum line with trap.
21. Ten milliliter solid phase peptide synthesis vessel (e.g., Chemglass).
22. Microcentrifuge.
23. Shaker.
24. High-performance liquid chromatography system (HPLC).
25. Lyophilizer.

2.3 Kinetic Analysis (K_M, k_{cat}/K_M, k_{cat}) of Caspase Fluorogenic Substrates

1. Individual tetrapeptide caspase substrate (e.g., Ac-DEVD-ACC for caspase 3 assay) dissolved in peptide grade DMSO to the final concentration of 10, 25, or 50 mM (*see* **Notes 1, 2, 8**).
2. Other materials as described in section 2.1 under points 2–10.

3 Methods

3.1 Caspase Profiling by SCL

1. To ensure that the caspase of interest will display high enough activity during kinetic assay, each of caspase activity (P4-P2) should be tested in the initial screening. To perform the screening, prepare several samples in caspase buffer with different enzyme concentrations ranging from 1 to 500 nM (e.g., 1, 5, 25, 100, and 500 nM). Next, select several substrates from one sublibrary (start from P4) that are known or expected to be good caspase substrates.
2. Perform the initial caspase activity screening (*see* **Notes 1, 9–11**). Caspase should be preactivated in the assay buffer for 10 min at 37°C [28]. The combinatorial fluorogenic substrates tailored for caspases should be tested at the final concentration of 50 μM. If library substrates are dissolved in DMSO to the final concentration of 5 mM, vortex and spot 1 μl of selected substrates in each well of a 96-well plate and to each well add 99 μl of caspase-containing buffer starting from the lowest caspase concentration (e.g., from 1 to 500 nM) using multichannel pipette. For this caspase, samples with different concentrations should be placed in five separate reagent reservoirs. The general outline of the initial screening is presented in Fig. 2.
3. Monitor proteolytic reaction on a plate reader. Read fluorescence every 15 or 30 s for 15–60 min with excitation 355 nm and emission 460 nm (the overall time of the assay depends on the caspase activity).
4. Appropriate caspase concentration is when the best substrate gives 50–100 relative fluorescence units (RFU) per sec (RFU/s). Under these conditions even very poor caspase substrate (around 1 % of the best one) can be detected during

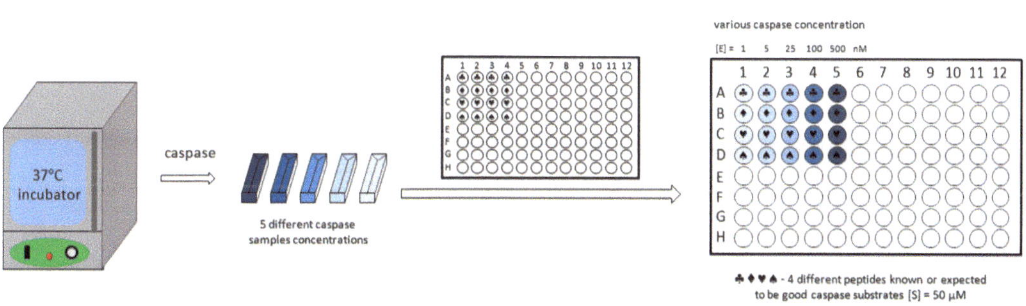

Fig. 2 The initial screening for determining optimal caspase concentration for a particular sublibrary screening. Several substrates from the combinatorial library have been selected, and their concentration was held constant (50 μM) during the whole assay. Caspase has been tested in five different concentrations (from 1 to 500 nM). Such a wide range of enzyme concentrations will enable the identification of the optimal enzyme concentration for each library assay. Characters *clubs*, *diamonds*, *hearts*, and *spades* represent four different substrates selected from the library

the assay (its RFU/s will be in the range 0.5–1.0). Substrates displaying RFU/s lower than 1 % of that of the best caspase substrate(s) cannot be considered to be caspase substrates. Note that RFU/s value depends on several factors, such as fluorescent tag, excitation and emission wavelengths, type of spectrofluorimeter, and software. The RFU/s values presented above were obtained with ACC tag, excitation 355 nm, emission 460 nm, Molecular Devices SpectraMax Gemini XPS, and Soft Max Pro 5 software.

5. Repeat the initial caspase activity screening for the P3 and P2 sublibraries separately (**steps 2–4**). In most cases all three sublibraries (P4, P3, and P2) can be screened with the same caspase concentration; however, there are some exceptions (e.g., caspase 9 screening).

6. After selecting the optimal caspase concentration for each sublibrary, perform the kinetic analysis of the whole CSL. Start with P4 sublibrary. One 96-well plate can accommodate four independent 19-membered sublibraries screenings (first set—columns 1–3, second set—columns 4–6, third set—columns 7–9, fourth set—columns 10–12). However, it is beneficial to perform only one experiment (one sublibrary) at a time. This is because pipetting takes a while and if we run too many columns in parallel, some very active substrates from the first columns might be already significantly hydrolyzed before we start an assay. To profile the caspase substrate specificity in P4 position, prepare the enzyme sample with the concentration established in **steps 2–4**: 3 columns × 8 wells × 99 μl = 2,378 μl (*see* **Note 12**). Incubate the enzyme sample in a 15 ml Falcon tube at 37 °C for 10 min (*see* **Note 11**).

7. In parallel, prepare the P4 sublibrary substrates. If the substrates are dissolved in DMSO to the final concentration of 5 mM, vortex and spot 1 μl of substrate into wells of the 96-well plate (columns 1–3). The pattern of spotting substrates from the P4 sublibrary on the 96-well plate is presented in Fig. 3.

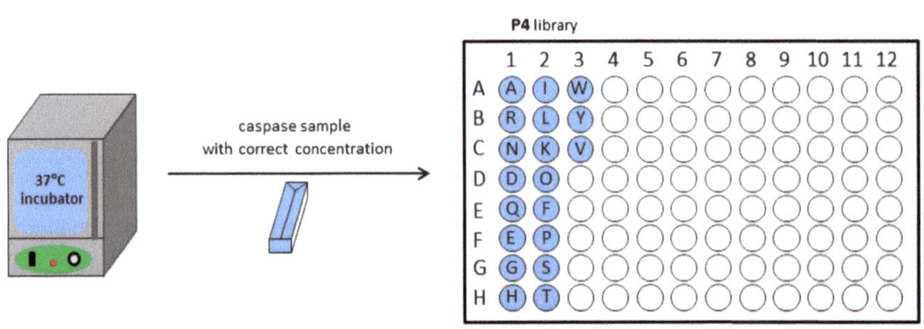

Fig. 3 The pattern of the P4 sublibrary substrates spotting on the 96-well plate. *Single letters* denote different amino acids

8. After 10 min of incubation, transfer the enzyme sample from the 15 ml Falcon tube to the reagent reservoir and add 99 μl in each well with substrates using 8-channel pipette. Mix well and run the experiment.
9. Monitor proteolytic reaction on a plate reader. Read fluorescence every 15 or 30 s for up to 60 min with excitation 355 nm and emission 460 nm. Stop reading when the very poor substrates will produce signal strong enough to be detected by the plate reader. For each substrate, extract only the linear part of the plot (*see* **Note 13**). In many cases it is possible to select the same time interval from hydrolysis curves for all substrates. However, sometimes each substrate hydrolysis should be considered separately, because the time range for linear plot differs among substrates and it is shorter for good substrates and longer for poor substrates. Figure 4 shows how to select the correct time range for various substrates.
10. To obtain the substrate specificity of a particular caspase in P4 position, analyze the data from kinetic assays. Find the best substrate with the highest RFU/s value and set this value as 100 %. Normalize cleavage rates of other substrates (in %) to that of the best substrate.
11. Repeat P4 sublibrary screening two more times (**steps 6–10**) and calculate the means and standard deviations for different substrates. If the standard deviation for any of the substrates is higher than 10 % of the mean, the assay must be repeated.
12. An alternative strategy to determine caspase substrate specificity is a simultaneous screening of P4, P3, and P2 sublibraries on a single 96-well plate. Note that for most caspases, the same enzyme concentration can be used for P4, P3, and P2 sublibraries profiling; however, there are some exceptions (e.g., caspase 9). The general idea of the approach with different enzyme concentrations is presented in Fig. 5. In any case, each experiment should be repeated at least two times.
13. To obtain substrate specificity profiles of particular caspase in P3 and P2 positions, repeat **steps 8–11**.
14. Select most active or specific amino acid in P4, P3, and P2 position (P1 is fixed with Asp) and proceed with synthesis of optimal tetrapeptide substrate (*see* **Note 14**) (Fig. 6).

3.2 Synthesis of ACC-Conjugated Tetrapeptide Substrates (See Fig. 7)

3.2.1 ACC-Resin Synthesis

1. Swell 0.048 mmol (100 mg) of Rink amide AM resin in a 5 ml DCM in a solid phase peptide synthesis vessel for 1 h with gentle stirring, once per 10 min to make functional group accessible [20] (*see* **Note 7**).
2. Remove DCM by vacuum filtration.
3. Wash the resin thoroughly with DMF, three times with 5 ml aliquots.

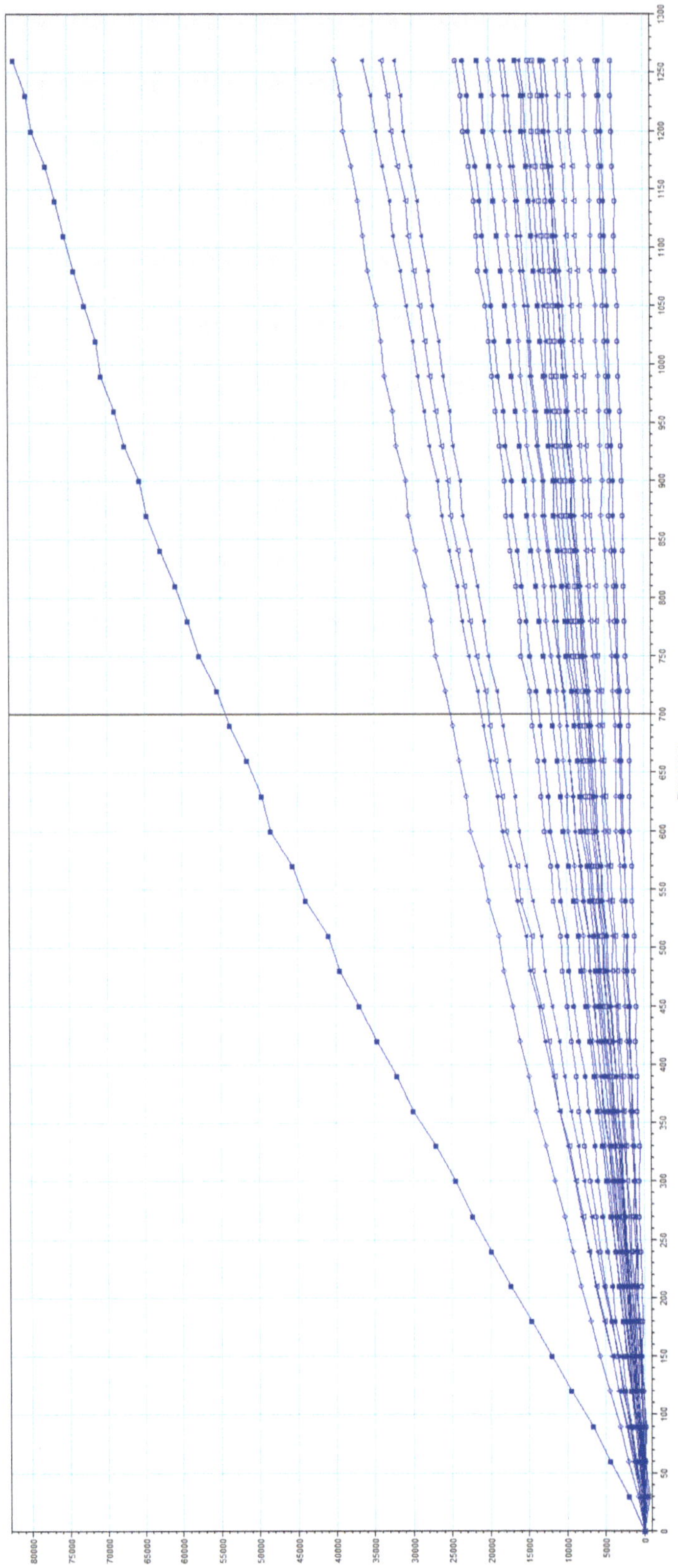

Fig. 4 The general scheme for how the linear part of the plot from library screening should be selected

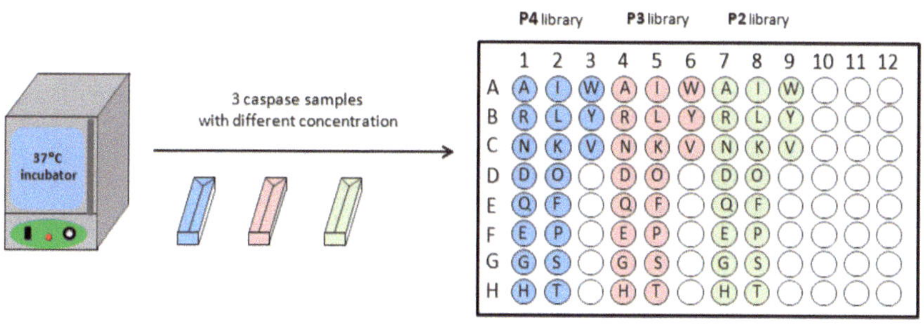

Fig. 5 The pattern of P4-P2 SCL spotting on the 96-well plate. This pattern is an alternative to the one presented in Fig. 3

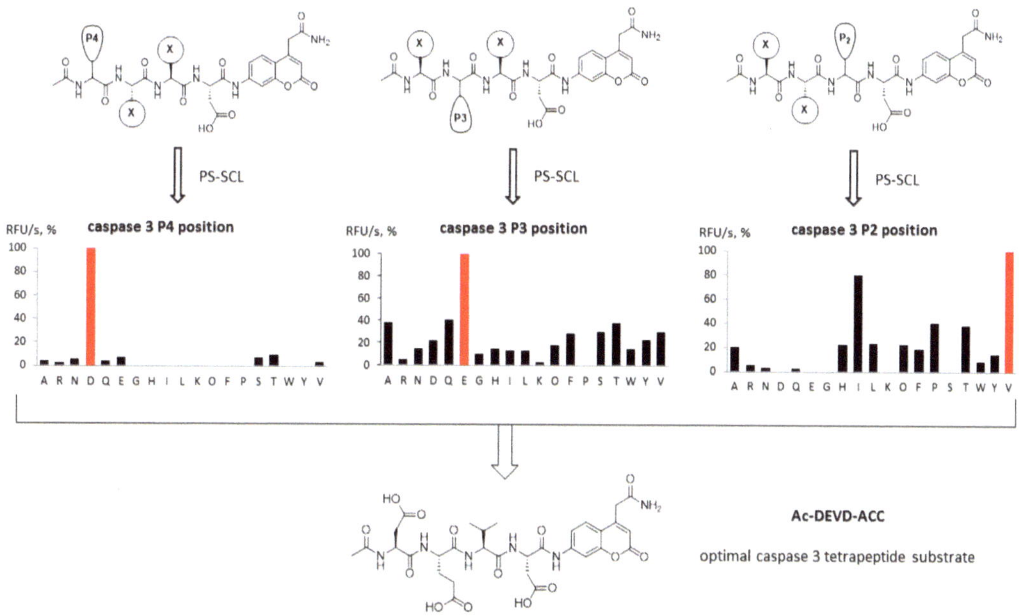

Fig. 6 Selection of optimal amino acids in P4-P2 positions from PS-SCL screening

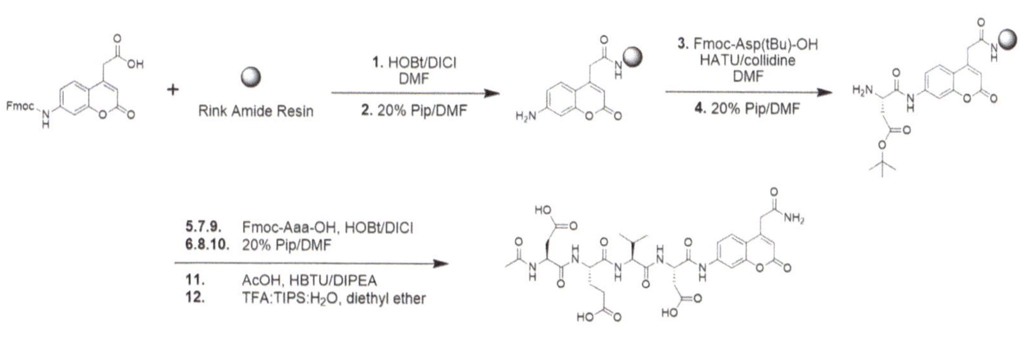

Fig. 7 Scheme for synthesis of Ac-DEVD-ACC. The numbers from *1* to *12* indicate successive steps of the synthesis. (*1*) Fmoc-ACC-OH coupling to the resin; (*2*, *4*, *6*, *8*, and *10*) Fmoc group deprotection; (*3*) Coupling the first amino acid to the NH_2-ACC-resin; (*5*, *7*, and *9*) peptide chain elongation; (*11*) acetylation of NH_2-tetrapeptide-ACC; (*12*) substrate cleavage from the resin (removing protecting groups from the substrate) and substrate precipitation

4. Remove *N*-Fmoc groups by incubating the resin in 5 ml of 20 % piperidine in DMF for 5, 5, and 30 min. Each time, agitate gently and remove solution by vacuum filtration [30] (*see* **Note 15**).
5. Wash the resin three times with 5 ml aliquots of DMF, then three times with 5 ml aliquots of DCM, and again three times with DMF. Agitate gently and remove solution each time.
6. Transfer a few resin beads to a glass test tube and perform a ninhydrin test to ensure that amine groups of resin are free (beads color should change from yellow to dark blue). If test is positive (beads are dark blue), proceed with synthesis, if negative (beads are yellow), repeat **steps 4–6** until the test is positive [31] (*see* **Note 16**).
7. Add Fmoc-ACC-OH (0.144 mmol, 63.5 mg), HOBt (0.144 mmol, 21.6 mg), and DICI (0.144 mmol, 22.55 μl) to 1 ml of DMF in an Eppendorf tube. After 5 min activation of functional groups, transfer solution to synthesis vessel with resin and incubate for 24 h with slow agitation [20] (*see* **Note 17**).
8. Filter the resin and wash three times with DMF. Then repeat **step** 7 with a half amount of reagents to increase coupling efficiency.
9. Remove reaction mixture by vacuum filtration and wash resin three times with DMF. Test for unreacted amine by ninhydrin test. If test is positive (beads are dark blue), repeat **step 8**, if negative (beads are yellow), proceed with **step 10** [31] (*see* **Note 16**).
10. To cap any unreacted resin, add 3 molar equivalents of acetic acid (0.144 mmol, 8.23 μl), DIPEA (0.144 mmol, 25.08 μl), and HBTU (0.144 mmol, 54 mg) in 1 ml of DMF in an Eppendorf tube. After 1 min of preactivation, transfer capping mixture to the resin and shake gently for 1 h [20].
11. Filter resin and wash three times in 5 ml aliquots of DMF.
12. To remove the *N*-Fmoc group from ACC-resin, repeat **steps 4–6** (positive ninhydrin test for free amine groups of ACC should yield dark red color; *see* **Note 16**).

3.2.2 NH_2-Asp-ACC-Resin Synthesis

1. Add 3 molar equivalents of each of Fmoc-*L*-Asp(O-*t*-Bu)-OH (0.144 mmol, 59.26 mg), HATU (0.144 mmol, 54.72 mg), and collidine (0.144 mmol, 18.74 μl) in 1 ml of DMF to an Eppendorf tube. After 1 min of preactivation, transfer mixture to the reaction vessel containing NH_2-ACC-resin and incubate for 24 h with gentle shaking (*see* **Note 17**).
2. Filter the resin and wash three times with 5 ml aliquots of DMF.
3. Repeat **step 1** using half amount of reagents to increase coupling efficiency and then repeat **step 2**.

4. Add 3 ml of 20 % piperidine in DMF to synthesis vessel and gently agitate for 5 min. Remove piperidine solution and then repeat deprotection treatment for 5 min and 30 min while agitating [32] (*see* **Note 15**).
5. Remove piperidine solution and wash three times with 5 ml aliquots of DMF, three times with 5 ml aliquots of DCM, and again three times with 5 ml aliquots of DMF.
6. Check for the presence of free amine group by ninhydrin test. If the test is negative (beads are yellow), repeat **steps 4–5**, if positive (beads are blue), continue with synthesis [31].

3.2.3 Peptide Chain Elongation: P2-P4 Positions Coupling [33, 34]

1. Add 3 molar equivalents of each Fmoc-*L*-Val-OH (0.144 mmol, 48.87 mg), HOBt (0.144 mmol, 22 mg), and DICI (0.144 mmol, 22 μl) in 1 ml to an Eppendorf tube. After 3 min of preactivation, transfer mixture to the reaction vessel containing NH_2-Asp-ACC-resin and incubate for 3 h with gentle shaking (*see* **Note 17**).
2. Filter the resin and wash three times with 5 ml aliquots of DMF.
3. Transfer a few resin beads to a test tube and perform ninhydrin test. If test is positive (beads are blue), repeat **steps 1–3**. If test is negative, proceed with **step 4** [31].
4. Remove Fmoc-protecting groups by incubating with three aliquots of 20 % piperidine in DMF for 5, 5, and 30 min removing piperidine solution each time [32] (*see* **Note 15**).
5. Wash resin three times with 5 ml aliquots of DMF, three times with 5 ml aliquots of DCM, and again three times with 5 ml aliquots of DMF.
6. Perform ninhydrin test. If test is positive (beads are blue), proceed with **step 7**. If test is negative (beads are yellow), repeat **steps 4–6** [31].
7. To substitute P3 position, repeat **steps 1–6**, but replace Fmoc-*L*-Val-OH by 3 equivalents of Fmoc-*L*-Glu(O-*t*-Bu)-OH (0.144 mmol, 64 mg).
8. To substitute P4 position, repeat **steps 1–6**, but replace Fmoc-*L*-Val-OH by 3 equivalents of Fmoc-*L*-Asp(O-*t*-Bu)-OH (0.144 mmol, 59 mg).
9. Protect free N-terminal amino group of resulting NH_2-DEVD-ACC-resin with acetyl group. Add 3 equivalents of acetic acid (0.144 mmol, 8.23 μl), DIPEA (0.144 mmol, 25.08 μl), and HBTU (0.144 mmol, 54 mg) in 1 ml of DMF to Eppendorf tube and activate for 1 min. Then transfer capping mixture to the vessel with resin and incubate with gentle agitation for 1 h.
10. Remove capping mixture and wash resin 3 times with 5 ml aliquots of DMF.

11. Transfer a few resin beads to a test tube and perform ninhydrin test. If test is positive (beads are blue), repeat **steps 9–11**. If test is negative (beads are yellow), proceed with **step 12** [31].
12. Wash resin three times with 5 ml aliquots of DCM and three times with 5 ml aliquots of methanol.
13. Dry resin over P_2O_5 under vacuum for 5–12 h.
14. Cleave peptide from resin by incubating in cleavage mixture with gentle mixing (*see* **Note 18**).
15. Collect the cleaved substrate in plastic tube and precipitate in 12 ml of cold diethyl ether for 30 min. The general scheme of Ac-DEVD-ACC synthesis is presented in Fig. 7.
16. Centrifuge precipitate at 3,000 × *g* for 5 min at 4 °C and remove supernatant. Wash pellet with 5 ml of diethyl ether, shake and centrifuge again at the same conditions. Remove supernatant. Dry crude peptide substrate on air.
17. Dissolve substrate in 1 ml of DMSO, purify on reverse phase HPLC (acetonitrile/water), and lyophilize.
18. Dissolve peptide in DMSO to final concentration of 50, 20, or 10 mM (*see* **Note 8**).
19. Store at −80 °C until use (*see* **Note 1**).

3.3 Kinetic Analysis (K_M, k_{cat}/K_M, k_{cat}) of Caspase Fluorogenic Substrates

1. The small molecule substrates of caspase are usually tetrapeptides conjugated with fluorophore [13, 14, 35]. To determine the kinetic parameters of a substrate, it is necessary to establish how much fluorescence signal (RFU) is produced by a certain amount of fluorophore. For this, vortex and spot 1 μl of 1 mM good caspase substrate in the well of the 96-well plate and add 99 μl of buffer containing active caspase (the final concentration of substrate in the well is 10 μM). Monitor proteolytic reaction on a plate reader. Read fluorescence every 10–15 s with excitation 355 nm and emission 460 nm. The time of measurement depends on the enzyme concentration, and the time after which all substrate is hydrolyzed can range from 30 min to 2 h. After all substrate is hydrolyzed, take the maximum RFU (not RFU/s) value and divide it by 10 to obtain the amount of fluorescence produced by 1 μM fluorophore.
2. To measure K_M for a selected substrate, prepare its serial dilution. Take the substrate stock in DMSO, vortex it, and dilute in assay buffer to known concentration (e.g., 500 μM for Ac-DEVD-ACC). Prepare at least 60 μl of this substrate. Next, vortex the sample, take 60 μl of this substrate, and transfer it into A1 well of 96-well plate. To the other wells (B1 to H1), add 20 μl of assay buffer. Next, transfer 40 μl of substrate from well A1 to B1 and mix it several times using single channel pipette. Then, transfer 40 μl of substrate from well B1 to C1

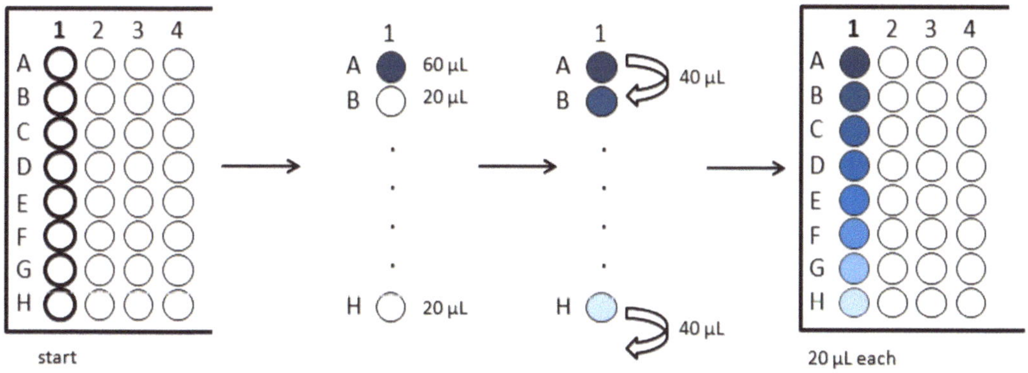

Fig. 8 The general outline of how serial dilution for K_m measurement should be prepared

and mix it several times. Repeat this procedure until the last well is filled with diluted substrate. Mix solution in H1 well and discard 40 μl. Now all wells in the first column contains 20 μl serial dilutions of Ac-DEVD-ACC in various concentration (from 500 μM to 29.3 μM; concentration of substrate in a given well is 2/3 of the concentration in the previous well) (*see* **Note 19**) (Fig. 8).

3. Prepare enzyme sample. For one column of wells, 640 μl of enzyme in buffer is required, but for comfortable pipetting, prepare around 700 μl (*see* **Note 12**). The enzyme concentration depends on the substrate (e.g., for assay with Ac-DEVD-ACC, the 0.5–1.0 nM solution of caspase 3 is usually used).
4. Transfer enzyme to a reagent reservoir and add 80 μl to each well using 8-channel pipette.
5. Monitor proteolytic reaction on a plate reader. Record fluorescence every 10–15 s with excitation 355 nm and emission 460 nm for 15–60 min. Calculate RFU/s for each well (from A1 to H1) using linear part of the plots.
6. Draw Michaelis-Menten plot: *y* axis is RFU/s or μM of ACC/s and *x* axis is substrate concentration. Calculate from the plot three kinetic parameters: K_M, k_{cat}/K_M, and k_{cat}. Substrate concentration range is often mismatched. When there are too many points on the plateau, decrease substrate concentration in the next experiment. On the other hand, if none of points achieve the V_{max}, the substrate concentration range should be shifted to higher values.
7. When the substrate concentration range is properly selected, repeat this experiment at least two times and present the results as average values with standard deviation. It is beneficial to carry out several experiments on one 96-well plate, but always calculate the total volume of caspase-containing buffer required for comfortable pipetting.

8. If K_M experiment has been carried out at least three times, it is possible to calculate three main substrate kinetic parameters K_M, k_{cat}/K_M, and k_{cat}. K_M can be determined directly by transferring obtained data into GraphPad Prism software (XY analyses → nonlinear regression (curve fit) → enzymes kinetic → Michaelis-Menten). In addition to the K_M value, one can also obtain a V_{max} value expressed in RFU/s. To calculate k_{cat}, divide V_{max} by the amount of fluorescence produced by 1 μM fluorophore (*see* **step 1**) and enzyme concentration used in the assay. Finally, calculate k_{cat}/K_M.

4 Notes

1. All substrates containing a fluorescent tag (e.g., ACC, AMC of AFC) are light sensitive so try to use dark-colored Eppendorfs. If you use transparent Eppendorfs, make sure that substrates are not exposed to UV light for a long time (use aluminum foil or hide tubes in a dark place when you are not using them at the moment).
2. P4, P3, and P2 sublibraries, as well as individual substrates (powders or DMSO solutions), should be stored at −80°C. Under such conditions they can be stored for up to several years.
3. Recombinant caspases can be stored at −80°C for several years.
4. Caspases are not very stable enzymes at higher temperatures, so after thawing they need to be stored at 4°C and used for kinetic assay within several hours. Therefore, it is a good practice to prepare many small (10–50 μl) aliquots of recombinant enzymes after expression.
5. Caspase reaction buffer contains DTT, so it needs to be prepared just before kinetic assays. We recommend not to store assay buffer at 4°C overnight. Freshly prepared buffer ensures good repeatability of results.
6. Different caspases have different optimal conditions for catalysis. The buffer presented here can be used for all caspases, but it is not the optimal one. Optimal buffers for individual caspases can be found elsewhere [14, 28, 36].
7. If the resin appears to be clammy, dry it overnight in the vacuum over P_2O_5.
8. Use only peptide grade DMSO to dissolve caspase substrates. Avoid long-stored open-bottle DMSO, because it is hygroscopic and this can lead to sample dilution.

9. Combinatorial libraries and individual substrates should be brought from −80°C to room temperature very slowly (over at least 3 h), without additional heating (i.e., in 37°C incubator).
10. All caspase substrates contain Asp residue in the P1 position and are well soluble in DMSO or water buffers at room temperature, so there is no need to warm them up.
11. For incubation, recombinant caspase should be added to the buffer heated up to 37°C, otherwise 10-min incubation of caspase in a large volume of buffers (10–20 ml) will not be sufficient to achieve 37°C.
12. Always prepare larger volume of enzyme in assay buffer than necessary to ensure comfortable pipetting.
13. It is very important to identify the linear part of the plot. However, there is no common range of time, which should be chosen. All sublibraries are scanned at the final concentration of 50 μM to ensure that velocity data are proportional to k_{cat}/K_m. Good substrates are hydrolyzed with a high rate (RFU/s = 50–100 units), so in this case only the first 5 min (or even shorter) period of reaction is enough for the kinetic analysis. Poor substrates are hydrolyzed much slower (RFU/s = 0.5–1.0) and require more time (>30 min) to produce a linear part of the plot.
14. SCL with Asp at the P1 position (Ac-X-X-X-Asp-ACC) can be used to determine substrate specificity profiles of almost all caspases from different organisms following above described protocol. However, some enzymes assigned to caspases family do not recognize aspartic acid in the P1 position of their substrates. A good example is paracaspase MALT1, which displays strict arginine specificity [24]. The P4-P2 specificity of MAL1 can be profiled using above described protocol; however, a different SCL is required (Ac-X-X-X-Arg-ACC).
15. Piperidine deprotection can lead to side reactions, such as diketopiperazine and aspartimide formation. Removing all piperidine solution is therefore critical. On the other hand, poor deprotection results in slow or incomplete coupling, thus repeated piperidine treatment helps to overcome the problem.
16. Primary amines condensed with ninhydrin can be detected by ninhydrin test (formed Schiff base, dark blue color, NH_2-amino acids). However, NH_2-ACC is an aromatic amine, which is detected by ninhydrin test by an appearance of orange to red color.
17. It is important to gently shake resin in vessel, because beads are susceptible to rubbing. Improper handling of resin during reaction can significantly decrease yield.

18. Silane derivative (TIPS) shows good efficacy in quenching carbocations. Cleavage should be completed in 2–3 h at room temperature to minimize side reactions.
19. Note that during K_M measurement experiments substrate concentration in particular well after serial dilution is not the final concentration during the assay, because substrate (20 μl) will be diluted 5 times with the enzyme-containing buffer (80 μl).

Acknowledgments

This work was supported by the National Science Centre grant 2011/03/B/ST5/01048 and the Foundation for Polish Science in Poland. This work is co-financed by the European Union as part of the European Social Fund.

References

1. Drag M, Salvesen GS (2010) Emerging principles in protease-based drug discovery. Nat Rev Drug Discov 9(9):690–701. doi:10.1038/nrd3053
2. Fuentes-Prior P, Salvesen GS (2004) The protein structures that shape caspase activity, specificity, activation and inhibition. Biochem J 384(Pt 2):201–232. doi:10.1042/BJ20041142
3. Salvesen GS, Dixit VM (1997) Caspases: intracellular signaling by proteolysis. Cell 91(4): 443–446
4. Pop C, Salvesen GS (2009) Human caspases: activation, specificity, and regulation. J Biol Chem 284(33):21777–21781. doi:10.1074/jbc.R800084200, R800084200
5. Thornberry NA, Lazebnik Y (1998) Caspases: enemies within. Science 281(5381):1312–1316
6. Poreba M, Drag M (2010) Current strategies for probing substrate specificity of proteases. Curr Med Chem 17(33):3968–3995
7. Ostresh JM, Winkle JH, Hamashin VT et al (1994) Peptide libraries: determination of relative reaction rates of protected amino acids in competitive couplings. Biopolymers 34(12): 1681–1689. doi:10.1002/bip.360341212
8. Rano TA, Timkey T, Peterson EP et al (1997) A combinatorial approach for determining protease specificities: application to interleukin-1beta converting enzyme (ICE). Chem Biol 4(2):149–155, S1074-5521(97)90258-1 [pii]
9. Sleath PR, Hendrickson RC, Kronheim SR et al (1990) Substrate specificity of the protease that processes human interleukin-1 beta. J Biol Chem 265(24):14526–14528
10. Howard AD, Kostura MJ, Thornberry N et al (1991) IL-1-converting enzyme requires aspartic acid residues for processing of the IL-1 beta precursor at two distinct sites and does not cleave 31-kDa IL-1 alpha. J Immunol 147(9): 2964–2969
11. Thornberry NA, Bull HG, Calaycay JR et al (1992) A novel heterodimeric cysteine protease is required for interleukin-1 beta processing in monocytes. Nature 356(6372):768–774. doi:10.1038/356768a0
12. Thornberry NA, Molineaux SM (1995) Interleukin-1 beta converting enzyme: a novel cysteine protease required for IL-1 beta production and implicated in programmed cell death. Protein Sci 4(1):3–12. doi:10.1002/pro.5560040102
13. Thornberry NA, Rano TA, Peterson EP et al (1997) A combinatorial approach defines specificities of members of the caspase family and granzyme B. Functional relationships established for key mediators of apoptosis. J Biol Chem 272(29):17907–17911
14. Garcia-Calvo M, Peterson EP, Rasper DM et al (1999) Purification and catalytic properties of human caspase family members. Cell Death Differ 6(4):362–369. doi:10.1038/sj.cdd.4400497
15. Wachmann K, Pop C, van Raam BJ et al (2010) Activation and specificity of human caspase-10. Biochemistry 49(38):8307–8315. doi:10.1021/bi100968m
16. Mikolajczyk J, Scott FL, Krajewski S et al (2004) Activation and substrate specificity of caspase-14. Biochemistry 43(32):10560–10569. doi:10.1021/bi0498048
17. Edwards PD, Mauger RC, Cottrell KM et al (2000) Synthesis and enzymatic evaluation of a

P1 arginine aminocoumarin substrate library for trypsin-like serine proteases. Bioorg Med Chem Lett 10(20):2291–2294

18. Backes BJ, Harris JL, Leonetti F et al (2000) Synthesis of positional-scanning libraries of fluorogenic peptide substrates to define the extended substrate specificity of plasmin and thrombin. Nat Biotechnol 18(2):187–193. doi:10.1038/72642
19. Harris JL, Backes BJ, Leonetti F et al (2000) Rapid and general profiling of protease specificity by using combinatorial fluorogenic substrate libraries. Proc Natl Acad Sci U S A 97(14):7754–7759. doi:10.1073/pnas.140132697, 140132697 [pii]
20. Maly DJ, Leonetti F, Backes BJ et al (2002) Expedient solid-phase synthesis of fluorogenic protease substrates using the 7-amino-4-carbamoylmethylcoumarin (ACC) fluorophore. J Org Chem 67(3):910–915, jo016140o [pii]
21. Walters J, Pop C, Scott FL et al (2009) A constitutively active and uninhibitable caspase-3 zymogen efficiently induces apoptosis. Biochem J 424(3):335–345. doi:10.1042/BJ20090825
22. Choe Y, Leonetti F, Greenbaum DC et al (2006) Substrate profiling of cysteine proteases using a combinatorial peptide library identifies functionally unique specificities. J Biol Chem 281(18):12824–12832. doi:10.1074/jbc.M513331200
23. Debela M, Magdolen V, Schechter N et al (2006) Specificity profiling of seven human tissue kallikreins reveals individual subsite preferences. J Biol Chem 281(35):25678–25688. doi:10.1074/jbc.M602372200, M602372200 [pii]
24. Hachmann J, Snipas SJ, van Raam BJ et al (2012) Mechanism and specificity of the human paracaspase MALT1. Biochem J 443(1): 287–295. doi:10.1042/BJ20120035
25. Drag M, Mikolajczyk J, Bekes M et al (2008) Positional-scanning fluorigenic substrate libraries reveal unexpected specificity determinants of DUBs (deubiquitinating enzymes). Biochem J 415(3):367–375. doi:10.1042/BJ20080779, BJ20080779 [pii]
26. Garcia-Calvo M, Peterson EP, Leiting B et al (1998) Inhibition of human caspases by peptide-based and macromolecular inhibitors. J Biol Chem 273(49):32608–32613
27. Stennicke HR, Salvesen GS (1999) Caspases: preparation and characterization. Methods 17(4):313–319. doi:10.1006/meth.1999.0745
28. Ekici OD, Li ZZ, Campbell AJ et al (2006) Design, synthesis, and evaluation of aza-peptide Michael acceptors as selective and potent inhibitors of caspases-2, -3, -6, -7, -8, -9, and -10. J Med Chem 49(19):5728–5749. doi:10.1021/jm0601405
29. Fu J, Yang Y, Zhang XW et al (2010) Discovery of 1H-benzo[d][1,2,3]triazol-1-yl 3,4,5-trimethoxybenzoate as a potential antiproliferative agent by inhibiting histone deacetylase. Bioorg Med Chem 18(24):8457–8462. doi:10.1016/j.bmc.2010.10.049
30. Carpino LA, Han GY (1972) 9-Fluorenylmethoxycarbonyl amino-protecting group. J Org Chem 37(22):3404–3409
31. Kaiser E, Colescott RL, Bossinger CD et al (1970) Color test for detection of free terminal amino groups in the solid-phase synthesis of peptides. Anal Biochem 34(2):595–598
32. Chang CD, Meienhofer J (1978) Solid-phase peptide synthesis using mild base cleavage of N alpha-fluorenylmethyloxycarbonylamino acids, exemplified by a synthesis of dihydrosomatostatin. Int J Pept Protein Res 11(3): 246–249
33. Chan WC, White PD (2000) Fmoc solid phase peptide synthesis, the practical approach series. Oxford University Press, New York, pp 1–74
34. Merrifield RB (1963) Solid phase peptide synthesis. I. The synthesis of a tetrapeptide. J Am Chem Soc 85(14):2149–2154
35. McStay GP, Salvesen GS, Green DR (2008) Overlapping cleavage motif selectivity of caspases: implications for analysis of apoptotic pathways. Cell Death Differ 15(2):322–331. doi:10.1038/sj.cdd.4402260
36. Stennicke HR, Salvesen GS (1997) Biochemical characteristics of caspases-3, -6, -7, and -8. J Biol Chem 272(41):25719–25723

Chapter 3

Global Identification of Caspase Substrates Using PROTOMAP (Protein Topography and Migration Analysis Platform)

Melissa M. Dix, Gabriel M. Simon, and Benjamin F. Cravatt

Abstract

Delineation of the natural substrate scope of proteases is important for understanding the functions of proteolytic pathways in physiology and disease. Herein we describe the protocol for PROTOMAP, a technique that combines SDS-PAGE with tandem mass spectrometry to globally identify shifts in protein migration indicative of proteolytic processing. When applied to cells undergoing apoptosis, this unbiased global method provides a snapshot of the topography and magnitude of proteolytic events associated with programmed cell death.

Key words SDS-PAGE, In-gel digestion, Mass spectrometry, Spectral counting, PROTOMAP, Peptograph, Substrate identification

1 Introduction

The key to elucidating the biological role of proteases is identification of their physiological substrates and the functional consequences of protein cleavage. For example, does a proteolytic cleavage event result in loss-of-function and rapid degradation of the substrate? Or, does cleavage produce a stable protein fragment that performs a new biochemical function? If so, what is the topography of the fragment, how large is it, and does it contain any known functional domains? The challenge of comprehensively defining the topography of protein cleavage in physiological systems and thereby predicting functional consequences of proteolysis has historically required generation of multiple antibodies that recognize epitopes throughout the length of proteins. In pursuit of a more general solution, we developed an unbiased proteomic platform that provides a global snapshot of proteolytic events occurring upon activation or inhibition of a protease(s). We have named this platform PROTOMAP for Protein Topography and Migration Analysis

Peter V. Bozhkov and Guy Salvesen (eds.), *Caspases, Paracaspases, and Metacaspases: Methods and Protocols*, Methods in Molecular Biology, vol. 1133, DOI 10.1007/978-1-4939-0357-3_3, © Springer Science+Business Media New York 2014

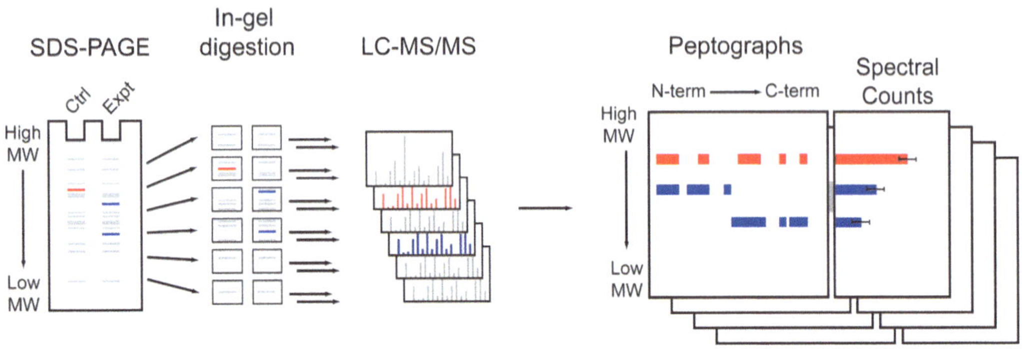

Fig. 1 Typical workflow for PROTOMAP experiment. Proteomes from control and experimental systems are separated by one-dimensional SDS-PAGE and the gel lanes cut into bands at defined intervals. Proteins are digested in-gel with trypsin, and the resulting peptides are analyzed by one-dimensional reverse-phase LC-MS/MS. The proteomic data are integrated into peptographs, which plot sequence coverage for a given protein versus SDS-PAGE migration (*left panel*). In the right panel, the peptograph also displays average spectral counts for each protein in each gel band. Proteins that undergo a proteolytic cleavage event are identified by a shift in migration from higher (*parental, red*) to lower (*fragments, blue*) MW species in control versus experimental samples. The sequence coverage shown in the left panel provides a topographical map of the protein fragments that persist following proteolysis. The magnitude of proteolysis is estimated by comparing spectral counts of the parental protein in the control versus experimental samples

Platform [1, 2]. In this platform, up-front fractionation of proteomic samples is performed by one-dimensional SDS-PAGE followed by shotgun liquid chromatography-electrospray tandem mass spectrometry (LC-MS/MS) analysis of proteins from individual gel bands. The resulting sequence and gel-migration information are then assembled into two-dimensional "peptographs," which, when combined with spectral counting data obtained through LC-MS/MS analysis, provide a semiquantitative topographical map for all detectable proteins in a sample.

In a typical PROTOMAP experiment (Fig. 1), control and experimental samples, one in which the protease or proteolytic pathway of interest is activated or inactivated, are run on SDS-PAGE. Each distinct gel lane is cut into bands of defined length intervals. The proteins in each band are then separately digested into peptides, analyzed, and sequenced using LC-MS/MS. Through the use of custom software detailed later in the protocol, the data are reassembled into a two-dimensional "peptograph" for each protein identified in the sample. The peptograph displays the sequence coverage for that protein in each band as well as the relative migration rate on the gel. Proteolyzed proteins are identified as proteins that experience a shift to a lower molecular weight in the protease-active sample. By comparing the spectral counts obtained for the full-length parental form of a substrate in the protease-active sample to the control sample, one can estimate the magnitude of cleavage event. This combined with near-complete topographical descriptions of cleavage patterns (often yielding precise sites-of-cleavage) enables prediction of the functional consequences of cleavage events.

Here we describe a detailed protocol for using PROTOMAP, which should be applicable to cell or tissue systems from a wide range of biological models and even primary human specimens.

2 Materials

2.1 SDS-PAGE

1. Tris-Glycine SDS-PAGE buffer (National Diagnostics, EC-870).
2. Loading dye.
3. 10 % acrylamide gel.
4. Ammonium bicarbonate.
5. Phosphate-buffered saline.
6. Prestained protein molecular weight marker.

2.2 In-Gel Digestion

1. Razor.
2. Clear glass plate.
3. Square glass container.
4. Formic acid.
5. Sequencing-grade modified trypsin (Promega).
6. Trypsin resuspension buffer (Promega).
7. Iodoacetamide.
8. Tris (2-carboxyethyl) phosphine.
9. High-purity water.
10. Optima-grade acetonitrile.
11. Ammonium bicarbonate.
12. Speed vacuum.
13. 1.5 mL Eppendorf tubes.
14. Incubator (37 °C).

2.3 Mass Spectrometric Analysis

1. MS buffer A: 95 % high-purity water, 5 % optima-grade acetonitrile, and 0.1 % formic acid.
2. MS buffer B: 20 % high-purity water, 80 % optima-grade acetonitrile, and 0.1 % formic acid.
3. 5 μm C18 reverse-phase resin (Phenomenex).
4. Model P-2000 CO_2 laser puller (Sutter Instrument).
5. Agilent 1100 series HPLC.
6. LTQ ion trap mass spectrometer (Thermo Electron).
7. High pressure bomb.
8. Fused silica capillary tubing (365 μm outer diameter, 100 μm inner diameter (ID); Agilent, cat. no. 160-2255-10).

2.4 Data Analysis and Generation of Peptographs

1. SEQUEST software.
2. DTASelect software.
3. Custom Perl scripts: coverage.pl, protomap.pl, and peptographer.pl (available at http://www.scripps.edu/chemphys/cravatt/protomap).

3 Methods

3.1 10 % SDS-PAGE and Cutting the Gel

1. Prepare 200 μg of each protein sample in 4× SDS-PAGE loading buffer (*see* **Note 1**).
2. Load sample flanked by prestained molecular weight markers onto a 10 % acrylamide gel (*see* **Note 2**).
3. Run the gel at constant 250 V/h until there are approximately 11 cm between the 250 kDa MW marker and the dye front.
4. Remove the gel from the plates. Cut off the stacking gel and any excess gel below the dye front (*see* **Note 3**).
5. Place the gel in a square glass dish and rinse in high-purity water for 10 min.
6. Clean a glass plate with water and then rinse with methanol and dry completely.
7. Place a paper grid with lines spaced 0.5 cm apart underneath the glass plate, and place the gel on top of the glass.
8. Line up the top molecular weight marker (250 kDa) in the first lane as well as the last lane on top of one of the lines.
9. Using a razor, cut one band above this marker and then cut across all gel lanes horizontally every 0.5 cm following the guide lines (*see* **Note 4**).
10. As soon as the gel is cut horizontally, cut the gel vertically in the middle of the MW markers to separate distinct samples.
11. One by one, take a single slice of gel and cut into small pieces (approx 1 mm^2) and place into microcentrifuge tubes containing 500 μL 100 mM ammonium bicarbonate.
12. Repeat for every gel slice for every sample (usually 22 bands per sample), placing each in a separate microcentrifuge tube with 500 μL 100 mM ammonium bicarbonate.

3.2 In-Gel Digestion (See Note 3)

Note: All washes are performed with slow vortexing.

1. Wash each gel band 2× for 15 min with 500 μL 100 mM ammonium bicarbonate.
2. After wash(es), remove supernatant and add enough 10 mM TCEP to completely cover gel band (approx 200 μL).
3. Place at 37 °C for 30 min.
4. Quick spin and remove TCEP with aspirator or pipetman.

5. Add enough 55 mM iodoacetamide (made fresh) to cover gel band (approx 200 μL).
6. Place tubes in the dark for 30 min at room temperature.
7. Remove supernatant and wash gel bands 2–3× for 15 min in 500 μL 50:50 acetonitrile:100 mM ammonium bicarbonate.
8. Remove supernatant and add 50 μL acetonitrile to completely dry gel band (should turn opaque).
9. Remove acetonitrile and speed vac for ~5 min or until slices are completely dry.
10. Resuspend a 20 μg bottle of trypsin in 200 μL of trypsin resuspension buffer.
11. Dilute 1:10 with 25 mM ammonium bicarbonate (10 ng/μL final concentration of trypsin).
12. Add 200 ng of trypsin (20 μL) to dehydrated gel slices (or if sure of exact protein concentration, use a 50:1 substrate/enzyme ratio).
13. Incubate gel slices in trypsin for 10 min to rehydrate.
14. Add enough 25 mM ammonium bicarbonate to cover gel bands and place at 37 °C overnight.
15. The next morning transfer supernatant into clean tube and add enough 5 % formic acid to cover gel bands and incubate at room temperature for 15 min.
16. Transfer supernatant to the tube with supernatant from the o/n digestion.
17. Add enough acetonitrile to cover gel bands (approx 100–200 μL) and incubate at room temperature for 15 min. Transfer supernatant to tube with digested peptides.
18. Repeat acetonitrile elutions until gel bands have become completely opaque (2–3 times).
19. Dry peptide sample to a final volume of <10 μL in speed vac.
20. At this stage peptides may be stored at −80 °C (*see* **Note 5**).

3.3 Preparing Columns for Mass Spectrometric Analysis

1. Cut 30 cm of 100 μm ID fused silica capillary tubing and burn off 2–3 cm of polyimide coating using a flame.
2. Remove the burnt coating with methanol and a Kimwipe. Place the tubing in the laser puller with the newly exposed silica in the laser's path and pull two 5 μm tips.
3. Create a methanol Aqua C18-reverse-phase slurry in an Eppendorf tube, and pressure-load the tips from the previous step with 11 cm of Aqua C18 reverse-phase slurry.
4. Equilibrate the tip on an Agilent 1100 series HPLC using a gradient of 40 % buffer A; 60 % buffer B to 100 % buffer A; 0 % buffer B over 30 min, followed by a 10 min wash with 100 %

buffer A with a flow rate of 0.1 mL/min with a tee splitter to reduce the flow rate to 300–400 nL/min.

5. Resuspend peptide samples from in-gel digestion in 10 μL of buffer A.
6. Pressure-load sample onto column equilibrated in **step 4**.
7. Connect column to instrument and ensure that it is actively flowing.
8. Create a method in Xcalibur with the following parameters: Mass spectrometer: scanning in data-dependent scanning mode with one full MS scan (400–1,800 m/z) followed by seven MS^2 scans of the *n*th most abundant ion.

   HPLC: Create a 125 min gradient with the following buffer concentrations:

0–10 min: 100 % A
10–20 min: 100–85 % A
20–80 min: 85–55 % A
80–90 min: 55–0 % A
100–105 min: 0–100 % A
105–125 min: 100 % A

9. Run control samples in chronological order.
10. Prepare new column for experimental samples and run in chronological order.

3.4 Data Analysis and Generation of Peptographs

The instructions below are intended to be used on the Microsoft Windows operating system. However, DTASelect and Perl are both platform independent and can be run successfully on most operating systems (e.g., OS X and Linux) although some details will differ slightly.

1. Raw mass spectrometry data must be searched with SEQUEST and filtered using DTASelect. Detailed instructions for using these programs are beyond the scope of this protocol, but details can be found in the literature [3]. A separate folder should be created for the DTASelect output files for each gel band and for each sample (Fig. 2).
2. DTASelect should be run with the following parameters: "y -1 –Stn RKD –Stc RKD –p 1." These parameters specify one peptide per protein (-p 1) and fully tryptic peptides or half tryptic ending on an arginine, lysine, or aspartic acid (indicative of caspase cleavage) (y -1 –Stn RKD –Stc RKD). These parameters are not required but are helpful for the identification of candidate sites of caspase cleavage.
3. Download the PROTOMAP software from http://www.scripps.edu/chemphys/cravatt/protomap

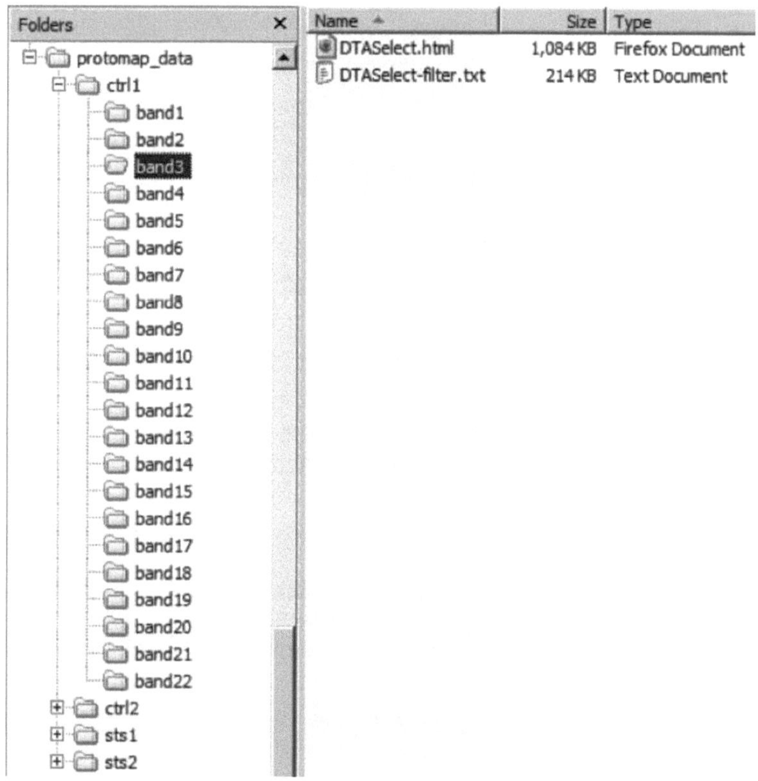

Fig. 2 Typical organization of DTASelect data. The only requirement is that the DTASelect data for each gel band are stored in separate folders

4. Install a Windows-compatible distribution of Perl. The ActivePerl distribution from ActiveState is free of charge and is available for download from the ActiveState website (http://www.activestate.com/activeperl).
5. Download the custom PROTOMAP Perl scripts from http://www.scripps.edu/cravatt/protomap/. Extract the scripts into the top-level folder in which your sub-folders containing DTASelect data reside. Briefly, the purposes of the three scripts are:
   (a) *coverage.pl* – reads each *DTASelect.html* file and stores the sequence covered by all of the peptides for each protein in files called *coverg.txt* in each folder.
   (b) *protomap.pl* – must be run after running *coverage.pl*. This script reads each *DTAselect-filter.txt* file and coverage maps generated by *coverage.pl* to create a single file called *all-data.pmap* that contains combined spectral-count and sequence-coverage information for the entire dataset.
   (c) *peptographer.pl* – must be run last. It takes a pmap file as input and produces all of the graphical peptographs and a web-browser interface to view them.

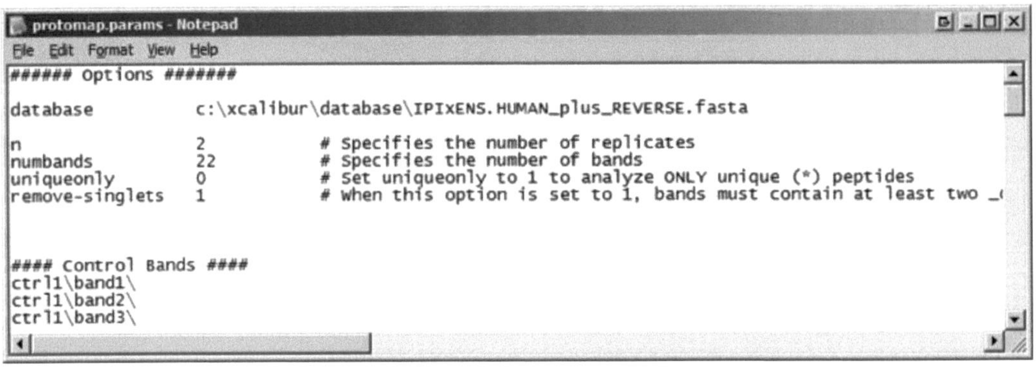

Fig. 3 The contents of a typical *protomap.params* file. Parameters specifying the location of the FASTA database and several optional switches for controlling the behavior of the script are specified in the "Options" section. Below this section are two sections that specify the relative paths of the control and experimental sections, respectively

6. Create a *protomap.params* file. An example of a *protomap.params* file is shown in Fig. 3. The *protomap.params* file tells the scripts how many samples there are, where they are, and in what order they should be organized. You can download a sample *protomap.params* file at http://www.scripps.edu/chemphys/cravatt/protomap. Place it in the folder containing the scripts (*coverage.pl*, *protomap.pl*, and *peptographer.pl*). Then, modify the params file for your own PROTOMAP datasets using a text editor. The first section of the protomap.params file, under the ###### Options ###### header, contains a set of parameters that you should modify depending on your experiment. Be sure that the *n* and *numbands* parameters correctly specify the number of replicates and bands, respectively. The next two sections (#### Control bands #### and #### Experimental bands #####) designate the locations of all of the folders containing your data from each gel band. There should be the same number of folders in each section, and each folder should contain *DTAselect.html* and *DTAselect-filter.txt* files from **step 3**. Other files in those folders will be ignored.
7. Run the scripts:
   (a) Open a *command prompt* and navigate to the folder that contains your scripts, data, and params file. Type coverage.pl and press <enter>.
   (b) When coverage.pl has finished, type protomap.pl and press <enter>. It may take some time to run, depending on the complexity of your data and the speed of your CPU.
   (c) When protomap.pl has finished, an output file called alldata.pmap is created. It is recommended that you rename this file with a more descriptive filename (but keep the .pmap extension).

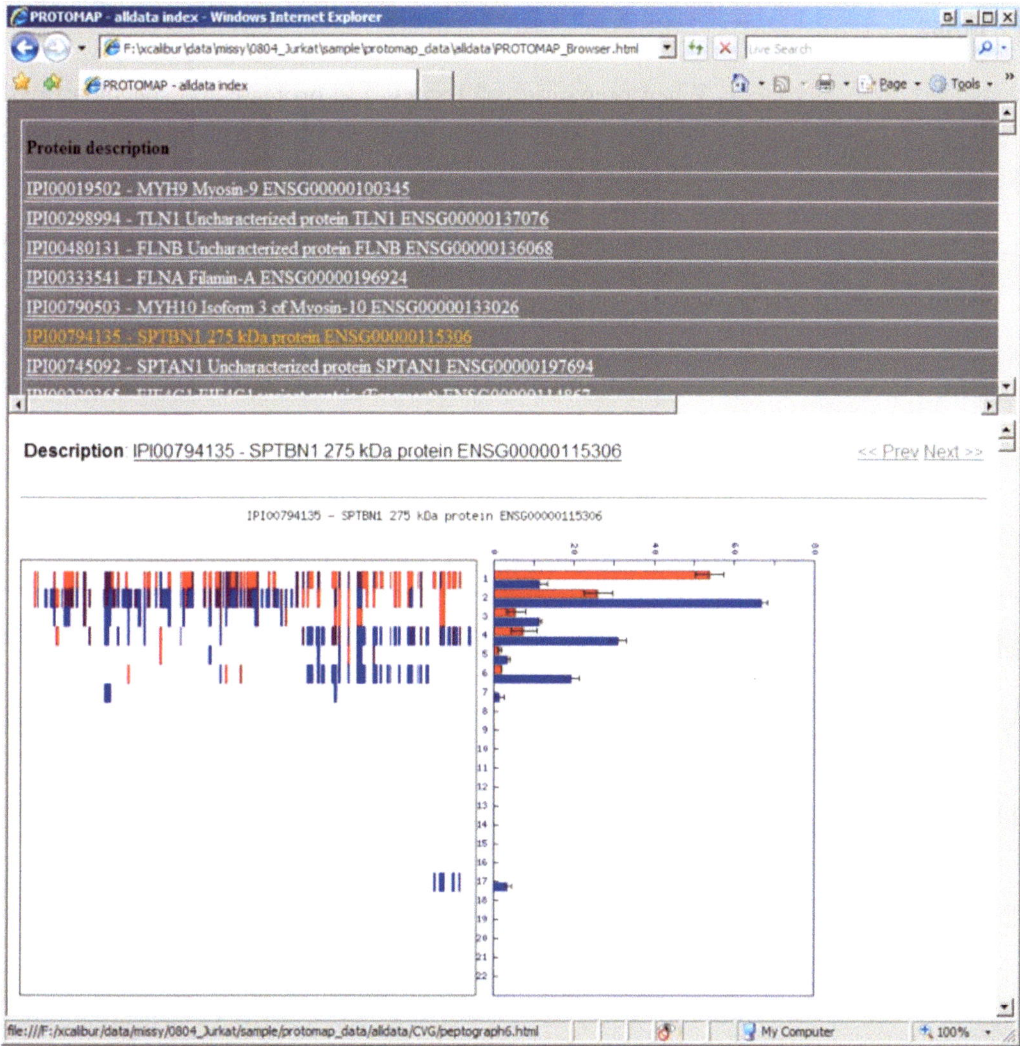

Fig. 4 The web-browser interface for viewing peptographs that is created upon successful completion of the *protomap.pl* script

(d) Type peptographer.pl alldata.pmap and press <enter> (or use the filename you renamed alldata.pmap to). It may take some time to run, depending on the complexity of your data and the speed of your CPU.

8. If peptographer.pl completed successfully, a new folder called alldata was generated (or whatever your .pmap file was called).
9. Open this folder in File Explorer and double-click on the PROTOMAP_Browser.html file. This will open a browser interface that should permit you to easily search or browse your peptographs (Fig. 4).

4 Notes

1. When preparing cell lysates, add protease and specific caspase inhibitors to eliminate any residual proteolytic activity. When possible prepare the samples and run the gel in the same day.
2. If enough lanes are available, run the sample in duplicate so you will have a back-up if something goes wrong during the mass spectrometric analysis.
3. When handling the gel and performing the in-gel digestions, make sure to always wear gloves and minimize exposure of the gel bands to possible keratin contamination.
4. When cutting the gel, make sure to keep the gel hydrated enough to prevent curling/shrinking, if you notice this happening splash a little high-purity water onto the gel.
5. Make sure all glassware is clean and free of any detergents, which could negatively impact the mass spectrometric analysis.

References

1. Dix MM, Simon GM, Cravatt BF (2008) Global mapping of the topography and magnitude of proteolytic events in apoptosis. Cell 134:679–691
2. Simon GM, Dix MM, Cravatt BF (2009) Comparative assessment of large-scale proteomic studies of apoptotic proteolysis. ACS Chem Biol 4:401–408
3. Cociorva D, Tabb DL, Yates JR (2006) Validation of tandem mass spectrometry database search results using DTASelect. Curr Protoc Bioinformatics 13.4.1–13.4.14, supplement 16, Johnwiley and Sons

Chapter 4

Caspase-2 Protocols

Loretta Dorstyn and Sharad Kumar

Abstract

Caspase-2 has been shown to function in apoptosis and in some non-apoptotic pathways, including tumor suppression and aging. Caspase-2 has some unique features and is the only caspase that constitutively localizes to the nucleus, although its nuclear function remains unknown. During apoptosis signaling, caspase-2 rapidly homodimerizes, which leads to its activation and proteolytic processing. The activation of caspase-2 can be measured by assessing its dimerization and/or cleavage of the caspase-2 zymogen and its substrates. This chapter outlines commonly used methods to purify recombinant caspase-2 and assess its activity and function in vitro and in cultured cells or tissue extracts.

Key words Apoptosis, Caspase-2, Caspase activity, Peptide substrates, Recombinant protein, Immunoblotting

1 Introduction

Caspase-2, one of the most conserved caspases in metazoans, has been shown to play redundant and non-redundant roles in apoptotic and non-apoptotic processes [1–3]. Caspase-2 contains a caspase activation and recruitment domain (CARD), which is a protein–protein interaction domain required for the dimerization of pro-caspase-2 molecules and for its activation [4–6]. The caspase-2 CARD also facilitates recruitment into large multiprotein complexes [7, 8]. For example, caspase-2 interacts with the CARD-containing proteins RAIDD (RIP-associated Ich-1/CED homologous protein with a death domain) and PIDD (p53-induced protein with a death domain) to form a protein complex called "PIDDosome," which is suggested to promote caspase-2 activation [8]. However, the role of the PIDDosome in cell death is unclear, and knockout studies suggest that caspase-2 activation can occur in the absence of PIDD [9, 10].

Caspase-2 zymogen contains basal enzyme activity that is enhanced by dimerization and further proteolytic processing into subunits (p19 and p12) (Fig. 1) [4, 7, 11–13]. Structural studies

Peter V. Bozhkov and Guy Salvesen (eds.), *Caspases, Paracaspases, and Metacaspases: Methods and Protocols*, Methods in Molecular Biology, vol. 1133, DOI 10.1007/978-1-4939-0357-3_4, © Springer Science+Business Media New York 2014

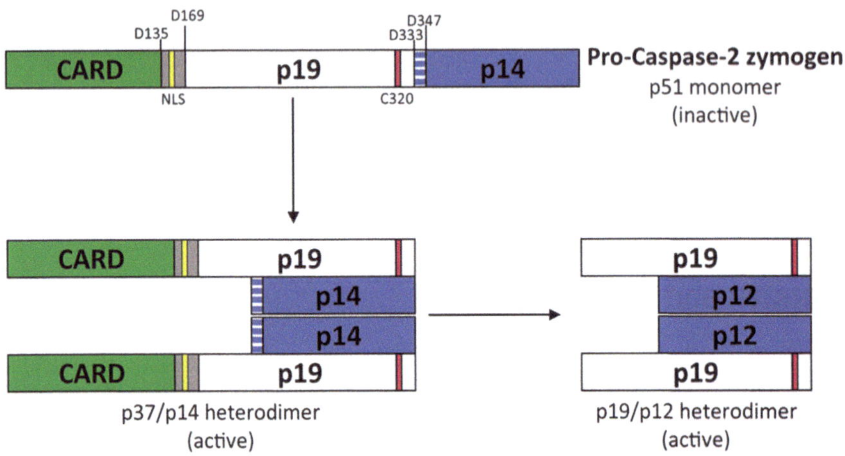

Fig. 1 Schematic representation of caspase-2 structure and activation. Caspase-2 zymogen exists as a monomer. In response to apoptotic stimuli, pro-caspase-2 molecules dimerize via their CARD and this leads to cleavage between the large (p19) and small (p14) enzyme subunits at D333. This cleavage event results in the formation of an active stable tetrameric complex. Further processing of this complex at D169 and D347 can also occur by caspase-3 to generate fully mature caspase-2 enzyme. NLS, nuclear localization sequence (*yellow line*); C320, catalytic cysteine residue (*red line*)

on human caspase-2 indicate that it forms a p19/p12 homodimer in solution with two active sites [14]. Interestingly the caspase-2 dimer has an intersubunit disulfide bridge between two adjacent cysteine residues (Cys^{329}–Cys^{390}), which is proposed to be important for caspase-2 dimerization and activation. As a consequence of its ability to readily dimerize in solution, caspase-2 can autoactivate when overexpressed in cultured cells [1, 4]. The unique structure of its active site confers somewhat atypical substrate specificity for caspase-2. Unlike other caspases, caspase-2 substrate recognition requires a pentapeptide sequence (XDxxD), with an additional small hydrophobic P5 residue (X) that accommodates a unique hydrophobic S5 binding pocket in the binding groove of the caspase-2 dimer [14, 15].

Caspase-2 is classed as an initiator caspase which acts upstream of mitochondrial outer membrane permeabilization (MOMP) [7, 16, 17]. When activated, caspase-2 cleaves several cellular targets including Bid, Mdm2, and ICAD [2]. Caspase-2-mediated cleavage of Bid induces MOMP and apoptosis [16–19]. Caspase-2 has been shown to be required for apoptosis induced by cytotoxic drugs and DNA-damaging agents [20, 21], TRAIL [22], heat shock [19], cytoskeletal disruption [20], and in oocyte cell death in response to nutrient deprivation [23–26]. Caspase-2 activation in cells is commonly monitored and detected following these apoptosis stimuli. However, its importance in apoptosis remains somewhat contentious as *Caspase-2*$^{-/-}$ (*Casp2*$^{-/-}$) mice are viable and fertile with only minor apoptotic defects in some cell types [20, 24, 25, 27].

These studies indicate that the role of caspase-2 in developmental cell death is redundant or can be compensated by other caspases and/or that caspase-2 has context and tissue specific roles [24]. Studies with *Casp2*$^{-/-}$ mice indicate that caspase-2 has important regulatory functions in non-apoptotic contexts [2, 28, 29]. For example, *Casp2*$^{-/-}$ mice display premature aging-related traits [30–32] and accumulation of oxidative stress [32]. More recently, our studies have shown that caspase-2 is important in DNA damage signaling, genomic stability, and tumor suppression [21, 29].

A unique feature of caspase-2 is its nuclear localization [5, 6, 33]. While caspase-2 has been shown to translocate from nucleus to cytosol during apoptosis [34, 35], it also constitutively localizes to the nucleus during cell death signaling [5, 6]. Therefore, the translocation of caspase-2 during cell death is not necessarily an indication of its activation and function. The nuclear function of caspase-2 remains poorly understood, although it has been suggested that nuclear localization is important for caspase-2 functions in the DNA damage signaling response and genomic stability [21].

The activation of caspases is required for the initiation and execution of apoptosis. As for other caspases, the most common method to assess caspase-2 activation is by measuring the cleavage of specific peptide substrates which are conjugated to a fluorogenic tag. Cleavage of the peptide releases the tag, and fluorescence emission is directly correlated to the amount of substrate cleaved and therefore an indication of the extent of caspase activity. The optimal substrate specificity for caspase-2 has been determined using peptide combinatorial libraries [15, 36], and the synthetic peptide VDVAD is commonly used to assess caspase-2 activity. However, it is important to note that other caspases, including caspase-3, the most abundant effector apoptotic caspase, are also able to cleave this peptide substrate [15, 36]. Therefore, when assaying caspase-2 activity from an apoptotic cell lysate, it is not exclusively indicative of caspase-2 activation alone. Unfortunately, there are no biochemical assays that uniquely determine caspase-2 enzyme activity in cells or tissue lysates.

Purified active recombinant caspase-2 protein is a useful biochemical tool to assess caspase-2-mediated substrate cleavage in vitro, using cell extracts or in vitro-synthesized proteins. There are currently only a few published substrates which are cleaved by caspase-2 in apoptotic cells [2], but there are no known specific substrates which can be used as a readout for caspase-2 specific activity and function in vivo. The use of recombinant caspase-2 can help to validate specific substrates in vitro. In addition to using peptide substrates, ^{35}S-labeled caspase-2 substrates can be synthesized in vitro and their cleavage by recombinant caspase-2 determined by fluorography. Similarly, auto-cleavage of caspase-2 can be assessed by incubation of recombinant protein with ^{35}S-labeled pro-caspase-2 (Fig. 2).

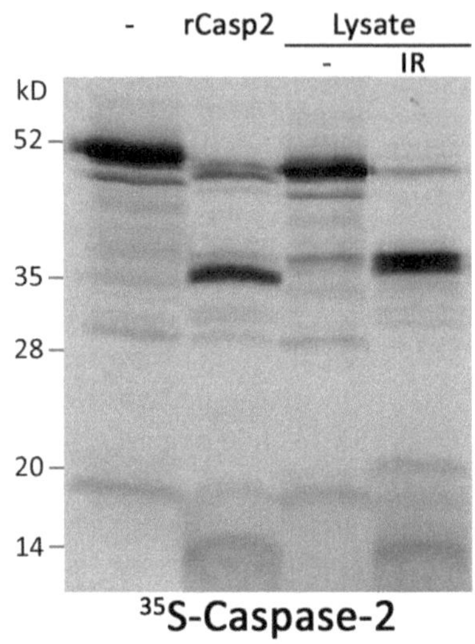

Fig. 2 In vitro processing of caspase-2. [^{35}S]-methionine-labeled caspase-2 was translated in vitro and incubated with recombinant caspase-2 protein (rCasp2) or with cell lysates from untreated or γ-irradiation (IR)-treated cells. Caspase-2 cleavage was assessed by autoradiography. Cleavage of caspase-2 by IR-treated cell lysate yields some differential products and is likely to represent cleavage by caspase-3, the predominant activated effector caspase in apoptotic cells. Caspase-2 major cleavage products are seen at 37, 19, and 14/12 kDa

Caspase-2 processing in apoptotic cells can be detected in apoptotic cell lysates by immunoblotting with specific caspase-2 antibodies. However, it is important to note that this approach is not necessarily a measure of caspase-2 activation, since the initial activation of caspase-2 occurs by dimerization and does not require proteolytic processing [12, 37]. Furthermore, in apoptotic cells, caspase-2 zymogen can be cleaved by active caspase-3, which contributes to the majority of the caspase activity present in cells undergoing apoptosis.

A robust method to detect active caspase-2 in apoptotic cells is the use of a biotin-tagged peptide inhibitor such as the pan-caspase inhibitor biotin-VAD-FMK, which acts to irreversibly bind to the active site of activated caspases [38–40]. Active caspase-2 can then be immunoprecipitated using immobilized streptavidin and detected by immunoblotting [41, 42]. This chapter describes some of these routinely used methods to assess caspase-2 activation, function, and activity during apoptosis.

2 Materials

2.1 Tissue Culture of Cell Lines

1. A humidified incubator at 37 °C with 5 % CO_2.
2. Cell culture medium: Dulbecco's Modified Eagle's Medium (DMEM) or RPMI (commercially available from Sigma or SAFC

Biosciences) supplemented with 10 mM 4-(2-hydroxyethyl)-1-piperazineethanesulfonic acid (HEPES), 100 mM penicillin/streptomycin (CSL Biosciences), and 10 % fetal bovine serum.

3. Trypsin (0.25 %) diluted in Hank's buffered salt solution (HBSS) (SAFC Biosciences). This can be aliquoted and stored at −20 °C.
4. Phosphate buffered saline (PBS): 2 mM KH_2PO_4, 10 mM Na_2HPO_4, 150 mM NaCl, pH 7.2.
5. Water bath at 37 °C.
6. Laminar flow hood.
7. Bench top centrifuge with swing-bucket rotor.

2.2 Protein Extraction

1. Cell extraction buffer: 50 mM HEPES-KOH pH 7.5, 50 mM NaCl, 10 mM dithiothreitol (DTT), 0.5 mM ethylenediaminetetraacetic acid (EDTA), 0.1 % 3-[(3-cholamidopropyl)-dimethylammonio]-1-propanesulfonate (CHAPS), 10 % sucrose, pH 7.5, and containing protease inhibitor cocktail such as Complete™ (Roche).
2. 2× Protein Loading Buffer (PLB): 100 mM Tris–HCl pH 6.8, 200 mM DTT, 20 % glycerol, 4 % SDS, 0.2 % bromophenol blue.
3. Liquid nitrogen.
4. Ice-cold water.
5. Water bath (with boiling water).
6. Bench top centrifuge.
7. Protein concentration assay reagent (such as BCA or Bradford reagent).
8. Plate reader equipped to measure absorbance between 540 and 595 nm.

2.3 Preparation of Recombinant Caspase-2

1. Competent *Escherichia coli* bacterial cells optimized for expression of recombinant proteins (such as BL21 (DE3) cells or BL21 star cells).
2. Luria broth (LB): 1 % Bacto-tryptone, 0.5 % Bacto-yeast extract, 1 % NaCl, pH 7.0.
3. 1 mM isopropyl β-D-thiogalactoside (IPTG).
4. Caspase-2 bacterial expression constructs (e.g., pET-Caspase-2 or pGEX-Caspase-2).
5. Antibiotics for selective growth of transformed bacteria (e.g., 100 μg/ml ampicillin; 34 μg/ml chloramphenicol).
6. Petri dishes (90 mm) containing LB/agar: 1 % Bacto-tryptone, 0.5 % Bacto-yeast extract, 1 % NaCl pH 7.0, 1.5 % Bacto-agar.
7. Orbital shaking incubator at 30 °C.
8. Orbital shaker at 4 °C.

9. Centrifuge.
10. PBS.
11. Chromatography columns (such as Bio-Rad Poly-prep chromatography columns).
12. Ice.
13. Sonicator.
14. Glutathione Sepharose for the purification of GST-tagged proteins or Ni-NTA beads for the purification of 6×His-tagged proteins.
15. Elution Buffer: GST—50 mM Tris–HCl pH 8.0, 10 mM reduced glutathione; Ni-NTA—50 mM sodium phosphate pH 7.0, 300 mM NaCl, 200 mM imidazole.

2.4 Fluorogenic Caspase-2 Activity Assays

1. Caspase-2 synthetic peptide substrate: peptide substrates comprising an *N*-acetyl (Ac)- and C-terminal-conjugated 7-amino-4-trifluoromethylcoumarin (AFC) or 7-amino-4-methylcoumarin (AMC) reporters. Ac-Val-Asp-Val-Ala-Asp-AMC/AFC (Ac-VDVAD-AMC/AFC) is commonly used (available from various commercial sources such as Calbiochem). Colorimetric substrates with *p*-nitroanilide (pNA) reporter are also available (but not as sensitive). AMC and AFC-conjugated substrates are made up as 5–10 mM stock solutions in dimethyl sulfoxide (DMSO) and can be stored in aliquots at −20 °C.
2. Caspase-2 peptide inhibitors: commonly conjugated to chloromethylketone (CMK), fluoromethyl ketone (FMK), or CHO-aldehyde and act as irreversible inhibitors with no added cytotoxic effects. Commonly used (semi-specific) caspase-2 inhibitors include z-VDVAD-FMK and Ac-VDVAD-CHO. Caspase inhibitors are made up at 5–10 mM stock in DMSO and stored in aliquots at −20 °C. Other commonly used pan-caspase inhibitors are z-valine-alanine-aspartate-fluoromethyl ketone (z-VAD-FMK) and quinolyl-valyl-*O*-methylaspartyl-[-2,6-difluorophenoxy]-methyl ketone (Q-VD-OPh). Avoid repeated freeze–thaw cycles.
3. Caspase-2 Assay Buffer: 0.1 M (2-[*N*-morpholino]ethanesulfonic acid) (MES), pH 6.5, 10 % sucrose, 0.1 % CHAPS, 0.5 mM EDTA, supplemented with 10 mM DTT and protease inhibitors (*see* **Notes 1** and **2**).
4. Liquid nitrogen.
5. Ice.
6. Bench top centrifuge.
7. 96-well plate.
8. Fluorometric plate reader equipped with a thermostated plate reader, such as a FLUOstar Optima Luminescence Spectrometer (BMG LabTech).

2.5 In Vitro Translation of Caspase Substrates

1. In vitro-coupled transcription/translation kit, such as the TNT™ Coupled Reticulocyte Lysate System (commercially available from Promega Corporation, USA).
2. ^{35}S-methionine (ICN Biochemicals), for the synthesis of radio-labeled proteins. Labeled proteins can be stored for up to 2 weeks at −70 °C.
3. Ribonuclease inhibitor, such as RNasin (Ambion) or RNAseOUT ribonuclease inhibitor (Invitrogen).
4. Heat block set to 30 °C.

2.6 Protein Electrophoresis

1. Gel precasting apparatus.
2. Resolving gel: 10–15 % acrylamide/bis solution (37.5:1 with 2.6 % C), 375 mM Tris–HCl, pH 8.8, 0.1 % SDS, 0.1 % ammonium persulfate (APS), 0.05 % *N,N,N,N'*-Tetramethyl-ethylenediamine (TEMED).
3. Stacking gel: 5 % acrylamide/bis solution (37.5:1 with 2.6 % C), 125 mM Tris–HCl, pH 6.8, 0.1 % SDS, 0.1 % APS, 0.1 % TEMED (note that acrylamide is a neurotoxin when unpolymerized and so care should be taken to avoid exposure). APS is prepared as a 10 % solution in water and frozen in aliquots at −20 °C.
4. Running Buffer (1×): 25 mM Tris-HCl, 192 mM glycine, 0.1 % SDS. A 5× buffer can be pre-made and stored at room temperature.
5. Pre-stained molecular weight markers such as Kaleidoscope markers (Bio-Rad), pre-stained protein markers (Invitrogen), or unstained protein markers (Invitrogen).
6. Protein electrophoresis apparatus, such as Hoefer minigel tank or Bio-Rad Mini-PROTEAN Electrophoresis system.
7. Power pack supply with constant voltage and constant current settings.
8. Protein stain such as Coomassie Brilliant Blue R-250 (Bio-Rad) made up as a 0.25 % (w/v) stock in H_2O/methanol/acetic acid (5:4:1).

2.7 Immunoblotting

1. Protein Transfer Buffer: 49 mM Tris-HCl pH 7.0, 39 mM glycine, 0.0375 % SDS, 20 % methanol.
2. Polyvinylidene difluoride (PVDF), such as low-fluorescence Immobilon-FL (Millipore).
3. Whatman filter paper.
4. Protein transfer apparatus; either semidry transfer system (Hoefer) or wet transfer system such as the Mini-Trans Blot from Bio-Rad.
5. Rocking platform or shaker.

6. Power pack supply with constant current optional settings.
7. Wash buffer: PBS with 0.05 % Tween-20 (PBS-T).
8. Membrane blocking buffer: 3 % nonfat milk or 3 % bovine serum albumin (BSA) in PBS.
9. Caspase-2 antibody: 1 μg/ml anti-caspase-2 rat monoclonal antibody (clone 11B4) diluted in 3 % skim milk in PBS-T (*see* **Note 3**).
10. Secondary antibody: anti-rat secondary antibody conjugated to alkaline phosphatase (AP) or horse radish peroxidase (HRP).
11. Protein detection solutions: enhanced chemifluorescence (ECF) (AttoPhos, GE Healthcare/Amersham) or enhanced chemiluminescence (ECL) (ECL Plus GE Healthcare/ Amersham).
12. X-ray film and an X-ray developing machine for ECL detection.
13. Laser scanner for fluorescent detection of proteins, such as the Typhoon FLA 7000 Image Scanner for ECF detection (GE Healthcare/Amersham).

2.8 Affinity Capture of Active Caspase-2

1. Affinity label biotin-VAD-FMK is prepared as 10 mM stock in DMSO, aliquoted and stored at −70 °C.
2. Affinity Label Dilution Buffer: 50 mM HEPES-KOH pH 7.0, 50 mM NaCl, 1 mM $MgCl_2$, 1 mM ethylene glycol-bis [β-aminoethyl ether]-*N*,*N*,*N*′,*N*′-tetraacetic acid (EGTA), 1 mM EDTA, 1 mM DTT, and supplemented with protease inhibitor cocktail.
3. Cell Lysis Buffer: 50 mM HEPES-KOH pH 7.0, 50 mM NaCl, 1 mM EDTA, 1 mM $MgCl_2$, 5 mM DTT, 0.1 % CHAPS, 1 mM phenylmethanesulfonyl fluoride (PMSF), and supplemented with protease inhibitor cocktail.
4. Streptavidin Sepharose.
5. 2× Protein Loading Buffer (PLB).
6. Rotating platform at 4 °C.
7. Bench top centrifuge.

3 Methods

3.1 Purification of Active Recombinant Caspase-2

1. A bacterial expression plasmid containing the caspase-2 coding region in frame of a C-terminal 6×His tag (pET-Caspase-2) or GST-tag (pGEX-Caspase-2) (*see* **Note 4**) is transformed into BL21 cells (typically cells that express lysozyme such as BL21-star or BL21-pLysE) using a standard heat shock method, and

cells are plated onto LB/agar containing the appropriate antibiotic (ampicillin, kanamycin, and/or chloramphenicol) for selection of transformed bacterial colonies. Bacterial colonies are grown on LB/agar petri dishes overnight at 37 °C.

2. Single colonies are inoculated into 5 ml LB (with antibiotic) and cultured overnight (O/N) with shaking at 37 °C.
3. 5 ml O/N culture is subcultured 1:20 into 200 ml LB (with antibiotic) and incubated at 30 °C for 3 h with shaking.
4. Add IPTG to a final concentration 1 mM, and incubate a further 4 h with shaking at 30 °C.
5. Pellet bacterial cells by centrifugation at 6,000 × *g* for 15 min at 4 °C. Store the bacterial pellet at −20 °C overnight.
6. Resuspend pellet in 10 ml PBS and lyse bacteria by sonication (5 × 20 s pulses)
7. Add Triton X-100 to final concentration of 0.1 %, and incubate 30 min at 4 °C.
8. Clarify by centrifugation at 15,000 × *g* for 30 min at 4 °C and transfer cleared supernatant to a new tube. At this stage, the lysate can be used in a caspase-2 activity assay (Subheading 3.2) to determine the activity level of recombinant protein.
9. Recombinant caspase-2 protein is now ready to purify on glutathione Sepharose beads (Caspase-2-GST) or Ni-NTA beads (Caspase-2-His_6) (*see* **Notes 5** and **6**).
10. Wash glutathione Sepharose beads or Ni-NTA beads in PBS.
11. Add 50 μl bead volume to the cleared lysate and incubate at 4 °C for 1 h with gentle rotation.
12. Transfer beads to a chromatography column (e.g., Poly-prep chromatography column from Bio-Rad) and wash beads three times with 10 ml of PBS.
13. Elute recombinant caspase-2 protein in 5 bead volumes of elution buffer in five separate fractions and quantitate protein concentration using a standard BCA or Bradford assay.
14. Purified protein can also be checked by resolving protein on SDS-PAGE and proteins then stained with Coomassie Brilliant Bluc.

3.2 Measuring Caspase-2 Activity

Caspase activity assays are typically carried out over a time course (i.e., every 10–15 min over a 1–2 h time course). The spectrophotometer should be equipped with a regulated temperature chamber that can accommodate 96-well plates. Alternatively the release of AFC, AMC fluorescent tags can be measured following different time points over a fixed period of incubation at 37 °C. The addition of a caspase-2 inhibitor (z-VDVAD-FMK or Ac-VDVAD-CHO) or pan-caspase inhibitors (z-VAD-FMK or Q-VD-OPh) should be used as controls. If assessing caspase-2 activity in protein extracts, controls should also include untreated cells and apoptotic cell extracts.

1. To assess the activity of recombinant caspase-2 protein, add 100 nM purified caspase-2 to Caspase-2 Assay Buffer supplemented with 100 μM VDVAD-AMC (or AFC-conjugated substrate) in a total volume 100 μl.
2. To monitor caspase-2-like activity in apoptotic protein extracts, add cell extract containing 20–50 μg total protein to Caspase-2 Assay Buffer supplemented with 100 μM VDVAD-AMC (or AFC-conjugated substrate) in a total volume 100 μl.
3. As a control, recombinant caspase-2 protein should be preincubated with inhibitor peptide for 30 min at room temperature, prior to the addition of peptide substrate.
4. Monitor cleavage by measurement of AMC/AFC fluorescence every 10–15 min over 1–2 h on a spectrophotometer at 37 °C. AMC fluorescence is measured at excitation wavelength 360 nm and emission wavelength 460 nm (AFC fluorescence: excitation 400 nm; emission 505 nm).
5. Plot the data as relative fluorescence units versus time (min) for each sample, and calculate the slope of the line from the linear part of the curve (*see* **Note 7**).
6. Caspase activity is expressed as picomole substrate hydrolyzed/min/μg protein. To do this, generate an AMC/AFC fluorescence calibration curve by measuring fluorescence of each of the AMC/AFC calibration standards (serial dilutions with concentration range 0–50 μM) made in assay buffer (final volume 100 μl). Plot relative fluorescence units (RFU) versus AMC/AFC concentration (μM) and calculate the slope of the linear region for the calibration standard.
7. Calculate

$$\begin{aligned} &\textit{Caspase activity} \\ &(\textit{pmol / min}) = [\textit{slope of sample } (RFU\ /\ min) \\ &\qquad \times [1\ /\ \textit{slope of calibration curve}](\mu M\ /\ RFU) \\ &\qquad \times \textit{assay volume}(\mu l)]/\textit{amount of proteinused}(\mu g) \end{aligned}$$

3.3 Assessing Caspase-2 Activity by Cleavage of Labeled Substrates

To validate caspase-2 substrates and/or determine caspase-2 activation, in vitro qualitative assays can be carried out with recombinant caspase-2 incubated with either in vitro-translated protein substrates or cell extracts.

3.3.1 Cleavage of ^{35}S-Met Labeled Caspase-2 Substrates In Vitro

1. The cDNA encoding the specific caspase-2 substrate is cloned in plasmid vectors that carry either SP6, T3, or T7 promoters (e.g., pBluescript, pGEM-T, or pcDNA3 vectors). If it is not possible to clone the entire protein coding region, truncated

coding regions containing the putative caspase-2 cleavage site can be used [43].

2. In vitro transcription/translation of each substrate is carried out using the Promega TNT-Rabbit Reticulocyte Translation System according to the instructions provided by the manufacturer. Reactions will contain 25 μl TNT lysate, 2 μl TNT reaction buffer, 1 μl RNA polymerase (T3, T7, or SP6), 1 μl amino acid mixture lacking methionine, 3 μl ^{35}S-methionine, 1 μl ribonuclease inhibitor, 1 μg plasmid DNA, and sterile RNase-free water to a final volume of 50 μl.
3. Incubate reactions at 30 °C for approximately 2 h.
4. Centrifuge samples at 10,000 × *g* for 5 min and transfer supernatant to a fresh tube. At this stage, the in vitro-translated ^{35}S-labeled protein can be stored at −70 °C for up to 2 weeks.
5. For cleavage assays, incubate 5 μl of ^{35}S-labeled protein at 37 °C for 2 h with 50–100 nM recombinant caspase-2 in caspase assay buffer in a total volume 50 μl.
6. To controls, preincubate recombinant caspase-2 with a caspase inhibitor (e.g., 50 μM Ac-VDVAD-CHO, z-VAD-FMK, or Q-VD-OPh) for 30 min prior to the addition of in vitro-translated ^{35}S-labeled protein substrate.
7. To stop reactions, add 50 μl of 2× PLB to each tube, boil for 5 min, and centrifuge at 10,000 × *g* for 5 min.
8. Check substrate cleavage products by resolving ^{35}S-labeled proteins by electrophoresis on a 10–15 % polyacrylamide/SDS gel.
9. Once resolved, transfer proteins to PVDF membranes using either wet or semidry transfer apparatus (for approximately 90 min at 130 mA).
10. Visualize ^{35}S-labeled protein bands by autoradiography, following overnight exposure to X-ray film or phosphor screen. Longer exposure time may be required if signal intensity is low. An example of autoradiograph showing caspase-2 cleavage by recombinant caspase-2 or apoptotic lysate is shown in Fig. 2.

3.3.2 Cleavage of Substrates in Cells Extracts by Recombinant Caspase-2

An alternative method to demonstrate caspase-2-mediated cleavage of substrate is to incubate recombinant caspase-2 with cell extracts and observe protein substrate cleavage by immunoblotting using specific antibodies against a known endogenous caspase-2 substrates.

1. To prepare protein cell extracts, cultured cells are harvested by gentle scraping into medium or by treatment with trypsin for 5 min at 37 °C (for adherent cells), or cell culture medium is simply collected (for suspension cells). For tissues, homogenize frozen tissue in cell extraction buffer using a tissue homogenizer.

2. Pellet cells by centrifugation at 200 × *g* for 5 min and wash in cold PBS.
3. Resuspend cell pellet in cell extraction buffer (at approximately 1×10^7 cells/ml).
4. Snap freeze cells in liquid nitrogen and then thaw in ice-cold water. Repeat this three times.
5. Protein extracts are cleared by centrifugation at 15,000 × *g* for 10 min at 4 °C and cleared supernatant (cytosolic extract) is transferred to a clean tube.
6. Protein concentrations are determined using standard BCA assays (Bio-Rad) and extracts are stored at −70 °C in small aliquots. Protein extracts and caspase activity are stable for several months at −70 °C.
7. To observe caspase-2-mediated cleavage of cellular protein substrates, incubate cell extract containing 20–30 μg total protein with 100 nM recombinant caspase-2 at 37 °C for 2–3 h in a total volume of 50 μl of Caspase-2 Assay Buffer.
8. To stop the reaction, add 50 μl of 2× PLB, boil for 5 min, and centrifuge at 10,000 × *g* for 5 min to remove any insoluble material.
9. Resolve proteins on a 12–15 % polyacrylamide/SDS gel, and transfer proteins to PVDF membrane using a wet or semidry protein transfer apparatus.
10. Incubate PVDF membrane in 3 % skim milk (or 3 % BSA) in PBST for 1 h at room temperature or overnight at 4 °C.
11. Dilute primary antibody as suggested by the manufacturer in 1–5 % skim milk/PBST (or 3 % BSA) and incubate the PVDF membrane with the antibody solution for 1 h at room temperature with gentle shaking. Incubation with caspase-2 antibody will show cleavage of caspase-2 and can be used as a control to demonstrate activity of the recombinant caspase-2 protein (*see* **Note 3**). Cellular proteins known to be cleaved by caspase-2 are Bid, Golgin-130, and Mdm2 and can be detected using antibodies specific to these proteins.
12. Wash the membrane twice for 5 min each and then twice for 10 min each and incubate with the appropriate secondary antibody diluted in 1–5 % skim milk/PBST (or 3 % BSA).
13. Protein detection can be carried out using ECL or ECF (as per instruction supplied by manufacturer) and visualized on X-ray film (for ECL) or for ECF imaging on a fluorescence scanner bed (such as a Typhoon FLA 7000 Image Scanner, GE Healthcare).

3.4 Assessing Active Caspase-2 In Vivo by Affinity Labeling

Affinity labeling of caspase-2 can be carried out in cells in culture following apoptotic stimulation or can be carried out using apoptotic cell extracts. The affinity label biotin-VAD-FMK is a pan-caspase inhibitor which will interact with all active caspases and has been used to efficiently detect active caspase-2 following immunoblotting with anti-caspase-2 antibody [41, 42] (*see* **Note 8**).

1. The affinity label can be added directly to cells in culture medium. In this case, approximately 1×10^7 cells are left untreated or treated with apoptotic stimuli and then incubated with 50 μM biotin-VAD-FMK for 2 h at 37 °C. As a negative control, include cells incubated in the same volume of DMSO.
2. Lyse cells in lysis buffer followed by freeze–thaw cycles in liquid nitrogen/ice-cold water as described in Subheading 3.3.2 (**steps 1–5**). Affinity label can also be added at this point to the cell extract at a final concentration of 50 μM. In this case, dilute the affinity label to 100 μM in Dilution Buffer, add an equal volume to cell extract containing 100–500 μg total protein, and incubate at room temperature for 1 h.
3. Increase the sample volume to 1 ml in lysis buffer and add 50 μl streptavidin-coupled Sepharose. Incubate extracts with the Sepharose on a rotating platform, overnight at 4 °C.
4. Precipitate the bound Sepharose by centrifugation at $2{,}000 \times g$ for 3 min and wash three times with 1 ml lysis buffer (*see* **Note 9**).
5. Add an equal bed volume of 2× PLB (50 μl) and boil samples for 5 min.
6. Resolve proteins on 15 % SDS-PAGE and transfer to PVDF membrane (*see* Subheading 3.3.2, **steps 9** and **10**).
7. Detect caspase-2 by immunoblotting with anti-caspase-2 rat monoclonal antibody (11B4) (*see* Subheading 3.3.2, **steps 11–13**).

4 Notes

1. Caspase-2 activity appears to be optimal at a slightly acidic pH [44], so assay buffer containing 0.1 M MES, pH 6.5, is commonly used. However, protein still demonstrates activity in buffer containing 0.1 M HEPES, pH 7.0. Caspase-2 Assay Buffer without DTT can be prepared in advance and stored at room temperature for several months, with DTT added fresh as required.
2. Caspase assay buffer and cell lysis buffer should contain freshly prepared protease inhibitor cocktail to prevent nonspecific hydrolysis of caspase substrates during protein extract preparation. Make sure that the protease inhibitor mix does not inhibit caspases.
3. There are numerous commercial sources of caspase-2 antibodies; however, many available antibodies are of poor quality. Specificity and affinity of the antibody can be established using

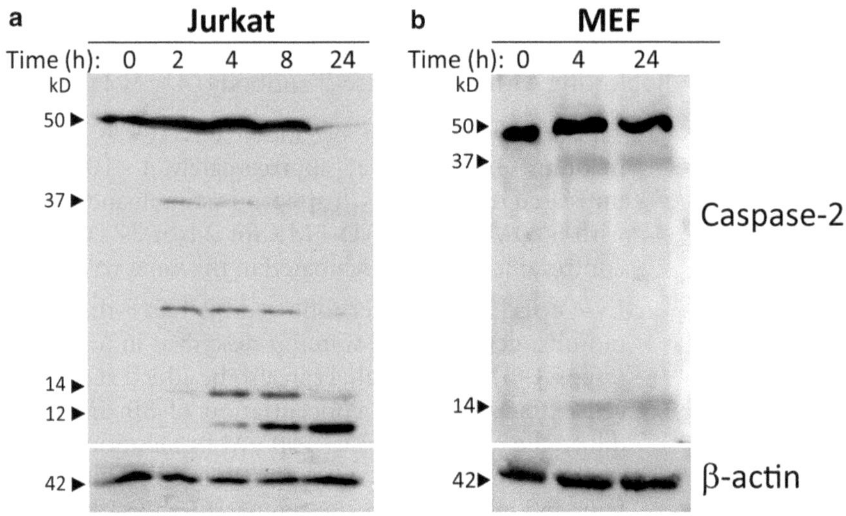

Fig. 3 Processing of caspase-2 in apoptotic cell lysates. Jurkat cells (**a**) or mouse embryonic fibroblasts (MEF) (**b**) were treated with etoposide (40 μM) and cell lysates prepared at the indicated time points over 24 h. Protein was electrophoresed through 15 % (**a**) or 4–15 % SDS-PAGE gradient gel (**b**) and transferred to PVDF membrane. Immunoblotting was carried out using anti-caspase-2 rabbit polyclonal antibody (SC-625; Santa Cruz) to detect human caspase-2 (**a**) or with anti-caspase-2 rat monoclonal antibody (11B4, Millipore) to detect mouse caspase-2 (**b**). β-Actin monoclonal antibody (Sigma) was used as a protein load control. Signals were detected by ECL

recombinant caspases. We have found that the rat anti-caspase-2 antibody clone (11B4) works best to detect full-length caspase-2 and cleaved 37 kDa caspase-2 (minus pro-domain) (Fig. 3). Some antibodies, such as rabbit anti-caspase-2 (C10, Santa Cruz Biotechnology, Santa Cruz, CA, USA) will detect both the precursor and one or more cleavage products (Fig. 3), while others only appear to recognize full-length caspase-2. Active caspase-2 antibody is available from some companies (Cell Signaling) and will detect cleaved small subunits (p19 and p14/p12). In some cell types, the half-life of processed caspase-2 subunits can be short, but a decrease in zymogen signal can generally be monitored to determine whether caspase-2 is cleaved following specific stimuli.

4. Expression of a caspase-2 protein without the pro-domain (i.e., minus CARD) in bacteria has been demonstrated to have higher activity than full-length caspase-2 protein.
5. GST or 6×His tags should be placed C-terminal to caspase-2 protein, since caspase-2 is rapidly processed following the CARD, when expressed in bacteria. We have not found that the GST or 6×His tag interferes with the activity of caspase-2; however, it is also possible to cleave these tags provided the tag is preceded by a thrombin cleavage site or TEV cleavage site.

6. There are also several commercial suppliers of purified recombinant caspase-2, such as Enzo Life Sciences (USA), Millipore (USA), and MP Biomedicals (USA).
7. For caspase activity assays, if sample is highly active, the substrate will be rapidly saturated. In this case, dilute out protein or lysate to get a linear response.
8. There are several different methods describing affinity capture of active caspases, and the protocol described here is a modification of the methods described in Faleiro et al. and in Dorstyn et al. [38, 40]. There are various caspase affinity labels available, and we have found that biotin-VAD-FMK labeling works best to detect active caspase-2. Incubating cells with biotin-VAD-FMK prior to cell lysis seems to be more effective in our hands.
9. If nonspecific binding is a problem, then this can be minimized by including a wash step in lysis buffer containing 0.5 M NaCl. However, this will need to be followed by additional 2–3 washes in lysis buffer (without NaCl) to reduce the high salt concentration that may interfere with electrophoresis of samples.

Acknowledgments

The financial support of the National Health and Medical Research Council through project grants 1021456 and 1043057 is gratefully acknowledged. LD is supported by a South Australian Cancer Research Collaborative Fellowship and SK by an NHMRC Senior Principal Research Fellowship (1002863).

References

1. Kumar S, Kinoshita M, Noda M et al (1994) Induction of apoptosis by the mouse Nedd2 gene, which encodes a protein similar to the product of the *Caenorhabditis elegans* cell death gene ced-3 and the mammalian IL-1 beta-converting enzyme. Genes Dev 8: 1613–1626
2. Kumar S (2009) Caspase 2 in apoptosis, the DNA damage response and tumour suppression: enigma no more? Nat Rev Cancer 9: 897–903
3. Lamkanfi M, Declercq W, Kalai M et al (2002) Alice in caspase land. A phylogenetic analysis of caspases from worm to man. Cell Death Differ 9:358–361
4. Butt AJ, Harvey NL, Parasivam G et al (1998) Dimerization and autoprocessing of the Nedd2 (caspase-2) precursor requires both the prodomain and the carboxyl-terminal regions. J Biol Chem 273:6763–6768
5. Colussi PA, Harvey NL, Kumar S (1998) Prodomain-dependent nuclear localization of the caspase-2 (Nedd2) precursor. A novel function for a caspase prodomain. J Biol Chem 273: 24535–24542
6. Baliga BC, Colussi PA, Read SH et al (2003) Role of prodomain in importin-mediated nuclear localization and activation of caspase-2. J Biol Chem 278:4899–4905
7. Baliga BC, Read SH, Kumar S (2004) The biochemical mechanism of caspase-2 activation. Cell Death Differ 11:1234–1241
8. Tinel A, Tschopp J (2004) The PIDDosome, a protein complex implicated in activation of caspase-2 in response to genotoxic stress. Science 304:843–846

9. Kim IR, Murakami K, Chen NJ et al (2009) DNA damage- and stress-induced apoptosis occurs independently of PIDD. Apoptosis 14: 1039–1049
10. Manzl C, Krumschnabel G, Bock F et al (2009) Caspase-2 activation in the absence of PIDDosome formation. J Cell Biol 185: 291–303
11. Kumar S, Colussi PA (1999) Prodomains–adaptors–oligomerization: the pursuit of caspase activation in apoptosis. Trends Biochem Sci 24:1–4
12. Shi Y (2004) Caspase activation, inhibition, and reactivation: a mechanistic view. Protein Sci 13:1979–1987
13. Kumar S (1999) Mechanisms mediating caspase activation in cell death. Cell Death Differ 6:1060–1066
14. Schweizer A, Briand C, Grutter MG (2003) Crystal structure of caspase-2, apical initiator of the intrinsic apoptotic pathway. J Biol Chem 278:42441–42447
15. Talanian RV, Quinlan C, Trautz S et al (1997) Substrate specificities of caspase family proteases. J Biol Chem 272:9677–9682
16. Guo Y, Srinivasula SM, Druilhe A et al (2002) Caspase-2 induces apoptosis by releasing proapoptotic proteins from mitochondria. J Biol Chem 277:13430–13437
17. Lassus P, Opitz-Araya X, Lazebnik Y (2002) Requirement for caspase-2 in stress-induced apoptosis before mitochondrial permeabilization. Science 297:1352–1354
18. Robertson JD, Enoksson M, Suomela M et al (2002) Caspase-2 acts upstream of mitochondria to promote cytochrome c release during etoposide-induced apoptosis. J Biol Chem 277: 29803–29809
19. Bonzon C, Bouchier-Hayes L, Pagliari LJ et al (2006) Caspase-2-induced apoptosis requires bid cleavage: a physiological role for bid in heat shock-induced death. Mol Biol Cell 17: 2150–2157
20. Ho LH, Read SH, Dorstyn L et al (2008) Caspase-2 is required for cell death induced by cytoskeletal disruption. Oncogene 27: 3393–3404
21. Dorstyn L, Puccini J, Wilson CH et al (2012) Caspase-2 deficiency promotes aberrant DNA-damage response and genetic instability. Cell Death Differ 19:1288–1298
22. Shin S, Lee Y, Kim W et al (2005) Caspase-2 primes cancer cells for TRAIL-mediated apoptosis by processing procaspase-8. EMBO J 24:3532–3542
23. Krumschnabel G, Sohm B, Bock F et al (2009) The enigma of caspase-2: the laymen's view. Cell Death Differ 16:195–207
24. Bergeron L, Perez GI, Macdonald G et al (1998) Defects in regulation of apoptosis in caspase-2-deficient mice. Genes Dev 12: 1304–1314
25. O'Reilly LA, Ekert P, Harvey N et al (2002) Caspase-2 is not required for thymocyte or neuronal apoptosis even though cleavage of caspase-2 is dependent on both Apaf-1 and caspase-9. Cell Death Differ 9:832–841
26. Nutt LK, Margolis SS, Jensen M et al (2005) Metabolic regulation of oocyte cell death through the CaMKII-mediated phosphorylation of caspase-2. Cell 123:89–103
27. Kumar S (2007) Caspase function in programmed cell death. Cell Death Differ 14: 32–43
28. Sidi S, Sanda T, Kennedy RD et al (2008) Chk1 suppresses a caspase-2 apoptotic response to DNA damage that bypasses p53, Bcl-2, and caspase-3. Cell 133:864–877
29. Ho LH, Taylor R, Dorstyn L et al (2009) A tumor suppressor function for caspase-2. Proc Natl Acad Sci U S A 106:5336–5341
30. Braga M, Sinha Hikim AP, Datta S et al (2008) Involvement of oxidative stress and caspase 2-mediated intrinsic pathway signaling in age-related increase in muscle cell apoptosis in mice. Apoptosis 13:822–832
31. Zhang Y, Padalecki SS, Chaudhuri AR et al (2007) Caspase-2 deficiency enhances aging-related traits in mice. Mech Ageing Dev 128:213–221
32. Shalini S, Dorstyn L, Wilson C et al (2012) Impaired antioxidant defence and accumulation of oxidative stress in caspase-2-deficient mice. Cell Death Differ 19:1370–1380
33. Paroni G, Henderson C, Schneider C et al (2002) Caspase-2 can trigger cytochrome C release and apoptosis from the nucleus. J Biol Chem 277:15147–15161
34. Bouchier-Hayes L, Oberst A, McStay GP et al (2009) Characterization of cytoplasmic caspase-2 activation by induced proximity. Mol Cell 35:830–840
35. Tinnikov AA, Samuels HH (2013) A novel cell lysis approach reveals that caspase-2 rapidly translocates from the nucleus to the cytoplasm in response to apoptotic stimuli. PLoS One 8:e61085
36. Thornberry NA, Rano TA, Peterson EP et al (1997) A combinatorial approach defines

specificities of members of the caspase family and granzyme B. Functional relationships established for key mediators of apoptosis. J Biol Chem 272:17907–17911
37. Stennicke HR, Deveraux QL, Humke EW et al (1999) Caspase-9 can be activated without proteolytic processing. J Biol Chem 274: 8359–8362
38. Dorstyn L, Kumar S (2008) A biochemical analysis of the activation of the *Drosophila* caspase DRONC. Cell Death Differ 15:461–470
39. Thornberry NA, Peterson EP, Zhao JJ et al (1994) Inactivation of interleukin-1 beta converting enzyme by peptide (acyloxy)methyl ketones. Biochemistry 33:3934–3940
40. Faleiro L, Kobayashi R, Fearnhead H et al (1997) Multiple species of CPP32 and Mch2 are the major active caspases present in apoptotic cells. EMBO J 16:2271–2281
41. Read SH, Baliga BC, Ekert PG et al (2002) A novel Apaf-1-independent putative caspase-2 activation complex. J Cell Biol 159:739–745
42. Tu S, McStay GP, Boucher LM et al (2006) In situ trapping of activated initiator caspases reveals a role for caspase-2 in heat shock-induced apoptosis. Nat Cell Biol 8:72–77
43. Harvey NL, Butt AJ, Kumar S (1997) Functional activation of Nedd2/ICH-1 (caspase-2) is an early process in apoptosis. J Biol Chem 272:13134–13139
44. Garcia-Calvo M, Peterson EP, Rasper DM et al (1999) Purification and catalytic properties of human caspase family members. Cell Death Differ 6:362–369

Chapter 5

Caspase-14 Protocols

Mami Yamamoto-Tanaka and Toshihiko Hibino

Abstract

Unlike other caspase family members, caspase-14 shows restricted expression, being found mostly in epidermis and its appendages. It has been suggested that caspase-14 is not involved in apoptosis or inflammation, but participates in keratinocyte terminal differentiation. Its activation occurs at the corneocyte formation. In previous work, we have purified active caspase-14 from human corneocyte extracts. In addition, we have clarified activation mechanism of caspase-14, where kallikrein-related peptidase 7 (KLK7) generates an intermediate form from procaspase-14 and this form finally converts procaspase-14 to active, mature caspase-14. Here we describe techniques for measurement of caspase-14 activity using synthetic substrate, purification of caspase-14 from corneocyte extract, preparation of constitutively active caspase-14 and specific antibody, quantification of total and active caspase-14 in corneocyte extracts using ELISA, as well as methods for caspase-14 activation and its visualization by immunohistochemistry.

Key words Caspase-14, Epidermis, Keratinocyte, Terminal differentiation, Corneocyte, Kallikrein-related peptidase

1 Introduction

Caspase-14 is a member of the caspase family, which is most exclusively expressed in epidermis and its appendages [1–3]. Keratinocytes are the major components of epidermis and they constitute 90 % of this protective tissue. Keratinocytes undergo terminal differentiation that is precisely controlled, starting from proliferating basal cells; progressing to the appearance of spinous, granular cells; and ultimately leading to cornified cells [4]. This process results in formation of the permeability barrier that protects the body against hazardous environmental factors. Epidermal barrier includes various components, such as natural moisturizing factor (NMF), extracellular lipid membranes, cornified envelope (CE), and internal structure of macrofibrils to support and to maintain flexible yet sturdy architecture. Dramatic changes occur at the transition phase from granular cells to cornified cells. Cellular organelles are digested and cytoplasm is occupied with keratin

Peter V. Bozhkov and Guy Salvesen (eds.), *Caspases, Paracaspases, and Metacaspases: Methods and Protocols*,
Methods in Molecular Biology, vol. 1133, DOI 10.1007/978-1-4939-0357-3_5,

fibers embedded in filaggrin to form macrofibrils. They also gain a reinforced, cytoplasmic membrane called CE, wherein omega-hydroxyceramides are covalently attached to cover the interstices of the cornified cells [5], and special structural proteins, such as involucrin and loricrin, are cross-linked to the membrane by transglutaminase [6, 7]. Internally, the cells lose their nuclei and cytoplasmic organelles, becoming filled with keratin filaments that are bound with an amorphous material, filaggrin. Caspase-14 is activated during this transition phase.

Previously we purified active caspase-14 from the extract of human cornified cells [8]. The primary structure of the purified caspase-14 consists of large and small subunits generated by the removal of a 6-amino-acid linker peptide. The C-terminal of the large subunit was identified as D^{146}, indicating that the activation-dependent cleavage occurs at this site. A cleavage-site-directed antibody, h14D146, was prepared and found to react only with active caspase-14. Immunohistochemical study using this antibody showed strong positive staining in the cornified layer, as well as of some of the granular cells.

We also constructed a constitutively active caspase-14 (revC14) based on the sequence information of purified human caspase-14. Large subunit and small subunit was reversely fused with an N-terminal His-tag. Recombinant revC14 demonstrated similar enzymatic activity with that of the purified caspase-14 from the corneocyte extracts.

For the quantitative analysis of caspase-14, we established two ELISA assays to quantify the total amount of caspase-14 and that of active caspase-14. We produced four clones of monoclonal antibodies to caspase-14 and used clone 3 as the immobilizing antibody. A commercial antibody, H99, and cleavage-site-directed antibody, h14D146 [8], were used for the quantification of total and active caspase-14 in extracts of tape-stripped corneocytes, respectively. Caspase-14 was found mostly in active form (71–94 %) in the extracts from healthy individuals. In patients with atopic dermatitis, active caspase-14 was markedly downregulated compared to age-matched controls in both lesional (7.5 %) and non-lesional skin (10.6 %) [9].

The maturation and activation mechanisms of caspases are generally well understood, except for those of caspase-14, which is activated at the onset of keratinocyte terminal differentiation. We found that the chymotrypsin-like serine protease kallikrein-related peptidase 7 (KLK7) cleaved procaspase-14 at Y^{178}, generating an intermediate form consisting of a large (20 kDa) and a small subunit (8 kDa) [10]. We confirmed the formation of the intermediate form by raising a cleavage-site-directed antibody, h14Y178. This antibody was prepared in the same manner as h14D146, an antibody specific to the active, mature form of caspase-14. We also constructed a constitutively active form of the intermediate, revC14-Y178. The substrate specificity of revC14-Y178 was completely different from that of caspase-14, showing broad specificity for various caspase substrates except WEHD-AFC, the preferred substrate of active, mature caspase-14. We propose the following mechanism of caspase-14 activation. Synthesis of procaspase-14 and preproKLK7 is initiated at the stratum spinosum (SS) and peaks at the stratum granulosum (SG). Activated KLK7 cleaves procaspase-14 at Y^{178}, generating the intermediate form (p20/p8), which acts on procaspase-14 leading to the production of mature form, p17/p11. The proposed mechanism of caspase-14 activation is thus distinct from canonical mechanisms of caspase activation mediated by apoptosome, DISC, or inflammasome [11, 12].

Here we describe techniques for measurement of caspase-14 activity using synthetic substrate, purification of caspase-14 from corneocyte extract, preparation of constitutively active caspase-14 and specific antibody, quantification of total and active caspase-14 in corneocyte extracts using ELISA, as well as methods for caspase-14 activation and its visualization by immunohistochemistry.

2 Materials

2.1 Equipment

1. Plate reader (Fluoroskan Ascent FL or similar).
2. Glass homogenizer (e.g., Wheaton Tenbroeck Tissue Grinder, 40 ml).
3. Protein concentrators (e.g., Amicon Ultra, Millipore).
4. Fast Desalting column HR10/10 (Amersham Biosciences) or equivalent.
5. HiPrep 16/10 Q XL column.
6. Mono Q column.
7. Mono S cation-exchange column.
8. Chromatofocusing Mono P column.
9. Superdex 75 gel column.
10. Spectrophotometer with UV.
11. PD-10 column (GE healthcare).
12. 96-well ELISA plate.
13. Microtome for paraffin sections.
14. Embedder Sakura ETP (Sakura Finetek) or similar.

2.2 Reagents

1. Ac-WEHD-7-amino-4-trifluoromethylcoumarin (AFC; e.g., from BioVision).
2. Caspase-14 assay buffer: 1.3 M tri-sodium citrate dihydrate, 0.1 M HEPES (pH 7.5), 60 mM NaCl, 0.01 % CHAPS, and 5 mM DTT (*see* **Note 1**).

3. Extraction buffer: 0.1 M Tris–HCl (pH 8.0), 0.14 M NaCl, and 1 mM phenylmethane sulfonyl fluoride.
4. 20 mM Tris–HCl (pH 8.0).
5. 1 M Tris–HCl (pH 8.0).
6. 0–1 M linear NaCl gradient.
7. 20 mM acetate buffer (pH 4.5).
8. 25 mM ethanolamine (pH 8.3).
9. Polybuffer: Polybuffer 74 (70 %) + Polybuffer 96 (30 %), pH adjusted to 5.0 with acetic acid.
10. Bio-Rad protein assay kit or similar.
11. Coomassie Brilliant Blue (CBB).
12. H99 antibody directed to the large subunit of human caspase-14 (Santa Cruz Biotechnology).
13. C20 antibody directed to the small subunit of caspase-14 (Santa Cruz Biotechnology).
14. pQE 30 vector (Qiagen).
15. pET15b vector.
16. BL21 (DE3) bacterial strain (Novagen).
17. Overnight Express Autoinduction System 1 (Novagen, Merck KGaA).
18. Ni-NTA His Bind Resin (Novagen) or similar.
19. Ni-NTA Agarose.
20. Active KLK7 (e.g., from R & D systems).
21. KLK7 assay buffer: 0.1 M HEPES or Tris–HCl (pH 7.5), 1 mM $CaCl_2$.
22. 0.1 M glycine-HCl buffer (pH 2.3).
23. ImmunoBlock (DS Pharma Biomedical Co) or similar blocking reagent.
24. PBS containing 0.1 % Tween 20 (PBST).
25. ECL™ Anti-rabbit IgG Horseradish Peroxidase linked $F(ab)'^2$ fragment (GE Healthcare) or similar.
26. TMB Peroxidase EIA Substrate Kit (Bio-Rad Laboratories).
27. 0.1 M sulfuric acid.
28. Acetone.
29. Methyl benzoate.
30. Xylene.
31. Paraffin (melting point 58–60 °C).
32. Goat serum.
33. VECTASHIELD Mounting Medium (VECTOR Laboratories).
34. 4′,6′-Diamidino-2-phenylindole (DAPI).

3 Methods

3.1 Measurement of Caspase-14 Activity

1. Mix 10 μl of corneocyte extract and 70 μl of caspase-14 assay buffer (*see* **Note 2**).
2. Incubate for 15 min at room temperature.
3. Add 10 μl of 0.2 mM Ac-WEHD-AFC (dissolved in DMSO) and incubate for 30 min at 37 °C (*see* **Note 3**).
4. Measure enzyme activity on plate reader with 355 nm excitation and 460 nm emission.

3.2 Purification of Caspase-14 from Corneocyte Extract

1. Prepare 10–15 g of cornified cells by scrapping them from heels of healthy individuals.
2. Homogenize ca. 0.5 g in 20 ml of the extraction buffer using glass homogenizer.
3. Centrifuge at 15,000 × *g* for 60 min. Remove supernatant and concentrate to ca. 2 ml using protein concentrator.
4. Desalt the sample using Fast Desalting column equilibrated with 20 mM Tris–HCl (pH 8.0), or dialyze against the same buffer.
5. Apply the crude extract to a HiPrep 16/10 Q XL column, wash the column with 20 mM Tris–HCl (pH 8.0) and elute with a 0–1 M linear NaCl gradient.
6. Monitor hydrolytic activity against WEHD-AFC in each fraction.
7. Concentrate active fraction(s) to 3.5 ml using protein concentrator.
8. Exchange buffer with 20 mM Tris–HCl (pH 8.0) equilibrated with 20 mM acetate buffer (pH 4.5).
9. Apply active fractions to a Mono Q column equilibrated with the same buffer and elute with 0–1 M NaCl gradient.
10. After concentration and buffer-exchange, apply caspase-14-containing fractions to Mono S cation-exchange column equilibrated with 20 mM acetate buffer (pH 4.5) and elute with 0–1 M NaCl gradient.
11. After concentration and buffer-exchange, apply active fractions to a chromatofocusing Mono P column equilibrated with 25 mM ethanolamine (pH 8.3).
12. Elute with 46 ml of Polybuffer (pH 5.0) forming a pH gradient from pH 8.0 to pH 5.0 and finally with 2 M NaCl.
13. Apply caspase-14 containing fraction (less than 1 ml volume) on Superdex 75 gel column equilibrated with PBS.
14. Determine protein concentration using, e.g., a Bio-Rad protein assay kit.

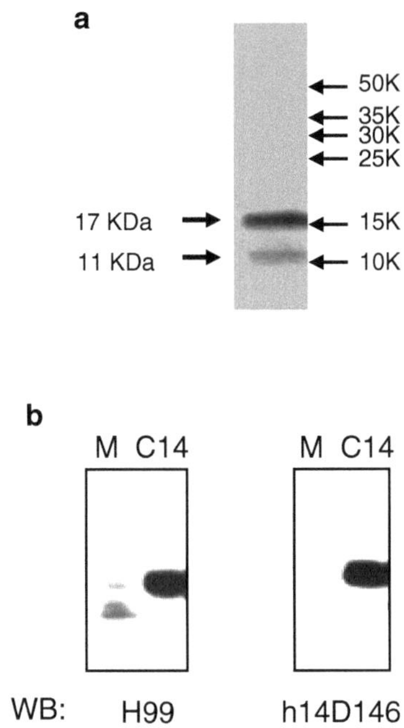

Fig. 1 Purification and characterization of human caspase-14. (**a**) SDS-PAGE analysis of purified caspase-14. Human caspase-14 was purified from corneocyte extract with successive chromatographic procedures. The caspase-14 containing fraction collected after Superdex gel chromatography was subjected to SDS-PAGE analysis. Gel was stained with CBB. The 17 kDa band corresponds to the large subunit and the 11 kDa bands to the small subunit of caspase-14. (**b**) Western blot analysis of the purified fraction. The purified fraction was verified with three different antibodies to caspase-14: H99 (recognizes the large subunit), h14D146 (recognizes cleavage site of the active caspase-14), and C20 (recognized the small subunit). *M* molecular weight marker, *C14* purified caspase-14 fraction

15. Check protein quality on SDS-PAGE (5–15 % gradient gel) after staining with CBB (*see* Fig. 1a).
16. For further characterization, Western blot can be performed using the cleavage-site-directed antibody (h14D146) (*see* Subheading 3.5), antibody to the large subunit (H99), and antibody to the small subunit of caspase-14 (C20) (*see* Fig. 1b).

3.3 Preparation of Constitutively Active Caspase-14 (revC14)

1. Procaspase-14 cDNA is PCR cloned into the Bam HI/Kpn I sites of pQE 30 vector using forward primer cgggatccatgagcaatccgcggtctttgg and reverse primer ggggtaccctagatgac catcacaatctc.
2. Clone large and small subunits using the procaspase-14 construct as a template DNA. The construct consisting of small

subunit fused to the N-terminal of the large subunit, ending at D^{146}, makes constitutively active caspase-14.

3. Insert revC14 cDNA into the pET15b vector at the Xho I/ Bam HI sites.
4. Restriction sites for revC14 consist of Xho I-small subunit-Sal I-large subunit-Bam HI. Primers used: forward small subunit with Xho I ccgctcgagaaagacagcccacaaaccatc, reverse primer with Sal I gaccctccggaaacggctgtatctgcaggtcgaccgca, forward large subunit primer with Sal I acgcgtcgacatgagcaatccgcg-gtctttgg, and reverse primer with Bam HI cgggattctaatctccacctact-gtttcaccggggt.
5. Express recombinant protein in the *E. coli* BL21 (DE3) using Overnight Express Autoinduction System 1 and purify with Ni-NTA resin.
6. Check proteolytic activity using WEHD-AFC as a substrate. revC14 has activity comparable with that of purified caspase-14 from corneocyte extract (*see* **Note 4**).

3.4 Procaspase-14 Activation by KLK7

1. Procaspase-14 cDNA is cloned into pQE 30 vector, expressed in *E. coli,* and recombinant protein is purified on Ni-NTA Agarose column.
2. Preincubate procaspase-14 (5 μg/μl, 10 μl) with 50–250 ng of active KLK7 (10 μl) in the KLK7 assay buffer at room temperature for 15 min (*see* **Note 5**).
3. Add 80 μl of the caspase-14 assay buffer and incubate at room temperature for 10 min.
4. Add 20 μl of 100 μM Ac-WEHD-AFC and incubate the mixture at 37 °C for 30 min.
5. Measure activity of caspase-14 on plate reader with 355 nm excitation and 460 nm emission.

3.5 Preparation of Cleavage-Site-Directed Antibody

3.5.1 Anti-mature Caspase-14 Antibody (h14D146)

1. Cleavage-site-directed antibody is prepared by immunizing rabbits with the synthetic hexapeptide, CTVGGD, in which TVGGD corresponds to the C-terminal end of the large subunit of the autoprocessed caspase-14. Cysteine residue is added for the coupling with career protein, such as keyhole limpet hemocyanin (KLH). h14D146 is purified with affinity chromatography as follows.
2. Dilute 10 ml of antiserum with PBS (1:1) and mix with 1 ml of TVGGD-coupled Agarose. Incubate at 4 °C for 3 h with gentle rotation.
3. Gel suspension is packed in a small column (PD-10 column is quite suitable for this purpose), wait until gel is settled, and discard supernatant.

4. After extensive washing with PBS, monospecific antibody is eluted with five 0.5 ml aliquots of 0.1 M glycine-HCl buffer (pH 2.3) and immediately neutralized with 0.5 ml of 1 M Tris–HCl (pH 8.0).
5. Measure absorbance at 280 nm in each fraction and calculate IgG concentration in mg/ml using formulae A280 × 0.714.
6. Exchange buffer to PBS using PD-10 column equilibrated with PBS or dialyze against PBS. Store in small aliquots at −80 °C.

3.5.2 Cleavage-Site-Directed Antibody (h14Y178)

Cleavage-site-directed antibody (h14Y178) is prepared by immunizing rabbits with synthetic hexapeptide CGYIAY, in which GYIAY corresponds to the C-terminal end of p20 cleaved by KLK7. h14Y178 can be purified using above described procedure for h14D146.

3.6 ELISA Assays for Caspase-14

It is important to distinguish zymogen from the active form of caspase-14, since proteolytic activation is a critical event for physiological action of caspase-14. Here we describe two kinds of ELISA quantification assays using H99 antibody directed to the large subunit of human caspase-14 (to measure total caspase-14) and cleavage-site-directed antibody, h14D146 (to measure active caspase-14). We used in-house monoclonal anti-caspase-14 antibody, clone 3, as an immobilized antibody that showed the highest sensitivity in the ELISA system. However, commercially available mouse monoclonal antibody to caspase-14 can also be used.

1. Immobilize 1 μg/ml of anti-caspase-14 (clone 3 or commercially available monoclonal antibody) in PBS by overnight incubation in a 96-well plate.
2. Tap off the solution from the plate.
3. Block by adding 200 μl/well of ImmunoBlock diluted with PBS and incubate for 1 h at room temperature.
4. Tap off the solution from the plate and wash each well with 200 μl of PBS containing 0.1 % Tween 20 (PBST).
5. Add 100 μl samples (cell extracts) and incubate at 37 °C for 1 h. Use recombinant procaspase-14 as a standard to quantify total caspase-14. Use revC14 as a standard to quantify active caspase-14.
6. Wash the wells three times with 200 μl of PBST.
7. For measurement of total caspase-14, add 2 μg/ml of H99 antibody in 1:20 diluted ImmunoBlock and incubate for 1 h at 37 °C. The active form of caspase-14 is measured in a similar way using h14D146 as a primary antibody instead of H99.

8. Wash the wells three times with 200 μl of PBST.
9. Add ECL™ Anti-rabbit IgG Horseradish Peroxidase linked F(ab)′2 fragment (GE Healthcare, UK) in PBST and 1:20 diluted ImmunoBlock (total volume 10 ml) and incubated for 1 h at 37 °C.
10. Wash the wells three times with 200 μl of PBST.
11. Develop color using TMB Peroxidase EIA Substrate Kit according to manufacturer's instructions.
12. Stop reaction by adding 0.1 M sulfuric acid and read absorbance at 450 nm (*see* Fig. 2).

3.7 Immunohistochemical Localization of Caspase-14

1. Fix skin tissues in acetone (the Amex method) to better preserve antigens as well as high molecular weight DNA and RNA that are normally destroyed by the routine formalin fixation and paraffin-embedding process [13] (*see* **Note 6**).
2. Briefly, immerse tissues in acetone at 4 °C, transfer to a freezer at −20 °C, and fix overnight.
3. Dehydrate tissues in acetone, first at 4 °C for 15 min and then for another 15 min at room temperature.
4. Clear tissues twice in methyl benzoate for 15 min, and then twice in xylene for 15 min at room temperature.
5. Embed in paraffin at 60 °C for 2–3 h in a vacuum-evaporating embedder.
6. Prepare thin sections using microtome and deparaffinize with xylene for 30 min.
7. Immerse in acetone 2–3 times.
8. Wash with PBS three times for 10 min.
9. Block sections with 10 % goat serum for 1 h at room temperature.
10. Wash sections with PBS for 5 min and incubate with appropriate primary antibody for 1 h at room temperature or at 4 °C overnight. Double staining can be done using anti-caspase-14 mAb and cleavage-site-directed antibody, h14D146 or h14Y178.
11. Wash with PBS three times for 10 min.
12. Incubate with secondary antibodies (e.g., Alexa Fluor 555 or 488; Molecular Probes) for 1 h at room temperature.
13. Wash with PBS three times for 10 min.
14. Immerse sections in VECTASHIELD and stain with DAPI (*see* Fig. 3; *see* **Note 7**).

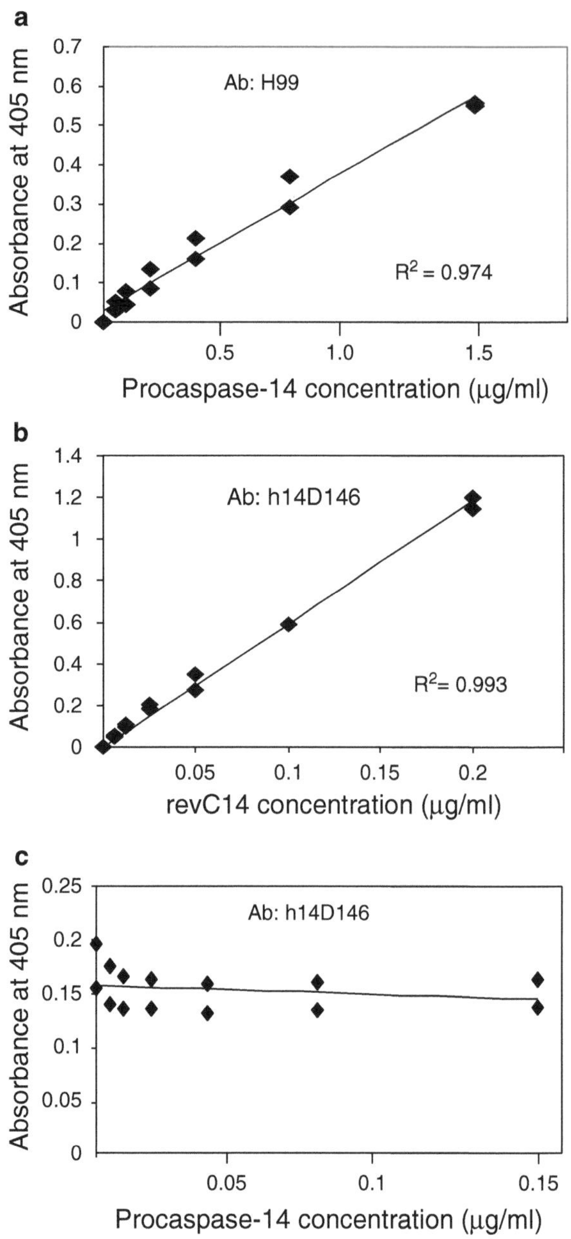

Fig. 2 Two types of caspase-14 ELISA assays. (**a**) Standard curve obtained from total caspase-14 measurement using H99 antibody. (**b**) Standard curve obtained from active caspase-14 (mature form) measurement using h14D146 antibody. (**c**) Proof of specificity for ELISA assay B showing that cleavage-site-directed antibody, h14D146, does not react with procaspase-14

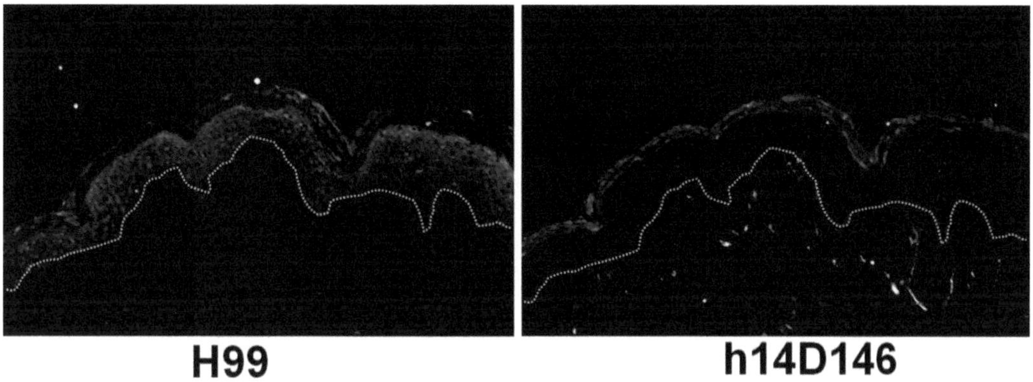

Fig. 3 Immunohistochemical localization of procaspase-14 and mature caspase-14. Procaspase-14 stained with H99 was present from spinous to granular layers. In contrast, the mature caspase-14, which was detected with h14D146, showed confined localization, being only found in the cornified layer

4 Notes

1. For the measurement of revC14-Y178, sodium citrate can be omitted.
2. Caspase-14 requires high concentration of kosmotropic salt for its activation. Usually 1–1.3 M sodium citrate is used since it shows neutral pH in the solution.
3. Substrates WEHD-AFC or WEHD-MCA are most suitable for caspase-14 activity measurement. Although substrate specificity of caspase-14 resembles that of caspase-1, YVAD-AFC (MCA) is not recommended, since it shows inhibitory effect.
4. It is quite hard to obtain purified caspase-14 from corneocyte extracts for the characterization because of very elaborate purification procedures. Chymotrypsin-activated caspase-14 or constitutively active revC14 should be used as an alternative source.
5. Activity of KLK7 is strongly suppressed in the presence of high concentrations of kosmotropic salts. For the activation assay, it is necessary to perform two-step procedure in different buffer systems.
6. When using the Amex method, preservation of antigenicity is almost equivalent to that of frozen sections. Thin sections of paraffin-embedded samples can be kept at 4 °C for 1 year.
7. Color development can be done using the 3,3′-diaminobenzidine (DAB) method, which produces an easily detectable brown color. The advantage of this method is that it enables permanent mounting of the sections. The disadvantage of the method is that it precludes double staining.

References

1. Eckhart L, Ban J, Fischer H, Tschachler E (2000) Caspase-14: analysis of gene structure and mRNA expression during keratinocyte differentiation. Biochem Biophys Res Commun 277:655–659
2. Van de Craen M, Van Loo G, Pype S, Van Criekinge W, Van den brande I, Molemans F, Fiers W, Declercq W, Vandenabeele P (1998) Identification of a new caspase homologue: caspase-14. Cell Death Differ 5:838–846
3. Eckhart L, Declercq W, Ban J, Rendl M, Lengauer B, Mayer C, Lippens S, Vandenabeele P, Tschachler E (2000) Terminal differentiation of human keratinocytes and stratum corneum formation is associated with caspase-14 activation. J Invest Dermatol 115:1148–1151
4. Watt FM (1983) Involucrin and other markers of keratinocyte terminal differentiation. J Invest Dermatol 81:100s–103s
5. Wertz PW (1997) Integral lipids of hair and stratum corneum. EXS 78:227–237
6. Hitomi K (2005) Transglutaminases in skin epidermis. Eur J Dermatol 15:313–319
7. Yoneda K, Demitsu T, Manabe M, Igarashi J, Kosaka H, Inagaki N, Takahashi H, Kon A, Kakurai M, Kubota Y (2010) Expression of wild-type, but not mutant, loricrin causes programmed cell death in HaCaT keratinocytes. J Dermatol 37:956–964
8. Hibino T, Fujita E, Tsuji Y, Nakanishi J, Iwaki H, Katagiri C, Momoi T (2010) Purification and characterization of active caspase-14 from human epidermis and development of the cleavage site-directed antibody. J Cell Biochem 109:487–497
9. Yamamoto M, Kamata Y, Iida T, Fukushima H, Nomura J, Saito M, Tajima M, Okubo Y, Momoi T, Tsuboi R, Hibino T (2011) Quantification of activated and total caspase-14 with newly developed ELISA systems in normal and atopic skin. J Dermatol Sci 61: 110–117
10. Yamamoto M, Miyai M, Matsumoto Y, Tsuboi R, Hibino T (2012) Kallikrein-related peptidase-7 regulates caspase-14 maturation during keratinocyte terminal differentiation by generating an intermediate form. J Biol Chem 287:32825–32834
11. Lamkanfi M, Festjens N, Declercq W, Vanden Berghe T, Vandenabeele P (2007) Caspases in cell survival, proliferation and differentiation. Cell Death Differ 14:44–55
12. Xu G, Shi Y (2007) Apoptosis signaling pathways and lymphocyte homeostasis. Cell Res 17:759–771
13. Sato Y, Mukai K, Furuya S, Kameya T, Hirohashi S (1992) The AMeX method: a multipurpose tissue-processing and paraffin-embedding method. Extraction of protein and application to immunoblotting. Am J Pathol 140:775–779

Chapter 6

Caspase Protocols in *Caenorhabditis elegans*

Eui Seung Lee and Ding Xue

Abstract

Caenorhabditis elegans genome has four genes (*ced-3*, *csp-1*, *csp-2*, and *csp-3*) encoding caspase-like proteins. Among these four proteins, CED-3 is the most well-known cell-killing caspase. Elucidation of the role of CED-3 as a central component of the apoptotic pathway in *C. elegans* has contributed to the understanding of the more complex apoptosis network in mammals and in other metazoa. In the highly conserved pathway of programmed cell death in *C. elegans*, CED-3 functions at the terminal step of this cell-killing pathway. Identification of CED-3 caspase substrates is essential for bridging the gaps between CED-3 activation and various downstream cell death execution events. If a protein is cleaved by CED-3 in vitro, this protein could be a potential CED-3 substrate in vivo. Here, we describe the method for purification of active CED-3 caspase. We will also describe in vitro assays for determining CED-3 proteolytic activity, CED-3 substrates, and CED-3 cleavage sites in the substrates.

Key words CED-3, CED-3 substrates, *C. elegans* apoptosis pathway, In vitro translation, In vitro cleavage assay

1 Introduction

Programmed cell death (PCD) or apoptosis was first described in *C. elegans* in the late 1970s [1–3], and many cell death genes identified in *C. elegans* have been shown to be conserved throughout metazoa [4]. Identification and analyses of the *C. elegans ced-3* gene and its homologues, caspases, in mammals and other organisms demonstrate that caspases are responsible for most apoptotic events [5]. Four genes encoding proteins with sequence similarity to caspases have been found in the *C. elegans* genome: *ced-3*, *csp-1*, *csp-2*, and *csp-3* [6, 7]. CED-3 is the key cell-killing caspase in *C. elegans* [6], whereas CSP-2 and CSP-3 do not display any caspase activity and, instead, function as caspase inhibitors [8, 9]. A recent study reports that *csp-1* also promotes apoptosis in *C. elegans* [10]. In this chapter, we will describe biochemical methods important for the characterization of the *C. elegans* CED-3 caspase and its substrates.

Peter V. Bozhkov and Guy Salvesen (eds.), *Caspases, Paracaspases, and Metacaspases: Methods and Protocols*, Methods in Molecular Biology, vol. 1133, DOI 10.1007/978-1-4939-0357-3_6, © Springer Science+Business Media New York 2014

The central cell-killing pathway in *C. elegans* (EGL-1 –| CED-9 –| CED-4 → CED-3) is well characterized [4]. CED-3 acts at the terminal step of this pathway and, as a protease, is expected to cleave numerous protein substrates to promote cell death execution [11, 12]. To identify CED-3 protease substrates, several different approaches have been employed [13–15]. Candidate CED-3 substrates identified from biochemical screens or genetic screens usually need to be confirmed as true in vitro substrates before time-consuming functional studies are performed. Therefore, a robust in vitro CED-3 caspase cleavage assay is critical for determining candidate CED-3 substrates [16–20]. To perform this experiment, successful purification of the CED-3 protease is essential.

Successful purification of active recombinant CED-3 can be achieved through bacterial expression [18]. In *C. elegans*, CED-3 is initially expressed as an inactive proenzyme of molecular weight 56 kDa. In dying cells, the CED-3 proenzyme is induced to undergo self-cleavage in a CED-4-dependent manner, leading to the generation of 24, 17, and 13 kDa cleavage products, and becomes self-activated [18, 21]. While many mammalian caspases are commercially available, CED-3 is not. In general, purification of a protease without losing its activity is challenging compared with preparation of other recombinant proteins. This has been a major issue in purification of CED-3, because CED-3 tends to lose its proteolytic activity during the purification process.

Here, we describe the method for purification of active CED-3 and further analytical assays using the purified CED-3 protease.

2 Materials

2.1 Purification of Recombinant CED-3

1. A pET-21 vector (Novagen) with a six-histidine affinity tag (His_6 tag) for C-terminal fusion can be used for expression of CED-3 (*see* **Notes 1–3**). Full-length CED-3 cDNA without a stop codon can be amplified by the polymerase chain reaction (PCR) and inserted into the vector via appropriate multiple cloning sites just before the His_6 coding region.
2. Bacterial strain BL21 (DE3).
3. 2 L of LB medium with an appropriate antibiotic in a 4 L flask.
4. 2 ml of IPTG stock (1 M).
5. 37 °C shaking incubator and 25 °C shaking incubator.
6. 4 °C high-speed centrifuge.
7. 250 ml or 500 ml centrifuge bottles for collecting bacterial cells.
8. Sonicator with a flat tip (*see* **Note 4**).

9. Syringes and 0.45 μm filters.
10. Ni-NTA beads (e.g., Qiagen catalog number 1018244).
11. Sonication buffer: 100 mM Tris–HCl pH 8.0, 250 mM NaCl, 0.5 mM sucrose, EDTA-free protease inhibitors cocktail (1 tablet/25 ml) (e.g., Roche catalog number 12483700).
12. Washing buffer 1: 100 mM Tris–HCl pH 8.0, 400 mM NaCl, 0.5 mM sucrose, 7.5 mM β-mercaptoethanol, 0.04 % NP-40, EDTA-free protease inhibitors cocktail (1 tablet/25 ml).
13. Washing buffer 2: washing buffer 1 + 25 mM imidazole.
14. Elution buffer 1: 100 mM Tris–HCl pH 8.0, 250 mM NaCl, 0.5 mM sucrose, 10 % glycerol, 7.5 mM β-mercaptoethanol, 0.04 % NP-40, 120 mM imidazole, EDTA-free protease inhibitors cocktail (1 tablet/25 ml).
15. Elution buffer 2: 100 mM Tris–HCl pH 8.0, 250 mM NaCl, 0.5 mM sucrose, 10 % glycerol, 7.5 mM β-mercaptoethanol, 0.04 % NP-40, 250 mM imidazole, EDTA-free protease inhibitors cocktail (1 tablet/25 ml).
16. Elution buffer 3: 100 mM Tris–HCl pH 8.0, 250 mM NaCl, 0.5 mM sucrose, 10 % glycerol, 7.5 mM β-mercaptoethanol, 0.04 % NP-40, 500 mM imidazole, EDTA-free protease inhibitors cocktail (1 tablet/25 ml).

2.2 Measuring the Activity of Recombinant CED-3

1. UV spectrophotometer and 96-well plate (*see* **Note 5**).
2. Fluorogenic caspase substrate: Ac-Asp-Glu-Val-Asp-AMC.
3. 2× CED-3 reaction buffer: 50 mM Tris–HCl pH 8.0, 0.5 mM EDTA, 0.5 mM sucrose, 5 % glycerol.

2.3 In Vitro Cleavage Assay Using Substrates Labeled by [^{35}S] Methionine

1. TNT transcription and translation-coupled rabbit reticulocyte system (Promega catalog number L4610).
2. [^{35}S]-Methionine.
3. RNase inhibitor.
4. SDS polyacrylamide gel (PAGE).
5. Fixation buffer: 50 % methanol, 10 % glacial acetic acid.
6. Gel drying buffer: 7 % methanol, 7 % glacial acetic acid, 1 % glycerol.
7. Phosphorimager.

2.4 Determination of the CED-3 Cleavage Site in the CED-3 Substrate

1. 30 °C incubator.
2. PVDF membrane.
3. Ponceau S solution: 2 g of Ponceau S, 30 g of trichloroacetic acid, 30 g of sulfosalicylic acid in 100 ml of H_2O.
4. Destaining solution: 5 % methanol, 7.5 % glacial acetic acid.

3 Methods

3.1 Purification of Recombinant CED-3

1. Transform BL21 (DE3) with the CED-3-expressing construct, inoculate a single transformed colony in 10 ml of LB medium containing 50 μg/ml ampicillin, and grow at 37 °C with shaking at 200 rpm overnight.
2. Transfer 10 ml LB culture into 2 L of LB medium with 50 μg/ml ampicillin and incubate at 37 °C with shaking at 200 rpm until OD_{600} of the culture reaches 0.5.
3. As soon as the OD_{600} of the bacterial culture reaches 0.5, place the culture on ice for 10 min.
4. Add IPTG to the bacterial culture to a final concentration of 1 mM.
5. Incubate the culture at room temperature with shaking at 200 rpm for 4 h.
6. Place the bacterial culture on ice for 10 min and spin down the cells at 4,500 × *g* for 10 min at 4 °C.
7. Discard the supernatant and move the pellet immediately to −80 °C freezer for at least 1 h (*see* **Note 6**).
8. Resuspend the frozen pellet in 10 ml of the cold sonication buffer. This should be performed at 4 °C.
9. Perform sonication to break up the resuspended cells until cleared lysate without viscosity is obtained (*see* **Note 7**). It is important to keep the lysate at low temperature.
10. Centrifuge the lysate at 18,000 × *g*, 4 °C, for 30 min and filter the supernatant using a syringe and a 0.45 μm filter.
11. Place 0.3 ml bed volume of Ni-NTA resin in a plastic column and wash the resin with 2 ml of the washing buffer. After this step, all procedures should be performed at 4 °C.
12. Load the filtered lysate onto the column and wait until all lysate has flown out of the resin.
13. Wash the column with 20 ml of the washing buffer (*see* **Note 8**).
14. Elute CED-3::His6 from the resin by adding 1 ml of elution buffer 1. Repeat this procedure by adding 1 ml of elution buffer 2 and 1 ml of elution buffer 3 each (*see* **Note 9**).

3.2 Measuring the Activity of Recombinant CED-3

1. Add 1 μl of purified CED-3 into a total 19 μl of reaction mixture, which contains 5 μM Ac-Asp-Glu-Val-Asp-AMC (*see* **Note 10**) and 10 μl of 2× CED-3 reaction buffer.
2. Incubate at 30 °C for 1 h.
3. Add 180 μl of distilled water.
4. Measure the fluorescent intensity using a UV spectrophotometer: Excitation 360, Emission 460.

5. As a negative control, the same concentration of DMSO as in the reaction solution can be used, as the substrate has been dissolved in DMSO. 5 μM AMC can be used as a positive control.

3.3 In Vitro Cleavage Assay Using Substrates Labeled by [^{35}S] Methionine

A candidate CED-3 substrate can be synthesized in vitro using the translation and transcription (TNT)-coupled reticulocyte lysate kit (Promega). Full-length cDNA encoding the candidate substrate can be cloned into a vector carrying the bacterial T7, T3, or SP6 promoter, such as pGEM-4Z, pET, or pBluescript vectors. To visualize the CED-3 cleavage products, the substrate protein should be labeled with [^{35}S] Methionine. The reaction mixture and procedure are as follows:

1. Prepare in vitro translation reaction mixture (10 μl reaction) (*see* **Note 11**):
   (a) 5 μl of TNT reticulocyte lysate.
   (b) 0.4 μl of TNT reaction buffer.
   (c) 0.2 μl of polymerase (T7, T3, or SP6 polymerase depending on the promoter).
   (d) 0.2 μl of amino acids mixture without Methionine.
   (e) 0.4 μl of [^{35}S] Methionine (1,000 Ci/mmol at 10 mCi/ml).
   (f) 0.2 μl of RNasin (ribonuclease inhibitor, 40 unit/μl).
   (g) 200 ng of the plasmid DNA template.
   (h) Add RNase-free water to a final volume of 10 μl.
2. Incubate the reaction at 30 °C for 90 min.
3. Add 5 μl of 2× CED-3 reaction buffer to 1–4 μl of the labeled product.
4. Add 1 μl of the purified CED-3.
5. Add distilled water to bring the final volume to 10 μl.
6. Incubate at 30 °C for 1–4 h.
7. Stop the reaction by adding SDS loading buffer and heating at 85 °C for 10 min.
8. Prepare a positive control: CED-9 is known to be one of the best CED-3 substrates and can be used as a positive control for the in vitro CED-3 cleavage assay [17]. If purified CED-3 has good activity, full-length CED-9 (31 kDa) should be completely digested to generate 7 and 24 kDa cleavage products (Fig. 1).

3.4 Determination of the CED-3 Cleavage Site in the CED-3 Substrate

CED-3 is known to recognize the amino acid sequence, Asp or Glu-X-X-Asp (also called P4-P3-P2-P1; X can be any amino acid) [18], and cleave right after the P1 Asp residue, which is absolutely conserved. Therefore, to identify the cleavage site(s) of a CED-3

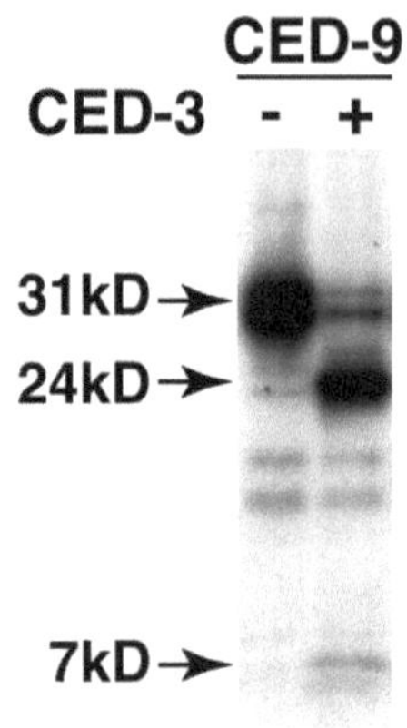

Fig. 1 In vitro cleavage assay. A pET-3a-CED-9 vector was used to synthesize [^{35}S] Methionine-labeled CED-9 using the TNT rabbit reticulocyte lysate system. The molecular weight of full-length CED-9 is 31 kDa. When incubated with purified CED-3, full-length CED-9 disappeared, and instead, 7 and 24 kDa fragments were observed

substrate, site-directed mutagenesis of the predicted CED-3 cleavage sequence is usually performed. Asp to Glu substitution or Asp to Ala substitution at the P1 position blocks CED-3 cleavage [16–20]. The mutated DNA template is used for the in vitro translation and cleavage assays, and the CED-3 cleavage pattern of the mutant protein is compared with that of the wild-type protein. If CED-3 cleavage of the mutant protein is blocked or greatly reduced, then the mutated site is likely a CED-3 cleavage site. However, if the substrate protein can be purified and is soluble, we would recommend to perform microsequencing analysis of the amino terminus of the cleavage product following in vitro cleavage with purified CED-3 to determine the caspase cleavage site.

1. Prepare at least 10 μg of the purified substrate protein from *E. coli.*
2. Incubate 10 μg of the purified protein with 1 μg of purified CED-3 in 50 μl CED-3 reaction buffer.
3. Incubate the reaction at 30 °C for 4–10 h.
4. For the negative control, use the elution buffer that was used for CED-3 purification.
5. Stop the reaction by adding SDS loading buffer and then heat the samples at 95 °C for 5 min.
6. Resolve the protein samples on SDS-PAGE. Load purified CED-3 alone to distinguish the CED-3 cleavage products from the 13 and 17 kDa CED-3 subunits.
7. Transfer the proteins on the SDS-PAGE gel to a PVDF membrane using a wet or semidry transfer apparatus.

8. Stain the membrane with the Ponceau S solution at room temperature for 5 min and wash with destaining buffer for 10 min.
9. Wash the membrane twice with distilled water for 5 min each.
10. Excise the predicted size of the stained protein band and use the membrane strip for N-terminal microsequencing (*see* **Note 12**).

4 Notes

1. Any pET vector carrying a six-histidine affinity tag can be used.
2. A His_6 affinity tag or a FLAG affinity tag can be used for the purification. But based on our experience, the activity of CED-3 purified using a FLAG tag was lower than that purified with a His_6 tag. The difference may be due to the concentration of elution in each eluting fraction. CED-3 with a His_6 tag can be eluted in high concentrations by controlling the concentrations of imidazole, while it is more difficult to do so with FLAG peptides.
3. CED-3 zymogen is auto-processed to generate 24, 17, and 13 kDa fragments [18]. Two of the fragments, 17 and 13 kDa, form a heterodimer that constitutes the catalytic protease and are positioned at the middle and at the C-terminus of the zymogen, respectively. Therefore, a C-terminal His_6 tag should be used for purifying CED-3 protease.
4. A micro-tip can also be used, but a flat tip is more efficient in this purification process, because the volume of CED-3 lysate is large.
5. For the detection of fluorogenic substrates, 96-well plates designed for UV light should be used.
6. Once the pellet has been frozen, it can be stored at −80 °C for up to several weeks.
7. The recommended procedure for sonication is 2 s on, 10 s off, and with sonicator at 40 % amplitude in a 4 °C room using a flat tip for a total of 5 min.
8. Do not allow the resin to dry, even for a short period of time.
9. Generally the highest concentration of CED-3 is eluted in elution buffer 2. Sometimes, a higher concentration of imidazole (1 M) may be applied to the column for the final elution.
10. Fluorogenic substrate stock solution can be made at 1 mM concentration in DMSO and stored at −20 °C.
11. Promega's manual recommends a 50 μl reaction. However, we found that a 10 μl reaction usually provides enough labeled product for the CED-3 cleavage assay.
12. At least 0.5–1 μg protein fragments are needed for successful microsequencing.

References

1. Sulston JE, Horvitz HR (1977) Post-embryonic cell lineages of the nematode, Caenorhabditis elegans. Dev Biol 56:110–156
2. Sulston JE, Schierenberg E, White JG, Thomson JN (1983) The embryonic cell lineage of the nematode Caenorhabditis elegans. Dev Biol 100:64–119
3. Hedgecock EM, Sulston JE, Thomson JN (1983) Mutations affecting programmed cell deaths in the nematode Caenorhabditis elegans. Science 220:1277–1279
4. Horvitz HR (1999) Genetic control of programmed cell death in the nematode Caenorhabditis elegans. Cancer Res 59: 1701s–1706s
5. Denault JB, Salvesen GS (2008) Apoptotic caspase activation and activity. Methods Mol Biol 414:191–220
6. Yuan J, Shaham S, Ledoux S, Ellis HM, Horvitz HR (1993) The C. elegans cell death gene ced-3 encodes a protein similar to mammalian interleukin-1 beta-converting enzyme. Cell 75:641–652
7. Shaham S (1998) Identification of multiple Caenorhabditis elegans caspases and their potential roles in proteolytic cascades. J Biol Chem 273:35109–35117
8. Geng X et al (2008) Inhibition of CED-3 zymogen activation and apoptosis in Caenorhabditis elegans by caspase homolog CSP-3. Nat Struct Mol Biol 15:1094–1101
9. Geng X et al (2009) Caenorhabditis elegans caspase homolog CSP-2 inhibits CED-3 autoactivation and apoptosis in germ cells. Cell Death Differ 16:1385–1394
10. Denning DP, Hatch V, Horvitz HR (2013) Both the caspase CSP-1 and a caspase-independent pathway promote programmed cell death in parallel to the canonical pathway for apoptosis in Caenorhabditis elegans. PLoS Genet 9:e1003341
11. Crawford ED, Wells JA (2011) Caspase substrates and cellular remodeling. Annu Rev Biochem 80:1055–1087
12. Conradt B, Xue D (2005) Programmed cell death. WormBook:1–13
13. Parrish J et al (2001) Mitochondrial endonuclease G is important for apoptosis in C. elegans. Nature 412:90–94
14. Taylor RC et al (2007) Establishing a blueprint for CED-3-dependent killing through identification of multiple substrates for this protease. J Biol Chem 282:15011–15021
15. Kim SY et al (2004) Identification of CED-3 substrates by a yeast-based screening method. Mol Biotechnol 27:1–6
16. Xue D, Horvitz HR (1995) Inhibition of the Caenorhabditis elegans cell-death protease CED-3 by a CED-3 cleavage site in baculovirus p35 protein. Nature 377:248–251
17. Xue D, Horvitz HR (1997) Caenorhabditis elegans CED-9 protein is a bifunctional cell-death inhibitor. Nature 390:305–308
18. Xue D, Shaham S, Horvitz HR (1996) The Caenorhabditis elegans cell-death protein CED-3 is a cysteine protease with substrate specificities similar to those of the human CPP32 protease. Genes Dev 10:1073–1083
19. Breckenridge DG et al (2008) Caenorhabditis elegans drp-1 and fis-2 regulate distinct cell-death execution pathways downstream of ced-3 and independent of ced-9. Mol Cell 31: 586–597
20. Nakagawa A, Shi Y, Kage-Nakadai E, Mitani S, Xue D (2010) Caspase-dependent conversion of Dicer ribonuclease into a death-promoting deoxyribonuclease. Science 328: 327–334
21. Yan N et al (2005) Structure of the CED-4-CED-9 complex provides insights into programmed cell death in Caenorhabditis elegans. Nature 437:831–837

Chapter 7

Detecting Caspase Activity in *Drosophila* Larval Imaginal Discs

Caitlin E. Fogarty and Andreas Bergmann

Abstract

Caspases are a highly specialized class of cell death proteases. Since they are synthesized as inactive full-length zymogens, activation—at least of effector caspases and to some extent also of initiator caspases—requires a proteolytic cleavage event, generating a large and a small subunit, two of each forming the active caspase. The proteolytic cleavage event generates neo-epitopes at both the C-terminus of the large subunit and the N-terminus of the small subunit. The cleaved Caspase-3 (CC3) antibody was raised against the neo-epitope of the large subunit and thus detects only cleaved, but not full-length, Caspase-3. Although raised against human cleaved Caspase-3, the CC3 antibody cross-reacts in other species and detects cleaved caspases, most notably DrICE and Dcp-1, in *Drosophila*. This protocol describes the procedure for use of the CC3 antibody to detect caspase activity in larval imaginal discs in *Drosophila*.

Key words *Drosophila*, Cleaved-Caspase-3, Dronc, DrICE, Dcp-1, Cell death, Non-apoptotic function, Imaginal Disc, Immunolabeling

1 Introduction

The genome of *Drosophila melanogaster* encodes for seven caspase genes (reviewed in [1, 2]). However, only three of these are involved in the execution of apoptosis: the initiator caspase Dronc and the effector caspases DrICE and Dcp-1 [1, 2]. The synthesis and activation of these three caspases is very similar to that of their human homologs [1]. Moreover, *Drosophila* is an ideal model system for studying the various functions of caspases. Overall, 50 % of caspase targets during apoptosis are conserved at the protein level between flies and humans [3]. This increases to 60 % when accounting for different protein targets within the same conserved pathway. Taken together with the fact that the intrinsic cell death cascade is remarkably well conserved in *Drosophila*, the fly is a very relevant model to study the roles of caspases in the initiation and execution of apoptotic cell death [2, 4]. Interestingly, recent studies have also shown that the roles of caspases in the fly go beyond

Peter V. Bozhkov and Guy Salvesen (eds.), *Caspases, Paracaspases, and Metacaspases: Methods and Protocols*,
Methods in Molecular Biology, vol. 1133, DOI 10.1007/978-1-4939-0357-3_7,

traditional cell death and include a plethora of non-apoptotic functions. Caspases have been implicated in cell migration and morphogenesis [5–7], compensatory cell proliferation [8–11], cell differentiation and maturation [12, 13], innate immunity [14, 15], and non-apoptotic alternative cell death [16].

Initially, to detect caspase activity in vivo, researchers employed indirect measures such as TUNEL, acridine orange, and propidium iodide [17–19]. However, these methods highlight cells in the late stages of programmed cell death and are not specific to the intrinsic cell death pathway [20]. More recently, direct methods to detect caspase activity have been developed, which take advantage of antibodies that specifically recognize neo-epitopes that are generated after the proteolytic processing of caspases. Ideally, these antibodies do not detect uncleaved, and therefore inactive, caspases. Typically, these antibodies are raised against the neo-epitope formed at the C-terminus of the large subunit of Caspase-3. The first antibody of its kind was the CM1 antibody, which is no longer available [21]. Currently, several antibodies are commercially available that are referred to as anti-Cleaved Caspase-3 (CC3 from Cell Signaling Technology) or anti-Active Caspase-3 (Abcam). Although raised against an epitope in human Caspase-3, these antibodies also cross-react with *Drosophila* caspases [22–24] which is surprising as only eight of 13 residues within the epitope are identical between human Caspase-3 and *Drosophila* effector caspases DrICE and Dcp-1. However, follow-up studies demonstrated that only the most C-terminal three residues (ETD), which are conserved between the three caspases, are required for detection by the CC3 antibody [23]. Because these three residues constitute part of the cleavage site of the initiator caspase Dronc, the specificity of the CC3 antibody comes with the caveat that it not only detects cleaved DrICE and Dcp-1, but also a currently unknown non-apoptotic protein (or proteins) bearing a similar epitope [23]. Because exposure of these sites is dependent on the proteolytic activity of the initiator caspase Dronc, we interpret the CC3 antibody as a marker of Dronc activity rather than of DrICE activity [23]. This becomes an especially important distinction in the study of non-apoptotic functions of caspases—processes that may be dependent solely on Dronc and not involve effector caspases at all.

In this chapter, we describe a method for using the CC3 antibody in *Drosophila* larval tissues, specifically in the epithelial layers found within the developing imaginal discs. The first part describes the technique for sample preparation and fixation. Proper sample handling at this point is critical to overall success of the method. The second and third components describe the immunolabeling and visualization of the CC3 antibody, the marker for Dronc activity. With slight modifications, one could apply similar methods to a vast array of embryonic, larval, and adult tissues to aid in the study of both the apoptotic and non-apoptotic functions of Dronc.

2 Materials

2.1 General Supplies and Reagents

1. Many fly stocks can be obtained from the Bloomington Drosophila Stock Center (http://flystocks.bio.indiana.edu/bloomhome.htm) (*see* **Note 1**).
2. Fly food and culture vials, incubator at 18 °C or 25 °C as needed, CO_2 source, humidifier, and anesthetizing pads.
3. Dissecting stereoscope with independent flexible light source (e.g., Fiber-Lite High Intensity Illuminator M-181-1 with dual gooseneck illuminator).
4. Nutator or rocking platform at room temperature and 4 °C.
5. Zeiss LSM 700 confocal microscope or similar compound microscope equipped with fluorescent lamp (e.g., X-Cite Series 120Q, Lumen Dynamics), CCD camera, and image processing software such as ImageJ (http://imagej.nih.gov/ij/) [25].

2.2 Sample Preparation

1. 1× Phosphate-buffered saline (PBS): 137 mM NaCl, 2.7 mM KCl, 8 mM Na_2HPO_4, 2 mM KH_2PO_4, pH 7.2. May be made at 10×, autoclaved, and stored at room temperature.
2. 1× Phosphate-buffered saline with Triton-X-100 (PBT): 1× PBS with 0.3 % (v/v) Triton X-100. May be made at 10×, sterile filtered with a 0.22 μm syringe filter, and stored at room temperature.
3. 16 % (w/v) Paraformaldehyde.
4. Glass multi-well spot plates.
5. Fine forceps (one #5 and two #55, Fine Science Tools by Dumont or similar).
6. Microcentrifuge tubes (0.5 mL recommended).
7. Ice bucket.

2.3 Immunolabeling and Visualization

1. Blocking solution: 1× PBT with 2 % (v/v) Normal Donkey Serum.
2. Rabbit anti-Cleaved-Caspase-3 (Asp175 Antibody, #9661, Cell Signaling Technologies) (*see* **Note 2**).
3. Appropriate fluorophore-conjugated anti-rabbit antibody.
4. VECTASHIELD Mounting Media with or without DAPI.
5. 0.15 mm black anodized steel insect pins and holders.
6. Glass slides and coverslips.
7. Clear nail polish.
8. Opaque freezer box.
9. Opaque slide folder, tray, or box.

3 Methods

3.1 Sample Preparation

This dissection technique is described assuming right hand dominance. Each person will likely find a variation of this method that works best for them.

1. Prepare a fresh aliquot of 4 % (v/v) paraformaldehyde in 1× PBS, place on ice.
2. Using #5 forceps, transfer larva to a glass well spot plate with drops of cold 1× PBS (*see* **Notes 3** and **4**).
3. Using the #55 forceps, firmly grasp the larva in the middle with right hand, and remove the posterior half of the larva with the left. Discard the posterior half.
4. Gently holding the cuticle with the left hand forceps, grasp the most anterior part of the mouth hooks with the right forceps and push the mouth hooks through cuticle, inverting the larva and exposing the brain lobes and imaginal discs.
5. Quickly remove the salivary gland and fat body, along with any components of the larval gut, as these structures may release degradative enzymes that could compromise imaginal disc integrity.
6. Grasping the cuticle, transfer to a microcentrifuge tube with fresh PBS on ice (*see* **Notes 5** and **6**).
7. After dissecting 10–15 larvae, remove PBS from the tube and add 4 % paraformaldehyde. Mix by pipetting until all samples are floating free from each other and the tube. Fix for 30 min at room temperature on a rotating platform (*see* **Note 7**).
8. Remove fixative and wash in 1× PBT, three times for 5 min each at room temperature (*see* **Note 8**).

3.2 Immunolabeling of Samples

1. Preincubate samples in fresh blocking solution for 15 min at room temperature on a rotating platform (*see* **Note 9**).
2. Incubate with anti-Cleaved-Caspase-3 antibody (1:100–1:500) diluted in fresh blocking solution, overnight at 4 °C on a rotating platform (*see* **Notes 10** and **11**).
3. Wash in PBT three times for 5 min each at room temperature.
4. Incubate with a fluorophore conjugated anti-rabbit antibody for 6 h at 4 °C on a rotating platform (*see* **Note 12**).
5. Wash in 1× PBT twice for 15 min each, followed by one wash in 1× PBS for 15 min, all at room temperature on a rotating platform.
6. Incubate in VECTASHIELD Mounting Media for at least 1 h at 4 °C on a rotating platform (*see* **Note 13**).

3.3 Mounting of Samples and Visualization of Cleaved-Caspase-3 Antibody

1. Using #55 forceps carefully transfer samples to a glass slide with a drop of VECTASHIELD (*see* **Note 14**).
2. Using the dissecting needles (insect pins in appropriate holders), separate the imaginal disc of interest from the surrounding tissue. For example, if removing the eye-antennal disc, sever the connection between the antennal tissue and the mouth hooks. Then carefully separate the eye tissue from the brain lobes and sever the connection at the optic stalk.
3. Remove the cuticle and other debris from the slide and arrange discs for imaging.
4. Carefully apply a coverslip over the samples and seal on at least two sides with clear nail polish (*see* **Note 15**).
5. View and image discs with a confocal microscope or compound microscope with fluorescent light source.
6. Store slides not currently being imaged in an opaque slide folder, tray, or after sufficient drying time in a slide box (*see* **Note 16**).

4 Notes

1. The Bloomington *Drosophila* Stock Center has a large collection of tissue-specific and developmentally regulated GAL4 drivers, which can be combined with any number of UAS linked transgenes or RNAi constructs, which could be useful in studying the apoptotic and non-apoptotic functions of Dronc. We encourage you to consult recent original works to determine the appropriate parental lines for use in your experiments.
2. The Cell Signaling Technologies CC3 antibody is a polyclonal antibody raised against the C-terminus of the large subunit of human Caspase-3. Some have noted varying levels of appropriate staining from lot to lot. If starting with a new lot of CC3, always compare to a lot previously determined to work in *Drosophila*. This will distinguish whether a failed procedure is due to a non-reactive antibody versus an improperly performed staining.
3. Take care to ensure that the vials do not become overcrowded. Crowding in the food can not only cause larvae to wander prematurely, potentially leading to incorrect developmental timing at dissection, but can also create a stressed environment, sometimes enhancing cell death phenotypes and confounding results. Depending on the overall health of the parental lines and the expected viability of the F1 progeny, we have found allowing a 24 h egg lay by eight virgin females mated with four to six males in narrow vials (25 mm O.D. × 99 mm H) yields sufficient progeny without crowding.

4. If the parental cross was set up such that dominant larval markers such as *Tubby* or a GFP reporter will positively or negatively mark the animals of interest, this is the appropriate time to sort them out. If sorting by fluorescence, keep the animals to be dissected on ice or transfer to new dishes with cold PBS immediately before dissection.
5. Do not attempt to fully dissect the imaginal discs from the brain lobes or cuticle at this point. The discs are very small and very delicate even after fixation. By leaving them attached to the bulkier structures during the staining process, it will be much easier to transfer, fix, wash, and stain the discs without damaging them. The excess structures will be removed later (*see* **step 2** in Subheading 3.3).
6. Someone who is proficient in this technique can dissect on average one larva every 30–45 s. However, a novice may find that it takes significantly longer. Exposed discs may be transferred to fresh PBS in a microcentrifuge tube and kept on ice until fixation, but they should never be left for more than 1 h. When beginning, dissect fewer animals at a time to reduce the time discs spend exposed in PBS before fixation.
7. Do not over fix samples. This protocol does not have a separate antigen retrieval step and excessive fixation can mask epitopes. Do not exceed 60 min of fixation at room temperature or 1.5 h on ice. Best results will come from the standard 30 min of fixation at room temperature.
8. Exposed discs may be stored in PBT at 4 °C overnight without any decrease in quality. Discs stored up to a week have been used successfully, but there can be noticeable decreases in quality of the samples.
9. All incubations from this point forward should be carried out in the dark to prevent degradation of the antibody or photobleaching of the sample. This is best achieved by keeping samples in a light-blocking freezer box. If this is not available, another option is to wrap sample racks in aluminum foil.
10. Primary antibody may be reused for subsequent samples. Users will even notice an improvement in quality of CC3 staining with a reduction in background after the first one or two trials.
11. It may be useful to include one or two additional antibodies at the same time to provide context to any CC3 positive cells—for example, co-staining third instar eye imaginal discs with the pan-neuronal marker ELAV would distinguish the developing photoreceptor neurons in the posterior from the as of yet undifferentiated proliferating cells in the anterior.

Take care to make sure that each antibody was raised in a different species which will prevent cross-reaction during the secondary incubation (e.g., rabbit anti-CC3 and rat anti-ELAV labeled with donkey anti-rabbit FITC and donkey anti-rat CY3 work very well together).

12. We routinely carry out our secondary incubations for 2.5 h at room temperature, but this does increase background fluorescence. When trying a new antibody (primary or secondary) or when in need of publication quality images, carry out the secondary incubation at 4 °C.
13. If using VECTASHIELD with DAPI, incubate for at least 2.5 h. DAPI staining will only improve with longer incubation prior to mounting. Samples may even be stored in VECTASHIELD for a few days at 4 °C in order to allow mounting and imaging to occur on the same day (*see* **Note 16**).
14. It is important to use sufficient VECTASHIELD so that samples do not dry out on the slide during dissection; however, too much can also cause problems. If the droplet is too deep, structures will be free floating and it will be difficult to hold them in place to make clean cuts. Also, if there is too much VECTASHIELD on the slide when placing the coverslip, samples will often move around and overlap, and any excess fluid outside the edges of the coverslip will prevent the nail polish from creating a secure seal.
15. This step is most essential if the slides will be imaged using oil immersion. If imaging with dry objectives, the slides can be imaged and then sealed prior to storage. If there is too much VECTASHIELD, the nail polish may pucker up or completely peel off within 24 h. Repeat sealing process as necessary.
16. We have found that when mounted properly, the slides may be stored at 4 °C overnight or at −20 °C for at least a week without significant reduction in quality of CC3 signal. However, other antibodies or fluorophores may not be as stable. Best results will come from the slides imaged on the same day they are mounted.

Acknowledgement

We would like to thank Yun Fan, Ernesto Perez, and Jillian Lindblad for their technical expertise and review of the manuscript. This work was supported by grants from the National Institutes of Health (GM068016 and GM107789).

References

1. Kumar S (2007) Caspase function in programmed cell death. Cell Death Differ 14(1):32–43. doi:10.1038/sj.cdd.4402060
2. Xu D, Woodfield SE, Lee TV, Fan Y, Antonio C, Bergmann A (2009) Genetic control of programmed cell death (apoptosis) in Drosophila. Fly (Austin) 3(1):78–90
3. Crawford ED, Seaman JE, Barber AE II, David DC, Babbitt PC, Burlingame AL, Wells JA (2012) Conservation of caspase substrates across metazoans suggests hierarchical importance of signaling pathways over specific targets and cleavage site motifs in apoptosis. Cell Death Differ 19(12):2040–2048. doi:10.1038/cdd.2012.99
4. Xu D, Li Y, Arcaro M, Lackey M, Bergmann A (2005) The CARD-carrying caspase Dronc is essential for most, but not all, developmental cell death in Drosophila. Development 132(9): 2125–2134
5. Geisbrecht ER, Montell DJ (2004) A role for Drosophila IAP1-mediated caspase inhibition in Rac-dependent cell migration. Cell 118(1): 111–125. doi:10.1016/j.cell.2004.06.020
6. Suzanne M, Petzoldt AG, Speder P, Coutelis JB, Steller H, Noselli S (2010) Coupling of apoptosis and L/R patterning controls stepwise organ looping. Curr Biol 20(19): 1773–1778. doi:10.1016/j.cub.2010.08.056
7. Kuranaga E, Matsunuma T, Kanuka H, Takemoto K, Koto A, Kimura K, Miura M (2011) Apoptosis controls the speed of looping morphogenesis in Drosophila male terminalia. Development 138(8):1493–1499. doi:10.1242/dev.058958
8. Fan Y, Bergmann A (2008) Distinct mechanisms of apoptosis-induced compensatory proliferation in proliferating and differentiating tissues in the Drosophila eye. Dev Cell 14(3):399–410. doi:10.1016/j.devcel.2008.01.003
9. Huh JR, Guo M, Hay BA (2004) Compensatory proliferation induced by cell death in the Drosophila wing disc requires activity of the apical cell death caspase Dronc in a nonapoptotic role. Curr Biol 14(14):1262–1266. doi:10.1016/j.cub.2004.06.015
10. Kondo S, Senoo-Matsuda N, Hiromi Y, Miura M (2006) DRONC coordinates cell death and compensatory proliferation. Mol Cell Biol 26(19):7258–7268. doi:10.1128/MCB.00183-06
11. Wells BS, Yoshida E, Johnston LA (2006) Compensatory proliferation in Drosophila imaginal discs requires Dronc-dependent p53 activity. Curr Biol 16(16):1606–1615. doi:10.1016/j.cub.2006.07.046
12. Arama E, Agapite J, Steller H (2003) Caspase activity and a specific cytochrome C are required for sperm differentiation in Drosophila. Dev Cell 4(5):687–697
13. Arama E, Bader M, Srivastava M, Bergmann A, Steller H (2006) The two Drosophila cytochrome C proteins can function in both respiration and caspase activation. EMBO J 25(1):232–243. doi:10.1038/sj.emboj.7600920
14. Stoven S, Silverman N, Junell A, Hedengren-Olcott M, Erturk D, Engstrom Y, Maniatis T, Hultmark D (2003) Caspase-mediated processing of the Drosophila NF-kappaB factor Relish. Proc Natl Acad Sci U S A 100(10):5991–5996. doi:10.1073/pnas.1035902100
15. Leulier F, Rodriguez A, Khush RS, Abrams JM, Lemaitre B (2000) The Drosophila caspase Dredd is required to resist gram-negative bacterial infection. EMBO Rep 1(4):353–358. doi:10.1093/embo-reports/kvd073
16. Yacobi-Sharon K, Namdar Y, Arama E (2013) Alternative germ cell death pathway in Drosophila involves HtrA2/Omi, lysosomes, and a caspase-9 counterpart. Dev Cell 25(1): 29–42. doi:10.1016/j.devcel.2013.02.002
17. Ciancio G, Pollack A, Taupier MA, Block NL, Irvin GL III (1988) Measurement of cell-cycle phase-specific cell death using Hoechst 33342 and propidium iodide: preservation by ethanol fixation. J Histochem Cytochem 36(9): 1147–1152
18. Abrams JM, White K, Fessler LI, Steller H (1993) Programmed cell death during Drosophila embryogenesis. Development 117(1):29–43
19. Gavrieli Y, Sherman Y, Ben-Sasson SA (1992) Identification of programmed cell death in situ via specific labeling of nuclear DNA fragmentation. J Cell Biol 119(3):493–501
20. Chen P, Abrams JM (2000) Analysis of programmed cell death and apoptosis in Drosophila. Methods Enzymol 322:65–76
21. Srinivasan A, Roth KA, Sayers RO, Shindler KS, Wong AM, Fritz LC, Tomaselli KJ (1998) In situ immunodetection of activated caspase-3 in apoptotic neurons in the developing nervous system. Cell Death Differ 5(12): 1004–1016. doi:10.1038/sj.cdd.4400449
22. Yu SY, Yoo SJ, Yang L, Zapata C, Srinivasan A, Hay BA, Baker NE (2002) A pathway of signals regulating effector and initiator caspases in the developing Drosophila eye. Development 129(13):3269–3278

23. Fan Y, Bergmann A (2010) The cleaved-Caspase-3 antibody is a marker of Caspase-9-like DRONC activity in Drosophila. Cell Death Differ 17(3):534–539. doi:10.1038/cdd.2009.185
24. Florentin A, Arama E (2012) Caspase levels and execution efficiencies determine the apoptotic potential of the cell. J Cell Biol 196(4): 513–527. doi:10.1083/jcb.201107133
25. Schneider CA, Rasband WS, Eliceiri KW (2012) NIH image to ImageJ: 25 years of image analysis. Nat Methods 9(7): 671–675

Chapter 8

Methods for the Study of Caspase Activation in the *Xenopus laevis* Oocyte and Egg Extract

Francis McCoy, Rashid Darbandi, and Leta K. Nutt

Abstract

The study of apoptosis and caspases has advanced greatly over recent decades. Studies conducted in the *Xenopus laevis* egg extract and oocyte model system have significantly contributed to these advances. Twenty years ago, Newmeyer and colleagues first showed that the *X. laevis* egg extract, when incubated at room temperature, reconstituted the key molecular events of cellular apoptosis including cytochrome *c* release, nuclear condensation, internucleosomal fragmentation, and caspase activation. The biochemical tractability of the egg extract system allows for robust study of apoptotic events and caspase activation. Its nature as a cell-free extract system allows substrates to be very simply added by pipette, and their effects on apoptosis and caspase activation and their placement in the apoptotic signaling pathway (e.g., pre- or post-mitochondrial) are subsequently very simply studied using the techniques described in this chapter. Also described in this chapter are assays that allow the study of caspase activation in intact oocytes, another valuable tool available when using the *X. laevis* model organism. Overall, the *X. laevis* egg extract/oocyte model is a robust, efficient, and biochemically tractable system that is ideal for the study of apoptosis and caspase activation.

Key words *Xenopus laevis*, Egg extract, Oocytes, Caspases, Apoptosis

1 Introduction

The *Xenopus laevis* egg extract is an ideal model system for the study of apoptosis. This cell-free in vitro system was first shown to recapitulate key apoptotic events nearly 20 years ago. After incubation at room temperature, these egg extracts underwent apoptosis, displaying nuclear condensation and internucleosomal fragmentation of synthetically formed nuclei. Apoptosis in these extracts was blocked by addition of BCL-2 and also required the presence of mitochondria [1]. Since this initial discovery, studies in the *X. laevis* egg extract have helped uncover some of the key mechanisms underlying the regulation of apoptosis: showing that BCL-2 inhibits apoptosis by blocking release of cytochrome *c* from mitochondria [2], identifying cytochrome *c*-induced activation of caspase-3 and the key structural features of cytochrome *c* involved in this activation [3],

Peter V. Bozhkov and Guy Salvesen (eds.), *Caspases, Paracaspases, and Metacaspases: Methods and Protocols*, Methods in Molecular Biology, vol. 1133, DOI 10.1007/978-1-4939-0357-3_8, © Springer Science+Business Media New York 2014

and discovering that the *Drosophila melanogaster* protein, reaper, could function in a vertebrate system to induce apoptosis upstream of mitochondria [4]. These studies and others have utilized the egg extract to study apoptosis in great detail [5–11].

Of all the caspases, caspase-2 is one of the least studied and least well understood. In the past this has likely been due to an inability to find defined contexts under which it is important for induction of apoptosis. However, caspase-2 knockout mice were shown to have an excess number of oocytes that were chemoresistive [12], implicating caspase-2 as being important for oocyte death. The use of various techniques to study caspase activation in the *X. laevis* oocyte and egg extract has since uncovered that caspase-2 appears to function under defined contexts and stimuli; in particular caspase-2 appears to be important for inducing apoptosis in response to metabolic cues [13–17].

These studies showed that oocyte cell death is metabolically regulated through CaMKII (Ca^{2+}/calmodulin-dependent protein kinase II)-mediated phosphorylation of caspase-2, resulting in inhibition of caspase-2 activation upstream of Bax/Bak activation and cytochrome *c* release. Specifically, the generation of NADPH through the pentose-phosphate pathway activates CaMKII that in turn phosphorylates and inactivates caspase-2 [13, 14]. Conversely, under metabolic deprivation, CaMKII is inactive, allowing for the dephosphorylation of caspase-2 by protein-phosphatase-1 (PP1). The model of metabolically regulated caspase-2-mediated oocyte apoptosis has been further expanded, with recent studies showing that metabolism acts to inhibit PP1, preventing dephosphorylation of CaMKII, and allowing for sustained phosphorylation of caspase-2 by CaMKII [17].

These and future findings rest on the techniques described below to properly measure caspase activity in the highly tractable *X. laevis* oocyte and egg extract. The process of collecting *X. laevis* eggs and the preparation of interphase and cytosolic egg extract are first covered (Subheading 3.1). As key markers of apoptosis, the assessment of caspase-3 and caspase-7 in extract (Subheadings 3.3 and 3.4) and in intact oocytes (Subheadings 3.10 and 3.11) is described. As discussed previously, caspase-2 is a critical point of control in the metabolic regulation of oocyte cell death, and the assessment of its activity in extract using a ^{35}S-labeled in vitro-translated protein is explained (Subheading 3.5).

The tractability of *X. laevis* egg extract has allowed for the development of techniques that enable complete dissection of each of the key steps in the apoptotic cascade. As a cell-free extract system, substrates can simply be added by pipette to the extract, and using the techniques described in this chapter, their effects on apoptosis and their placement as pre- or post-mitochondrial in the apoptotic cascade can be identified (Subheadings 3.6 and 3.7).

Exogenous sperm nuclei can even be added to *X. laevis* egg extract, and nuclear morphology can be studied as a technique for measuring apoptosis (Subheading 3.9). Caspase activation can also be measured by analyzing cleavage of caspase substrates, e.g., PARP, in the extract by measuring PARP cleavage (Subheading 3.8).

2 Materials

2.1 Collection of X. laevis Eggs and Preparation of Interphase Egg Extract

1. Mature female frogs (e.g., Nasco; http://www.enasco.com/page/xenopus).
2. Frog aquarium.
3. Pregnant mare serum gonadotropin (PMSG, Calbiochem). Reconstituted in autoclaved, sterile filtered (0.22 μm filter) ddH_2O, to a stock concentration of 200 U/mL. Store at −20 °C.
4. Human chorionic gonadotropin (hCG, Sigma-Aldrich). Reconstituted in autoclaved, sterile filtered (0.22 μm filter) ddH_2O, to a stock concentration of 1,000 U/mL. Store at 4 °C.
5. 2 % L-cysteine (w/v), pH 8.0.
6. 40× Marc's Modified Ringer's solution (MMR): 50 mM HEPES pH 7.7, 1 M NaCl, 20 mM KCl, 10 mM $MgSO_47H_2O$, 25 mM $CaCl_2$, and 2 mM EDTA.
7. 10× egg lysis buffer (ELB): 100 mM HEPES pH 7.7, 25 mM $MgCl_2$, and 500 mM KCl.
8. Cycloheximide. Reconstituted in ddH_2O to a stock concentration of 10 mg/mL.
9. Aprotinin. Reconstituted in ddH_2O to a stock concentration of 5 mg/mL.
10. Leupeptin. Reconstituted in ddH_2O to a stock concentration of 5 mg/mL.
11. Cytochalasin B. Reconstituted in dimethyl sulfoxide (DMSO) to a stock concentration of 5 mg/mL.
12. 200-mL glass beaker.
13. 14-mL polypropylene round-bottom tube.
14. IEC clinical centrifuge (or equivalent).
15. Avanti J-26 XPI centrifuge (Beckman Coulter), with a JS 13.1 swinging bucket rotor (or equivalent).
16. 10-mL syringe.
17. 16-gauge needle.
18. 15-mL polypropylene conical tubes.

2.2 Fractionation of Egg Cytosols

1. *X. laevis* egg extract.
2. 2.5-mL (11 × 34 mm) polycarbonate centrifuge tubes (e.g., Beckman).
3. TLS ultracentrifuge and TLS-55 swinging bucket rotor (Beckman) or equivalent.

2.3 Assessment of Caspase-3/7 Activity in X. laevis Egg Extract Using a Luminescent Caspase Substrate

1. *X. laevis* egg extract.
2. 20× energy mix (EM): 7.5 mM Phosphocreatine, 1 mM ATP, 0.1 mM EGTA, and 1 mM $MgCl_2$.
3. White 96-well, flat-bottomed plate.
4. DEVDase buffer: 50 mM HEPES pH 7.5, 100 mM NaCl, 0.1 % CHAPS, 10 mM DTT, 1 mM EDTA, and 10 % glycerol.
5. Caspase-Glo 3/7 assay substrate (Promega).
6. SpectraMax M3 microplate reader (Molecular Devices) or equivalent.

2.4 Assessment of Caspase-3/7 Activity in X. laevis Egg Extract Using a Chromophore-Linked Caspase Substrate

1. *X. laevis* egg extract.
2. 20× energy mix (EM): 7.5 mM Phosphocreatine, 1 mM ATP, 0.1 mM EGTA, and 1 mM $MgCl_2$.
3. Clear 96-well, flat-bottomed plate.
4. DEVDase buffer: 50 mM HEPES pH 7.5, 100 mM NaCl, 0.1 % CHAPS, 10 mM DTT, 1 mM EDTA, and 10 % glycerol.
5. Ac-DEVD-pNA (e.g., from Enzo Life Sciences).
6. SpectraMax M3 microplate reader (Molecular Devices) or equivalent.

2.5 Assessment of Caspase-2 Processing in X. laevis Egg Extract Using a ^{35}S-Labeled In Vitro-Translated Protein

1. *X. laevis* caspase-2 DNA (pSP64T).
2. TNT-coupled reticulocyte lysate system (Promega).
   (a) Rabbit reticulocyte lysate.
   (b) Reaction buffer.
   (c) SP6 RNA polymerase.
   (d) Amino acid mixture (minus methionine).
   (e) [^{35}S]methionine.
   (f) RNasin ribonuclease inhibitor.
3. Nuclease free water.
4. *X. laevis* egg extract.
5. SDS sample buffer: 150 mM Tris–HCl pH 6.8, 10 % SDS, 30 % Glycerol, and 5 % β-mercaptoethanol.
6. SDS-PAGE apparatus.
7. Gel dryer.
8. X-ray film and cassette.

2.5.1 Assessment of Cytochrome c Release from Mitochondria in X. laevis Egg Extract

1. *X. laevis* egg extract.
2. 20× EM.
3. Centrifuge filter, PVDF 0.1 μm (Millipore, cat. # UFC30VVNB or similar).
4. SDS sample buffer.
5. SDS-PAGE and Western blot apparatus.
6. Cytochrome *c* antibody (BD Pharmingen, cat # 556432 or similar).

2.5.2 Induction of Apoptosis in Cytosolic Fractions of X. laevis Egg Extract Using Cytochrome c

1. *X. laevis* egg extract cytosolic fraction (from Subheading 3.2).
2. 20× EM.
3. Materials from Subheading 2.3 to analyze caspase-3/7 activity.
4. Recombinant cytochrome *c* (*Equine*, Sigma-Aldrich or similar).
5. SDS sample buffer.
6. SDS-PAGE and Western blot apparatus.

2.6 Using Recombinant tBID and BCL-xL to Determine if Inhibition of Caspase Activation Lies Upstream of Mitochondria and BCL-2 Family Proteins

1. *X. laevis* egg extract.
2. 20× EM.
3. Recombinant mouse tBID (ProSpec, cat # PRO-644 or similar).
4. Recombinant human BCL-xL (Sigma-Aldrich, cat # SRP0187 or similar).
5. Glucose 6-phosphate sodium salt.
6. Materials from Subheading 2.3 to analyze caspase-3/7 activity.
7. Materials from Subheading 2.6 to analyze cytochrome *c* release.
8. Materials from Subheading 2.5 to analyze caspase-2 processing.

2.7 Assessment of PARP Cleavage by Western Blot

1. *X. laevis* egg extract.
2. 20× EM.
3. SDS sample buffer.
4. SDS-PAGE and Western blot apparatus.
5. PARP antibody (Novus Bio cat #NB001-64828 or similar).

2.8 Assessment of Nuclear Morphology in X. laevis Egg Extracts

1. *X. laevis* egg extract.
2. 20× EM.
3. Sperm chromatin. *See* Deming and Kornbluth for protocol [18].
4. Hoechst stain.
5. Microscope slide and cover slip.
6. Fluorescent microscope.

2.9 Isolation of Stage VI Oocytes from Mature, Female X. laevis Frogs

1. Mature female primed frogs.
2. 10 % benzocaine.
3. Dissection equipment.
4. Liberase blendzyme 3 (Roche, Switzerland) or similar.
5. OR-2 buffer: 5 mM HEPES pH 7.5, 82.5 mM NaCl, and 2 mM KCl.
6. Fetal bovine serum.
7. Gentamicin.
8. Glass petri dish.
9. Low temperature incubator.
10. Dissecting (stereo) microscope.

2.10 Assessment of Caspase-3 Activity in Intact X. laevis Oocytes Using Near-Infrared Caspase Substrate

1. Stage VI oocytes (*see* Subheading 2.10).
2. OR-2 buffer (*see* Subheading 2.10).
3. Black 96-well plate.
4. Model P-97 flaming/brown micropipette puller (Sutter Instrument Co.) or equivalent.
5. Glass capillary tubing: borosilicate glass, 10 cm length, OD 1.0 mm, ID 0.5 mm.
6. Ficoll PM 400.
7. Clean, glass microscope slide.
8. 40× MMR (*see* Subheading 2.1 for recipe).
9. Microinjection apparatus.
10. Dissecting (stereo) microscope.
11. IRDye 800CW/QC-1 CSP-3 (LI-COR).
12. Recombinant cytochrome *c* (*Equine*, Sigma-Aldrich).
13. Odyssey Infrared Imaging system (LI-COR) or equivalent.

3 Methods

The *X. laevis* egg/oocyte system provides a quick high-throughput model for studying the biochemical events of apoptosis and caspase activation. It recapitulates all the classical hallmarks of apoptosis and caspase activation. Using the extract model system, it is also possible to interrogate the intrinsic apoptotic pathway and dissect each stage. The protocols described below will provide examples of assessment of in vitro-translated caspase activation, how to determine if the initiating signal for induction of apoptosis is pre-mitochondrial, assessing cytochrome *c* release from mitochondria, assessment of executioner caspase (3/7) activation in the egg extract and whole oocytes, and how to analyze targets of executioner caspase activation including DNA fragmentation and poly(ADP-ribose) polymerase (PARP) cleavage.

3.1 Collection of X. laevis Eggs and Preparation of Interphase Egg Extract

Mature female *X. laevis* should be housed in appropriate conditions with tightly controlled water parameters. We house our frogs in a multi-rack system from Aquatic Habitats (http://www.aquatichabitats.com). The water must be maintained at 18 °C, at a pH 7.2, and a conductivity 2 mS. This system is expensive; however, less sophisticated and cheaper methods of housing *X. laevis* have also been used with success (*see* **Note 1**). There are two injections that must be performed to induce egg laying by *X. laevis* frogs. The first is an injection of 0.6 mL PMSG (200 U/mL stock), 3–7 days prior to the second injection, and the desired time for eggs to be laid. This "primes" the frog, inducing production and maturation of more eggs (*see* **Note 2**). The second injection is an injection of 0.6 mL of hCG (1,000 U/mL stock), the day before you wish the eggs to be laid (*see* **Note 3**). This injection will induce the frog to lay eggs, which can be collected the next morning. We inject intramuscularly in the upper quadriceps of the frog's leg.

Mature *X. laevis* eggs are arrested in metaphase of meiosis II, due to high cyclin B/cdc2 activity. Lysis of the eggs by centrifugation induces a release of calcium from intracellular stores, which in addition to the $CaCl_2$ in the egg lysis buffer induces activation of calcium/calmodulin regulated protein kinase (CaMKII) which leads to the degradation of cyclin B and entry into interphase of mitosis. The egg extract is then maintained in interphase by the addition of cycloheximide to the egg lysis buffer, preventing the synthesis of new proteins, namely, cyclin B (*see* **Note 4**).

Wash and de-jelly eggs

1. Collect the freshly laid eggs in a 200-mL glass beaker, and pour off any excess water.
2. Prepare solutions of 2 % L-cysteine pH 8, 1× MMR, and 1× ELB. For 1× ELB, include DTT at a final concentration of 1 mM and sucrose to a final concentration of 250 mM.
3. Wash and incubate the eggs in 2 % L-cysteine, pH 8 for approximately 5 min, swirling the eggs periodically. Use approximately 50 mL of 2 % cysteine per frog. *Xenopus* eggs are surrounded by a transparent jelly coat, which prevents the eggs from contacting each other, that is removed by cysteine. When the jelly coat is removed, the eggs will pack closely together (*see* **Note 5**).
4. When the eggs are de-jellied, pour off the cysteine, avoiding pouring off any eggs at this stage (*see* **Note 6**).
5. Wash eggs two times in 1× MMR. For washes, briefly swirl the eggs and then pour off as much MMR as possible without pouring off any of the eggs.

6. Wash eggs three times in 1× ELB. At this stage, bad or poor quality eggs can be removed. Mature, good quality, *X. laevis* eggs have a very distinct appearance. They have a dark half (animal pole) on top and a light half (vegetable pole) on bottom. Bad eggs vary in appearance, but any eggs that are large, puffy, white, and/or not uniformly pigmented with a distinct dark and light half should be removed. We remove bad eggs by essentially pouring off as many bad eggs as possible. The bad eggs are less dense than the good eggs, and so due to the sucrose in ELB these bad eggs will float closer to the top of the beaker when the ELB is first poured in, allowing them to be poured off while retaining the good eggs (*see* **Note 7**).

Lyse eggs

7. Pour the eggs into a 14-mL polypropylene round-bottom tube. From this point on, the eggs should be kept on ice as much as possible.
8. Centrifuge the eggs in a clinical centrifuge at approximately 1,500 × *g* for 25–30 s and remove any excess buffer from the packed eggs.
9. Add the following inhibitors from stocks based on the volume of packed eggs (*see* **Note 8**):
   (a) 5 μg/mL aprotinin—1 μL/mL of packed eggs
   (b) 5 μg/mL leupeptin—1 μL/mL of packed eggs
   (c) 5 μg/mL cytochalasin B—1 μL/mL of packed eggs
   (d) 50 μg/mL cycloheximide—5 μL/mL of packed eggs
10. Lyse the eggs by centrifugation at 15,600 × *g* for 10 min at 4 °C. We use an Avanti J-26 XPI centrifuge (Beckman Coulter), with a JS 13.1 swinging bucket rotor.
11. Following centrifugation, the eggs will separate into three distinct layers: a yellow layer on top that consists of lipids, the crude egg extract layer in the middle (this can vary in color from a pale to dark brown color), and a dark bottom layer consisting mainly of yolk proteins, glycogen, and pigment granules. The nuclei will also be in this bottom layer (*see* Fig. 1a, b).
12. Remove the crude egg extract (the middle layer) using a 16-G needle and 10-mL syringe. Firmly yet carefully pierce the side of the tube, taking care not to puncture the opposite side of the tube. Slowly draw out the crude extract (*see* **Note 9**). Transfer the extract to a 15-mL polypropylene conical tube on ice.

3.2 Fractionation of Egg Cytosols

Using the following protocol, it will be possible to separate the cytosolic fraction from crude extract. This cytosolic fraction will be competent to undergo apoptosis with the addition of cytochrome *c* as described later (Subheading 3.6.2).

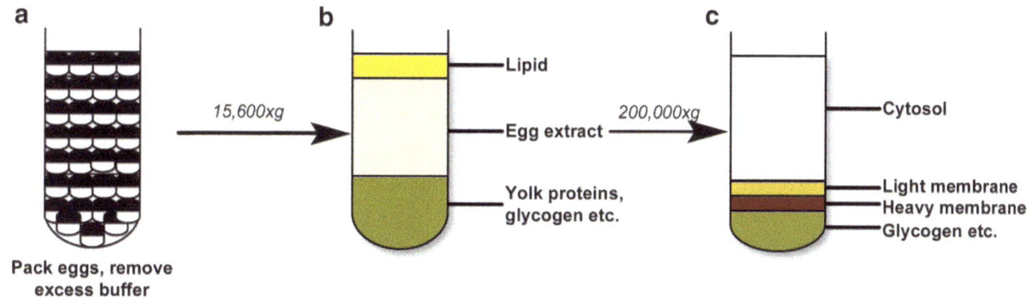

Fig. 1 Collection of *X. laevis* eggs and preparation of interphase egg extract and fraction of egg extract cytosol. (**a**) Eggs are packed by low-speed centrifugation and excess buffer is removed. Following the addition of cycloheximide, cytochalasin B, aprotinin, and leupeptin, the eggs are lysed by centrifugation at 15,600 × *g* for 10 min. (**b**) The eggs are fractionated into distinct layers following high-speed centrifugation: a lipid layer on top; a pellet of yolk proteins, glycogen, pigment granules, and nuclei; and a middle layer of crude egg extract. (**c**) Fractionation of crude egg extract is achieved by ultracentrifugation at 200,000 × *g*. This separates the crude extract into a clear cytosolic fraction on top, a pale light membrane fraction just below, a darker heavy membrane fraction, and a pellet consisting mainly of glycogen and ribosomes

1. Transfer the crude egg extract prepared as described in the previous section to a 2.5-mL polycarbonate centrifuge tube. All the steps in this procedure should be performed on ice (*see* **Note 10**).
2. Centrifuge the crude extract at 200,000 × *g* for 70 min at 4 °C.
3. Following this spin the extract will separate into a clear cytoplasmic layer, with two membrane layers below: a pale, light membrane fraction and a dark, heavy membrane fraction directly below this. There will be a final very dark layer at the bottom of the tube, consisting of mainly glycogen, ribosomes, etc. (Fig. 1c) (*see* **Note 11**).
4. Carefully remove the clear cytoplasmic layer, taking care not to disturb the underlying membrane layers, and transfer it to a fresh 2.5-mL centrifuge tube.
5. Centrifuge the cytoplasmic fraction again at 200,000 × *g* for 25 min. This step removes any remaining membrane contaminants.
6. Remove the clear cytoplasmic fraction, aliquot it, and store at −80 °C until ready to use.

3.3 Assessment of Caspase-3/7 Activity in X. laevis Egg Extract Using a Luminescent Caspase Substrate

Activation of caspases, responsible for the ordered dismantling of the cell, lies at the apex of the apoptotic program. Apoptosis generally culminates in the activation of the executioner caspase-3 and caspase-7 (*see* **Note 12**). These caspases have a very defined and specific cleavage motif, cleaving after the second aspartic acid residue in the DEVD motif. This specificity allows for the activation of caspase-3/7 to be studied very easily in vitro. The following protocol utilizes a pro-luminescent caspase-3/7 substrate (Promega) containing the DEVD motif. In the presence of active caspase-3/7, this substrate will be cleaved and release aminoluciferin, a substrate of luciferase that will result in the production of light. When this substrate is incubated in the egg extract undergoing apoptosis (and active caspase-3 and/or caspase-7), the substrate is cleaved, and a luminescent signal is produced proportional to the level of caspase-3/7 activity.

1. Pipette 85 μL of DEVDase buffer per well of a 96-well plate for each sample and/or time point to be assayed.
2. Supplement egg extract from Subheading 3.1 with 1:20 dilution of 20× EM. Add any treatments that you wish to examine (*see* **Note 13**).
3. Incubate the egg extract at room temperature for 6 h. A good quality extract will undergo caspase activation at approximately 3 or 4 h (*see* Fig. 2).
4. Add 3 μL of egg extract per well for each time point and/or sample. The plate should be kept on ice while samples are being taken, while the extract should remain at room temperature (*see* **Note 14**).
5. Add 5 μL of the Caspase-Glo 3/7 assay substrate (Promega) to each well. Mix briefly and gently, and then incubate the plate at 37 °C for 15 min.
6. Analyze the luminescent output of the plate using 3 s integrated readings on a microplate reader.

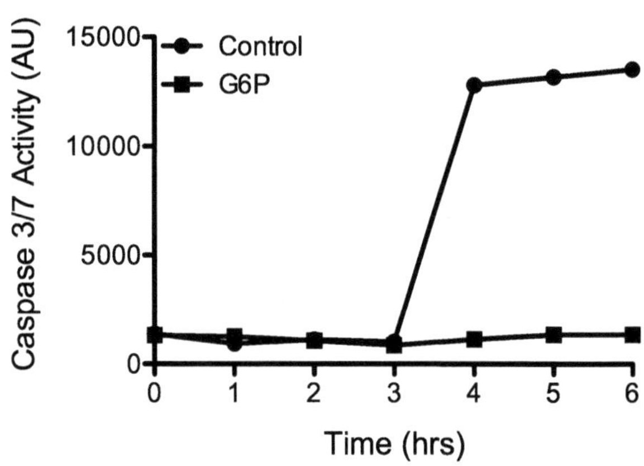

Fig. 2 Spontaneous apoptosis in *X. laevis* egg extracts. Egg extract was incubated at room temperature, and aliquots were taken at indicated time points and analyzed for caspase-3/7 activity as described in Subheading 3.3. Note that addition of G6P to a separate aliquot of extract inhibited this spontaneous apoptosis (McCoy and Nutt, unpublished data)

3.4 Assessment of Caspase-3/7 Activity in X. laevis Egg Extract Using a Chromophore-Linked Caspase Substrate

This protocol is almost identical to that described in the previous section. The only difference being that this protocol utilizes a chromophore-linked caspase substrate (Ac-DEVD-pNA). It must be pointed out that this method is less sensitive and more time consuming than the pro-luminescent caspase-3/7 substrate-based method. However, this method is cheaper; the substrate is less expensive and requires only the ability to read absorbance of a 96-well plate.

1. Pipette 85 μL of DEVDase buffer per well of a 96-well plate for each sample and/or time point to be assayed.
2. Supplement egg extract from Subheading 3.1 with 1:20 dilution of 20× EM. Add any treatments that you wish to examine.
3. Incubate the egg extract at room temperature for 6 h.
4. Add 3 μL of egg extract per well for each time point and/or sample. The plate should be kept on ice while samples are being taken, while the extract remains at room temperature (*see* **Note 14**).
5. Add 10 μL of Ac-DEVD-pNA substrate to each well. Mix briefly and gently, and then incubate the plate at 37 °C for 1 h.
6. Read the plate at 405 nm on a microplate reader.

3.5 Assessment of Caspase-2 Processing in X. laevis Egg Extract Using a ^{35}S-Labeled In Vitro-Translated Protein

Aside from the caspase-3/7 activity assays, there are currently no means by which to directly measure endogenous caspase activation in *X. laevis* egg extract. No commercial antibody is available at present that recognizes *Xenopus* caspases. Assays similar to those described in Subheadings 3.3 and 3.4 exist for other caspases, such as caspase-2, caspase-8, and caspase-9; however, these assays are not specific. Caspase-3 has been shown to more efficiently cleave most substrates than the caspases that the substrates are reportedly specific for [19]. Therefore, we use the protocol described below to quickly and simply analyze the processing of caspases other than 3 and 7 in the egg extract, caspase-2 in this case. This protocol in vitro translates caspase-2 using a rabbit reticulocyte system (Promega). This exogenous caspase-2 is then added to the egg extract, and its processing is monitored over time by autoradiography and used as a marker for caspase-2 activation with a characteristic cleavage pattern [13]. This protocol can also be adapted for use with any other caspase; all that is needed is plasmid DNA in a vector that is responsive to the appropriate promoter (T7 or SP6 in the case of the rabbit reticulocyte system).

1. Prepare the following reaction, using the components of the TNT-coupled reticulocyte lysate system (Promega), in a RNase/DNase-free Eppendorf:
   (a) 25 μL rabbit reticulocyte lysate.
   (b) 2 μL TNT buffer.

(c) 1 μL SP6 polymerase.

(d) 1 μL amino acid mixture, minus methionine (20 μM final concentration).

(e) 2 μL [^{35}S]methionine (20 μCi).

(f) 1 μL RNasin (40 μM final concentration).

(g) 2 μL pSP64T caspase-2 DNA (0.4 μg final concentration).

(h) 16 μL nuclease free water, to a final volume of 50 μL.

2. Incubate this reaction at 30 °C for 2 h.
3. Aliquot and store at −20 °C if not using immediately.
4. Add 8 μL of this TNT reaction to 50 μL of egg extract from Subheading 3.1 supplemented with EM, and incubate at room temperature for ~6 h. For each time point (either every hour or every 2 h), take 5 μL of this reaction and add it to 30 μL of SDS sample buffer.
5. Heat these samples for 3 min at 80 °C.
6. Analyze these samples by SDS-PAGE, loading 15 μL of sample/well (*see* **Note 15**).
7. Dry the gel using an appropriate gel-drying apparatus (we use a model 583 gel dryer from Bio-Rad), and expose to X-ray film (*see* **Note 16**).

3.6 Assessment of Cytochrome c Release from Mitochondria in X. laevis Egg Extract, and Induction of Apoptosis in Cytosolic Fractions of Egg Extract Using Cytochrome c

The *X. laevis* egg extract recapitulates all the key hallmarks of apoptosis, including cytochrome *c* release. Cytochrome *c* release, which follows mitochondrial outer membrane permeabilization (MOMP), is a key event in the intrinsic apoptotic pathway. Once released into the cytosol, cytochrome *c* forms the apoptosome; a caspase-9-activating platform composed of cytochrome *c*, APAF-1, dATP, and procaspase-9. Monitoring release of cytochrome *c* is a very useful tool in dissecting the apoptotic cascade under varying treatment conditions. The first protocol presented below (Subheading 3.6.1) describes how to very simply, and quickly, monitor cytochrome *c* release by removing mitochondria via filtration and subsequently immunoblotting for cytochrome *c* in the cytosolic fraction.

The *X. laevis egg* extract requires the presence of mitochondria to initiate apoptosis and caspase activation [1]. Subheading 3.2 described how to isolate the cytosolic fraction of egg extract, fractionating heavy and light membranes. Lacking mitochondria, this cytosolic fraction will not initiate apoptosis; however, it is still competent to undergo apoptosis, utilizing the post-mitochondrial machinery (caspase-9, apoptosome, etc.). The second protocol below (Subheading 3.6.2) describes how to initiate caspase activation in this cytosolic fraction, by addition of exogenous cytochrome *c* and utilizing the protocol described in Subheading 3.3. This cytosolic fraction can provide a simplified, tractable system in which to study the mechanisms of caspase activation under many different settings.

3.6.1 Assessment of Cytochrome c Release from Mitochondria in X. laevis Egg Extract

1. Supplement egg extract from Subheading 3.1 with 1:20 dilution of 20× EM (and any treatments you wish you to examine). Incubate this extract at room temperature for 6 h.
2. Every 30 or 60 min, collect 20 μL of extract and apply to a 0.1-μM filter tube. Centrifuge at 9,300 × *g* for 10 min at 4 °C. The filter will trap the mitochondria and any cytochrome *c* that has been released will be within the flow through.
3. Add 10 μL of SDS sample buffer (*see* Subheading 2.5) to the flow through, and subject to SDS PAGE gel (either 4 20 % gradient or 15 % gel).
4. Immunoblot with an anti-cytochrome *c* antibody. We recommend a 1:2,000 dilution of the primary antibody into 5 % fat-free milk and incubation of the membrane overnight at 4 °C. 1–2 h incubation with the secondary antibody, at a dilution of 1:10,000 should also be sufficient.

3.6.2 Induction of Apoptosis in Cytosolic Fractions of X. laevis Egg Extract Using Cytochrome c

1. Supplement cytosolic fraction of egg extract from Subheading 3.2 with 1:20 dilution of 20× EM.
2. Supplement cytosolic fractions with cytochrome *c*, at concentrations between 0.6 and 1 ng/μL.
3. Incubate at room temperature for 120 min. Take samples at 15, 30, 60, 90, and 120 min to assess caspase activity as described in Subheading 3.3.

3.7 Assessing the Contribution of a Pre-mitochondrial Signal to Caspase Activation

Mitochondria play a key role in the apoptotic signaling cascade, responsible for releasing pro-apoptotic factors such as cytochrome *c* following MOMP. The BLC-2 family of proteins is the gatekeeper of this mitochondrial permeabilization, with pro-apoptotic BCL-2 family proteins inducing MOMP and anti-apoptotic BCL-2 family proteins inhibiting induction of MOMP. In the *X. laevis* egg extract, caspase-2 activation lies upstream of mitochondria [13], while executioner caspase-3 and caspase-7 lie downstream of mitochondria and cytochrome *c* release. An important aspect of studying apoptosis and caspase activation is determining if the initiating stimulus lies upstream or downstream of mitochondria. Using the *X. laevis* egg extract system, this is easily achieved using techniques already described in combination with the simple addition of either a pro-apoptotic or anti-apoptotic BCL-2 family member.

3.7.1 Using Recombinant tBID to Determine if Inhibition of Caspase Activation Lies Upstream of Mitochondria and BCL-2 Family Proteins

An important pro-apoptotic BCL-2 family member is BID. This protein belongs to a subfamily of BCL-2 family proteins known as BH3-only proteins. BID is cleaved by the initiator caspases, caspase-8 and caspase-2, to form tBID. This truncated BID is the fully active form of BID and is a potent inducer of MOMP and cytochrome *c* release. This property allows it to be used as a tool to dissect if there are key signals upstream of mitochondria and the BCL-2 family of proteins important in inducing caspase activation and apoptosis. The protocol below is adapted from Nutt et al. (2005) [13]. In this study, we found that glucose-6-phosphate (G6P) was capable of inhibiting caspase-3/7 activity, using the assay described in Subheading 3.4. The initiating signal for induction of the apoptotic pathway was believed to be upstream of mitochondria, and G6P was believed to be imposing a block on this pre-mitochondrial signal. To test this hypothesis, we tested the activation of caspase-3/7 in the presence of G6P and tBID. If G6P was inhibiting apoptosis upstream of mitochondria, then tBID would be able to override and bypass this inhibition and induce cytochrome *c* release and subsequent caspase-3/7 activation, which is what we observed. This indicated that there was a signal upstream of mitochondria and BCL-2 family proteins that was responsible for initiating apoptosis.

1. Supplement egg extract from Subheading 3.1 with 1:20 dilution of 20× EM. Divide this extract into two separate aliquots. Treat one of these samples with 20 mM G6P, and treat the other sample with 20 mM G6P and 25 nM tBID.
2. Incubate the extract at room temperature. Every 30–60 min, take a 3 μL aliquot to perform the caspase-3/7 assay from Subheading 3.3 and 20 μL to perform the cytochrome *c* release assay from Subheading 3.6.

3.7.2 Using Recombinant BCL-xL to Determine if Inhibition of Caspase Activation Lies Upstream of Mitochondria and BCL-2 Family Proteins

An important anti-apoptotic BCL-2 family member is BCL-xL. This protein belongs to a subfamily of anti-apoptotic BCL-2 family proteins that prevent proteins like tBID and other members of the BH3-only protein family from inducing MOMP, cytochrome *c* release, and caspase activation. This property allows it to be used in a similar fashion to tBID in Subheading 3.7.1, in determining if the signals initiating apoptosis are upstream of mitochondria and BCL-2 family proteins. The protocol below is also adapted from Nutt et al. (2005) [13]. We hypothesized that the pre-mitochondrial signal responsible for inducing apoptosis was caspase-2. Caspase-2 was known to be important in regulating apoptosis in oocytes, and it had been shown to be capable of cleaving BID to its active tBID form. To test this hypothesis, we assessed the processing of caspase-2 in *X. laevis* egg extract, using the protocol described in Subheading 3.5, in the absence and presence of BCL-xL. We also examined cytochrome *c* release in these same extracts, in the absence and presence of BCL-xL, using the protocol described in Subheading 3.6. If caspase-2 activation was the pre-mitochondrial signal initiating apoptosis, then the addition of BCL-xL would inhibit cytochrome *c* release but not affect caspase-2 activation. We observed that in the presence of BCL-xL, caspase-2 was still processed, while cytochrome *c* release was completely inhibited. This indicated that caspase-2 was indeed an upstream of cytochrome *c* release.

1. Supplement egg extract from Subheading 3.1 with 1:20 dilution of 20× EM. Divide this extract into two separate aliquots, one for analyzing cytochrome *c* release and one for analyzing caspase-2 processing (*see* **Note 17**).
2. Add 8 μL of the TNT reaction from Subheading 3.5 per 50 μL of egg extract from Subheading 3.1 for analysis of caspase-2 processing.
3. Divide these two sets of samples (one for cytochrome *c* release and one for caspase 2 processing) into two separate aliquots. Supplement one of these aliquots with 500 nM recombinant BCL-xL, and leave the other untreated to serve as control.
4. Incubate the samples at room temperature for ~6 h. Every 30–60 min, take 20 μL to perform the cytochrome *c* release assay from Subheading 3.6, and every 1–2 h take 5 μL of the extract supplemented with TNT caspase-2 and add it to 30 μL of SDS sample buffer and process as described in Subheading 3.5.

3.8 Assessment of PARP Cleavage by Western Blot

Poly(ADP-ribose) polymerase (PARP) is a family of proteins that play an important role in DNA repair. PARP detects single-strand breaks in DNA, binds to the DNA, and initiates the repair process. PARP is a 116 kDa protein that during apoptosis is cleaved by caspase-3/7 thus preventing it from initiating DNA repair. This cleavage event can be detected very simply by immunoblotting and has been used extensively in mammalian systems to detect induction of apoptosis. We show here that it can also be used in *X. laevis* egg extracts as a robust marker for induction of apoptosis and caspase activation.

1. Supplement egg extract from Subheading 3.1 with 1:20 dilution of 20× EM.
2. Incubate this egg extract at room temperature for 6 h. Every hour, take a 4 μL aliquot of extract and add to 30 μL of SDS sample buffer.
3. Subject 10 μL of each sample to SDS-PAGE gel (using a 4–20 % gradient gel).
4. Immunoblot with an anti-PARP antibody. A 1:1,000 primary antibody concentration and a 1:2,000 secondary antibody concentration are recommended (*see* **Note 18**). PARP has a very distinctive cleavage pattern, with full-length PARP at ~116 kDa and a cleavage product at ~86 kDa (Fig. 3).

3.9 Assessment of Nuclear Morphology in X. laevis Egg Extracts

Another key hallmark of apoptosis and caspase activation is nuclear fragmentation and morphology. Although *X. laevis* egg extracts lack nuclei, addition of exogenous sperm chromatin results in formation of synthetic nuclei that appear condensed and fragmented in apoptotic extracts. This protocol has been used since the

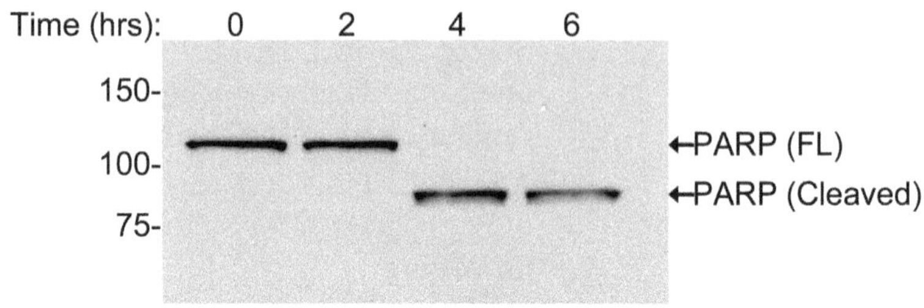

Fig. 3 PARP cleavage in *X. laevis* egg extracts. Egg extract was incubated at room temperature, and samples were taken at indicated time points and analyzed for PARP cleavage by immunoblotting as described in Subheading 3.8 (McCoy and Nutt, unpublished data)

spontaneous apoptosis of *X. laevis* egg extracts was first described [1]; however, it is not used as often today in favor of faster, more quantitative assays.

1. Prepare egg extract from Subheading 3.1 with EM.
2. Add 100× sperm chromatin at a 1:100 dilution, and incubate at room temperature. Nuclei should begin to form after 20–30 min and be fully formed soon after. Well-formed nuclei will appear round and homogenous.
3. After 2 h of incubation in egg extract, these synthetic nuclei should be observed at 30 min intervals for a further 4–6 h.
4. To obtain images of nuclei at desired time points, take an aliquot of extract and mix 1:1 with Hoechst stain on a microscope slide. Add a cover slip and observe the nuclei on a fluorescent microscope (*see* **Note 19**).

3.10 Isolation of Stage VI Oocytes from Mature, Female X. laevis Frogs

Stage VI *X. laevis* oocytes are arrested in prophase of meiosis 1, undergoing a G2-like growth phase. Priming of the frogs with PMSG (*see* Subheading 3.1) induces production of more stage VI oocytes. This protocol will describe how to harvest and isolate stage VI oocytes from female *X. laevis* frogs (*see* **Note 20**). Induction of apoptosis can also be studied in these whole oocytes, providing a primary, whole-cell model to validate results from experiments conducted in the egg extract.

1. Euthanize one mature female frog in 1 L of aquarium water containing 10 mL of 10 % benzocaine (0.1 % final concentration).
2. A secondary method of euthanasia should also be used, such as cervical dislocation.
3. Place the frog with its abdomen facing up, and make a vertical incision along the midline of the frog's abdomen. Peel back the skin and underlying tissue, and excise the ovaries and oocytes.

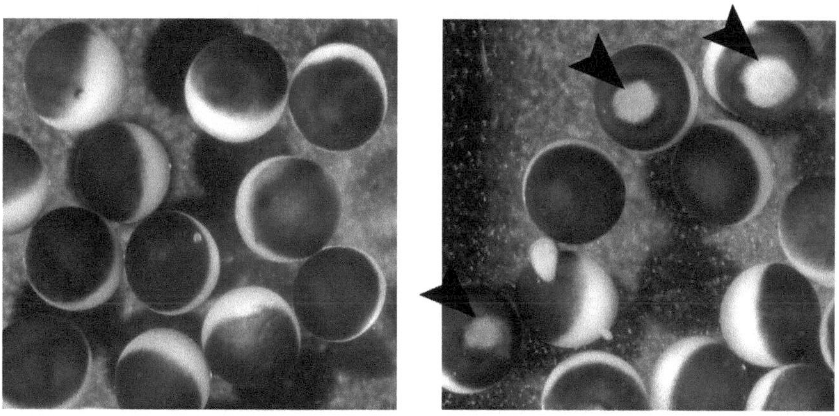

Fig. 4 Healthy stage VI oocytes, in comparison to stage VI oocytes undergoing apoptosis. Healthy stage VI *X. laevis* oocytes were harvested from mature, female frogs as described in Subheading 3.10. Oocytes were untreated as control (*left panel*) or treated with trans-dehydroandrosterone (DHEA) (*right panel*), incubated at room temperature, and monitored for induction of apoptosis. Note the distinctive "white dot" on the oocytes undergoing apoptosis in the *right panel* (*black arrowheads*; McCoy and Nutt, unpublished data)

4. Place the oocytes in a 50-mL conical tube containing OR-2 buffer. Wash the oocytes several times with OR-2 buffer, removing as much blood and extraneous tissue as possible.
5. Incubate the oocytes with agitation in OR-2 buffer and 2.8 U of Liberase blendzyme for ~2 h at room temperature.
6. When completely digested (*see* **Note 21**), wash the oocytes several more times with OR-2 buffer, pouring off as much debris and tissue as possible, leaving only oocytes behind.
7. Place the oocytes in a glass petri dish containing OR-2 buffer supplemented with 10 % fetal bovine serum (FBS) and 0.2 % gentamicin, and store at 18 °C.
8. Using a dissecting (stereo) microscope, sort as many stage VI oocytes as are required for the experiment. Stage VI oocytes have a very distinctive appearance, very similar to mature, laid eggs (*see* Fig. 4). A characteristic that can help differentiate stage VI oocytes is a "white band" around the midline of the oocyte. Store stage VI oocytes in OR-2 buffer supplemented with FBS and gentamicin at 18 °C until ready to use (*see* **Note 22**).

3.11 Assessment of Caspase-3/7 Activity in Intact X. laevis Oocytes Using a Near-Infrared Caspase Substrate

The large size and physical robustness of *X. laevis* oocytes makes them very amenable to study and microinjection. This means studying caspase activation directly in intact oocytes is also possible. The principle of the technique is similar to that used to study caspase-3/7 activity in egg extract; utilizing a fluorogenic, near-infrared caspase-3 substrate (LI-COR). Upon induction of apoptosis and activation of caspase-3, this substrate fluoresces, displaying a peak emission at 789 nm. While this technique does require more specialized equipment and knowledge of how to microinject oocytes, it is not difficult providing the equipment and tools are available. This protocol is adapted from one first published by Johnson et al. (2010) [20].

1. Prepare a solution of 3 % Ficoll in 0.4× MMR (*see* **Note 23**).
2. Prepare and calibrate needles of appropriate size for microinjection (*see* **Note 24**). The cytosolic volume of *X. laevis* oocytes is ~500 nL, and this volume should be used to calculate the final intracellular concentration of the injected solutions. The injection volume should be kept below 2 % of the total egg volume (20 nL in the case of the *X. laevis* oocytes).
3. Separate the required number of healthy, stage VI oocytes for the experiment.
4. Carefully and slowly coat a clean, glass microscope slide with the 3 % Ficoll solution. Ensure that none runs over the edges of the slide.
5. Transfer the oocytes to be injected to the Ficoll solution (*see* **Note 25**).
6. Inject oocytes with recombinant cytochrome *c*, to a final concentration of 200 nM, and with IRDye 800CW/QC-1 CSP-3 to a final concentration of 200–400 nM (*see* **Note 26**). Oocytes should be injected with the needle at a slight (~30–40°) angle on the dark half (animal pole) of the oocyte, just above the border of the dark and light halves.
7. Transfer injected oocytes to a black 96-well plate (3–5 oocytes per well), and incubate for 30 min—2 h at room temperature in OR-2 buffer.
8. Scan the plate on an Odyssey Infrared Imaging system (LI-COR), with a focus offset of 3 mm, a resolution of 84 μm, and an intensity setting of 5 in the 700 nm channel and 0.5 in the 800 nm channel (*see* **Note 27**).

4 Notes

1. Frogs can also be housed in dechlorinated tap water. However, the temperature still needs to be maintained at 18 °C and the water at an appropriate level of conductivity [18].
2. Ovulated females should have ideally a 3–6-month break in between cycles of priming and laying.
3. For our system, injecting frogs with hCG during the afternoon (~2 pm) results in consistently better quality eggs. Female frogs will usually begin laying eggs 10–12 h following injection; we believe the better quality of the eggs results from less time spent in the tanks prior to collection.

4. By lysing the eggs in the presence of EGTA, the effect of calcium will be blocked, and the eggs will remain arrested in metaphase of meiosis II. This extract is very resistant to induction of apoptosis [15].
5. *X. laevis* eggs are surrounded by a transparent jelly coat, which is removed by cysteine. Leaving the eggs in cysteine for too long can cause the eggs to lyse. When the eggs pack closely together, this is a sign that they are de-jellied. This will usually occur before the 5 min period is complete.
6. Once the jelly coat is removed, the following steps need to be done without pause. The eggs are more vulnerable without their jelly coat and taking too long can decrease the overall quality of the final extract.
7. This method may leave behind some bad eggs; however, the number of bad eggs is typically very small and not enough to affect the functionality of the final extract. If you require all bad eggs to be removed, an alternative method is to pour the eggs into a petri dish and remove the bad eggs with a Pasteur pipette. The pipette should be cut to provide a wider tip, so as to avoid lysing the eggs.
8. Aprotinin and leupeptin are protease inhibitors; aprotinin inhibits trypsin-, chymotrypsin-, and plasmin-like proteases, while leupeptin inhibits lysosomal proteases. Cytochalasin B inhibits actin polymerization and cycloheximide inhibits protein synthesis.
9. Be sure not to puncture too close to the dark bottom layer, and ensure the needle bevel is pointing up at first. As the lipid layer moves closer, slowly turn the bevel 90° so that it is facing away from the lipid layer. It is important to not take up any of the lipid layer or bottom pigmented layer. We typically do not attempt to remove all of the crude extract as this will more likely lead to contamination. Instead, by slowly and carefully removing all but a small portion of the crude extract layer, we avoid any contamination by the other layers. It is also common for the needle to be clogged (by the plastic of the tube during penetration) resulting in extremely slow removal of the extract. When this occurs, we insert a second needle perpendicular to the first needle and repeat the process. Do not remove the first needle as this will lead to loss of extract.
10. Be sure to fill the tube at least ¾ full to avoid the tube collapsing during centrifugation.
11. There may also be a thin pale/yellow layer of lipid above the clear cytoplasmic layer. This is normal and can be easily removed or avoided.
12. There are also caspase-independent pathways of apoptosis; however, they will not be discussed here.

13. It is best to prepare a "stock" egg extract supplemented with EM and then divide this between each sample you wish to assay.
14. The temperature in our laboratory is typically around 18–21 °C.
15. Take care not to run the gel too far as one of the cleavage products of caspase-2 is low molecular weight and can easily be lost off the bottom of the gel.
16. The length of exposure will depend on the age of the [^{35}S] methionine. With fresh [^{35}S]methionine, typically an overnight or one day exposure should be enough to give a reasonably strong signal.
17. An aliquot of extract could also be taken at each time point and used to assess caspase-3/7 activity using the assay described in Subheading 3.3.
18. We recommend cutting the membrane at or below ~75 kDa mark, as the antibody recognizes a nonspecific protein that can reduce overall signal intensity of the bands of interest.
19. Apoptotic nuclei should appear condensed and fragmented. The difference with early time point nuclei, or nuclei in extracts where apoptosis has been inhibited, should be very apparent.
20. This method will only summarize the procedure for physical removal of oocytes from female frogs. A more detailed description of *Xenopus* anatomy and dissection procedure can be found in Ref. [21].
21. When digestion is complete, the oocytes will be detached from each other and any connective tissue.
22. The oocytes should be sorted the same day they are harvested from the frog and stored for no more than one night at 18 °C, and the experiment performed the next day. It is also a good idea to quickly sort the oocytes again the next morning to remove any that died overnight.
23. The Ficoll will need to be dissolved in dH_2O first, by briefly microwaving until dissolved. This needs to be done prior to adding the MMR solution, as the heating will destroy the EDTA in MMR.
24. The settings we use on our micropipette puller are as follows: $P = 500$, heat = 350, pull = 200, $V = 100$, and time = 150. These needles then need to be cut, beveled, and calibrated to the point where a single injection will inject the desired volume (10—25 nL).
25. Care must be taken when physically handling the oocytes, as they are easily damaged when too much force is applied. We use a plastic Pasteur pipette, with the end cut off to create a wider opening for the oocytes, preventing them from becoming damaged.

26. When injecting two solutions, they can be injected separately. However, care must be taken when injecting the same oocyte twice not to cause too much damage to the oocyte. One sign of this would be excessive leaking from the oocyte. Performing the same number of injections in control oocytes will control for any effect of the injections alone.
27. It is also possible to observe apoptotic activity in oocytes visually. Upon induction of apoptosis, *X. laevis* oocytes develop a characteristic large, white spot in their dark half (animal pole) (Fig. 4).

References

1. Newmeyer DD, Farschon DM, Reed JC (1994) Cell-free apoptosis in Xenopus egg extracts: inhibition by Bcl-2 and requirement for an organelle fraction enriched in mitochondria. Cell 79(2):353–364
2. Kluck RM, Bossy-Wetzel E, Green DR, Newmeyer DD (1997) The release of cytochrome c from mitochondria: a primary site for Bcl-2 regulation of apoptosis. Science 275(5303):1132–1136
3. Kluck RM, Martin SJ, Hoffman BM, Zhou JS, Green DR, Newmeyer DD (1997) Cytochrome c activation of CPP32-like proteolysis plays a critical role in a Xenopus cell-free apoptosis system. EMBO J 16(15):4639–4649
4. Evans EK, Kuwana T, Strum SL, Smith JJ, Newmeyer DD, Kornbluth S (1997) Reaper-induced apoptosis in a vertebrate system. EMBO J 16(24):7372–7381
5. Evans EK, Lu W, Strum SL, Mayer BJ, Kornbluth S (1997) Crk is required for apoptosis in Xenopus egg extracts. EMBO J 16(2):230–241
6. Kuwana T, Smith JJ, Muzio M, Dixit V, Newmeyer DD, Kornbluth S (1998) Apoptosis induction by caspase-8 is amplified through the mitochondrial release of cytochrome c. J Biol Chem 273(26):16589–16594
7. Thress K, Evans EK, Kornbluth S (1999) Reaper-induced dissociation of a Scythe-sequestered cytochrome c-releasing activity. EMBO J 18(20):5486–5493
8. Kluck RM, Esposti MD, Perkins G, Renken C, Kuwana T, Bossy-Wetzel E et al (1999) The pro-apoptotic proteins, Bid and Bax, cause a limited permeabilization of the mitochondrial outer membrane that is enhanced by cytosol. J Cell Biol 147(4):809–822
9. Kluck RM, Ellerby LM, Ellerby HM, Naiem S, Yaffe MP, Margoliash E, Bredesen D, Mauk AG, Sherman F, Newmeyer DD (2000) Determinants of cytochrome c pro-apoptotic activity. The role of lysine 72 trimethylation. J Biol Chem 275(21):16127–16133
10. Tashker JS, Olson M, Kornbluth S (2002) Post-cytochrome C protection from apoptosis conferred by a MAPK pathway in Xenopus egg extracts. Mol Biol Cell 13(2):393–401
11. Deming PB, Schafer ZT, Tashker JS, Potts MB, Deshmukh M, Kornbluth S (2004) Bcr-Abl-mediated protection from apoptosis downstream of mitochondrial cytochrome c release. Mol Cell Biol 24(23):10289–10299
12. Bergeron L, Perez GI, Macdonald G, Shi L, Sun Y, Jurisicova A et al (1998) Defects in regulation of apoptosis in caspase-2-deficient mice. Genes Dev 12(9):1304–1314
13. Nutt LK, Margolis SS, Jensen M, Herman CE, Dunphy WG, Rathmell JC et al (2005) Metabolic regulation of oocyte cell death through the CaMKII-mediated phosphorylation of caspase-2. Cell 123(1):89–103
14. Nutt LK, Buchakjian MR, Gan E, Darbandi R, Yoon S-Y, Wu JQ et al (2009) Metabolic control of oocyte apoptosis mediated by 14-3-3ζ-regulated dephosphorylation of caspase-2. Dev Cell 16(6):856–866
15. Andersen JL, Johnson CE, Freel CD, Parrish AB, Day JL, Buchakjian MR et al (2009) Restraint of apoptosis during mitosis through interdomain phosphorylation of caspase-2. EMBO J 28(20):3216–3227
16. Andersen JL, Thompson JW, Lindblom KR, Johnson ES, Yang C-S, Lilley LR et al (2011) A biotin switch-based proteomics approach identifies 14-3-3ζ as a target of Sirt1 in the metabolic regulation of caspase-2. Mol Cell 43(5):834–842
17. McCoy F, Darbandi R, Chen SI, Eckard L, Dodd K, Jones K et al (2013) Metabolic regulation of CaMKII protein and caspases in Xenopus laevis egg extracts. J Biol Chem 288(13):8838–8848

18. Deming P, Kornbluth S (2006) Study of apoptosis in vitro using the Xenopus egg extract reconstitution system. Methods Mol Biol 332:379–393
19. McStay GP, Salvesen GS, Green DR (2007) Overlapping cleavage motif selectivity of caspases: implications for analysis of apoptotic pathways. Cell Death Differ 15(2): 322–331
20. Johnson CE, Freel CD, Kornbluth S (2009) Features of programmed cell death in intact Xenopus oocytes and early embryos revealed by near-infrared fluorescence and real-time monitoring. Cell Death Differ 17(1):170–179
21. Sive HL, Grainger RM, Harland RM (2000) Early development of Xenopus laevis: a laboratory manual. Cold Spring Harbor Laboratory Press. ISBN 978-087969942-0

Chapter 9

Caspase Protocols in Mice

Varsha Kaushal, Christian Herzog, Randy S. Haun, and Gur P. Kaushal

Abstract

Members of the caspase family of proteases are evolutionarily conserved cysteine proteases that play a crucial role as the central executioners of the apoptotic pathway. Since the discovery of caspases, many methods have been developed to detect their activation and are widely used in basic and clinical studies. In a mouse tissue, caspase activation can be monitored by cleavage of caspase-specific synthetic substrates and by detecting cleaved caspase by western blot analysis of the tissue extract. In tissue sections, active caspase can be detected by immunostaining using specific antibodies to the active caspase. In addition, among the myriads of caspase-specific substrates known so far, cleaved fragments produced by caspases from the substrates such as PARP, lamin A, and cytokeratin-18 can be monitored in tissue sections by immunostaining as well as western blots of tissue extracts. In general, more than one method should be used to ascertain detection of activation of caspases in a mouse tissue.

Key words Caspases, Immunoblot, Immunostaining, Antibodies, Tetrapeptide substrates, Chromophore, Homogenization, Deparaffinization

1 Introduction

Members of the caspase family of proteases are evolutionarily conserved cysteine proteases that play a crucial role as the central executioners of the apoptotic pathway. In addition to their apoptotic role in embryogenesis and maintaining cellular homeostasis, caspases play non-apoptotic roles in inflammation, cell differentiation, proliferation, motility, and neuronal functions including neural axon pruning and synapse elimination [1–3]. However, most of the characteristic biochemical and morphological features occurring in apoptotic cells that were originally observed by Kerr et al. [4] are generally attributed to the action of caspases on the specific cellular substrates [5–8].

Since the discovery of caspases in the 1990s many methods designed to detect activation of caspases have been developed and are widely used in basic and clinical studies. These methods include (1) determination of caspase activity using specific synthetic substrates designed to contain fluorogenic or chromogenic leaving

Peter V. Bozhkov and Guy Salvesen (eds.), *Caspases, Paracaspases, and Metacaspases: Methods and Protocols*, Methods in Molecular Biology, vol. 1133, DOI 10.1007/978-1-4939-0357-3_9, © Springer Science+Business Media New York 2014

groups after the aspartate residue; (2) detection of the specific active caspase by western blot using an antibody specific to active caspase; (3) detection of caspase-specific cleaved products of PARP, cytokeratin-18, and lamin A by western blot; (4) immunocytochemical localization of an activated caspase (cleaved caspase) using antibodies specific for active (cleaved) caspase; (5) detection of active caspases by use of *f*luorochrome-*l*abeled *i*nhibitors of *ca*spases (FLICA) as affinity ligands that bind to the active center of the active caspase [9]; and (6) use of tandem molecules of green-, blue-, cyan-, or yellow-fluorescent protein covalently linked to a small peptide that target a caspase [10]. These methods are routinely and widely used in in vitro cell culture studies. However, only some of these methods can be successfully used to detect caspase activation in in vivo situations. This chapter provides details of the following methods that can be used to determine caspase activation in a mouse tissue: (1) caspase enzyme assay in mouse tissue homogenates using synthetic peptide substrates, (2) detection of cleaved caspases in mouse tissue homogenates by western blot analysis, (3) immunostaining of caspases in tissue sections with antibodies specific to full-length and cleaved (active) caspases, and (4) detection of caspase-specific cleaved products of poly (ADP-ribose) polymerase (PARP), cytokeratin-18, and lamin A in tissue homogenates and tissue sections as a measure of caspase activation.

2 Materials

All animal protocols should be approved and conducted in accordance with the Institutional Animal Care and Use Committee.

2.1 Caspase Enzyme Assay in Mouse Tissue Homogenates Using Synthetic Peptide Substrates

1. Lysis buffer for mouse tissues: 50 mM HEPES, pH 7.5, 0.1 % CHAPS, 2 mM dithiothreitol, 0.1 % Nonidet P-40, 1 mM EDTA, 1 mM phenylmethylsulfonyl fluoride, 2 μg/ml leupeptin, and 2 μg/ml pepstatin A at 4 °C.
2. Thermo Scientific Pierce BCA (bicinchoninic acid) Protein Assay Kit.
3. Caspase assay buffer: 100 mM HEPES, pH 7.2, 10 % sucrose, 0.1 % CHAPS, 1 mM Na-EDTA, and 2 mM dithiothreitol.
4. Stock solution for caspase substrate: 20 mM caspase peptide substrates in dimethyl sulfoxide (DMSO). DEVD-AMC or DEVD -AFC for caspase-3/7, VEID-AMC or VEID-AFC for caspase-6, YVAD-AMC or YVAD-AFC for caspase-1, VDVAD-AMC or VDVAD-AFC for caspase-2, LEHD-AMC for caspase-9, and IETD-AMC or IETD-AFC for caspase-8.
5. Dounce homogenizer.
6. Microplate reader (Molecular Devices, SpectraMaxM5 or equivalent).

2.2 Detection of Cleaved Caspases in Mouse Tissue Homogenates by Western Blot Analysis

1. Lysis buffer: 50 mM HEPES, pH 7.5, 0.1 % CHAPS, 2 mM dithiothreitol, 0.1 % Nonidet P-40, 1 mM EDTA, 1 mM phenylmethylsulfonyl fluoride, 2 μg/ml leupeptin, and 2 μg/ml pepstatin A at 4 °C.
2. Thermo Scientific Pierce BCA Protein Assay Kit.
3. Dounce homogenizer.
4. MiniProtean II (Bio-Rad) protein gel apparatus.
5. TEMED.
6. Solution A: 30 % acrylamide, 0.8 % bisacrylamide (75 g acrylamide and 2 g *N*,*N*-methylene-bisacrylamide in 250 ml distilled water).
7. Solution B: 1.5 M Tris–HCl, pH 8.8. Dissolve 45.5 g Tris in 200 ml distilled water. Adjust pH with HCl to pH 8.8. Add distilled water to 250 ml.
8. Solution C: 0.5 M Tris–HCl, pH 6.8. Dissolve 15.1 g Tris in 200 ml distilled water. Adjust pH with HCl to 6.8. Add distilled water to 250 ml.
9. Solution D: 10 % SDS.
10. Solution E: 10 % (w/v) ammonium persulfate (APS; add 100 mg in 1 ml of distilled water, prepare at the time of gel casting).
11. 2× SDS-sample buffer: 0.76 g Tris, 10 ml of glycerol, 5 ml of 2-mercaptoethanol, 2 mg bromophenol-blue, 5 g SDS. Add distilled water to make 100 ml and adjust pH to 6.8 with 1 N HCl.
12. 10× SDS-running buffer: 60 g Tris, 288 g glycine, 20 g SDS in 2 L of distilled water. Adjust pH to 8.3.
13. 10× transfer buffer: 30 g Tris, 144 g glycine in 20 % MeOH to 1 L.
14. PVDF membrane.
15. 10× PBS: 160 g NaCl, 4 g KCl, 28.8 g Na_2HPO_4, 4.8 g KH_2PO_4. Add distilled water to 2 L.
16. PBS-Tween-20 buffer (1× PBS-T): 1× PBS, 0.05 % Tween-20.
17. Dry milk.
18. 1 M DTT stock solution in distilled water stored at −20 °C.
19. Primary antibodies to procaspases and cleaved (active) caspases (e.g., from Cell Signaling Technology Inc.).
20. Anti-GAPDH antibody (e.g., sc-47724, Santa Cruz Biotechnology).
21. Appropriate HRP-conjugated secondary antibodies.
22. Chemiluminescence reagent (e.g., Super Signal WestPico Chemiluminescence Reagent from Thermo Scientific).

23. Stripping solution for western blot membranes: 62.5 mM Tris–HCl, pH 6.7, 2 % SDS, and 100 mM 2-mercaptoethanol (should be added just before use).
24. Gel imaging system (optional; e.g., Bio-Rad Chemidoc-System).

2.3 Detection of Caspases by Immunostaining with Antibodies Specific to Full-Length and Cleaved (Active) Caspases

1. Tissue fixative: 10 % neutral-buffered formalin (e.g., Sigma HT501128).
2. Paraffin: Use paraffin heated to 56–58 °C.
3. PBS-T: 1× phosphate-buffered saline, pH 7.4, 0.05 % Tween-20.
4. Graded ethanol and xylene: 70, 80, 95, and 100 % ethanol and xylene.
5. Antigen retrieval buffer: 10 mM sodium citrate pH 6.0, 0.05 % Tween-20.
6. 1 % H_2O_2 in PBS containing 0.1 % sodium azide.
7. Blocking buffer: 5 % BSA in PBS-T with up to 0.1 % Triton X-100 as needed.
8. Appropriate primary and secondary antibodies.
9. Antibody dilution buffer: 5 % BSA in PBS-T.
10. HRP substrate solution: 0.2 mg/ml 3,3′-diaminobenzidine (DAB) tetrahydrochloride, 0.012 % H_2O_2, 20 mM citric acid monohydrate, 100 mM imidazole, 100 mM NaCl, pH 7.0.
11. 4′,6′-Diamidino-2-phenylindole (DAPI).
12. Eukitt mounting medium (Electron Microscopy Sciences) or other permanent mounting medium.
13. VECTASHIELD mounting medium.
14. Tissue processor, optional (e.g., Tissue Tek VIP E300 Tissue Processor from Sakura).
15. Rotary microtome (e.g., Microm HM325).
16. Slide warmer.

2.4 Detection of Caspase-Specific Cleaved Products of PARP, Cytokeratin-18, and Lamin A in Tissue Homogenates and Tissue Sections as a Measure of Caspase Activation

1. All reagents described in Subheading 2.2 (except for caspase antibodies) for western blot analysis and Subheading 2.3 for immunodetection.
2. Antibodies: PARP antibodies (e.g., from Cell Signaling Technology Inc.), anti-M30 CytoDEATH (cytokeratin-18 epitope) monoclonal antibody from Roche Applied Science, and antibody to cleaved lamin A (small subunit, e.g., from Cell Signaling Inc.). Cleaved lamin A antibody detects the fragments of lamin A/C resulting from cleavage at aspartic acid 230 by active caspase. It does not react with mature lamin A/C protein.

3 Methods

3.1 Caspase Enzyme Activity Assay in Mouse Tissue Homogenate Using Synthetic Peptide Substrates

Proteins susceptible to caspase cleavage possess a tetrapeptide recognition sequence containing the scissile P1 aspartate residue at its C-terminus. These proteins are recognized by the caspases and are cleaved following the essential aspartic acid residue. Thornberry et al. [11] applied a positional scanning synthetic combinatorial library approach with the tetrapeptide recognition sequence containing a 7-amino-4-methylcoumarin (AMC) fluorochrome (Ac-X-X-X-Asp-AMC) and determined cleavage specificities of different caspases in vitro. Based on the specific cleavage sites of various caspases, synthetic small peptide substrates containing a chromophore after the aspartate group have been synthesized that are commercially available (e.g., Peptide International Inc., Louisville, Kentucky; Bachem AG, Bubendorf, Switzerland) and are commonly used as synthetic substrates to determine the caspase activity. The peptides YVAD, VDVAD, DEVD, VEID, IETD, and LEHD are considered to be specific for caspase-1, -2, -3, -6, -8, and -9, respectively (*see* **Note 1**). The peptides are conjugated to AMC (detected by excitation at 380 nm and emission at 460 nm) or 7-amino-4-trifluoromethylcoumarin (AFC; detected by excitation at 405 nm and emission at 500 nm), or to the chromophore, *p*-nitroanilide (pNA: detected colorimetrically by absorbance at 400 nm). Upon cleavage by a specific caspase, the liberated fluorochrome or chromophore is detected by a fluorometer or spectrophotometer, respectively. Caspase activation using synthetic substrates has been previously determined in diabetic mouse liver [12], renal ischemia–reperfusion injury [13], cerebral ischemia [14], and lungs during renal ischemia [15]. Thus, the activity of caspases in crude homogenates of tissues can be determined using the synthetic substrates (*see* **Note 2**).

3.1.1 Preparation of Tissue Homogenate

1. Harvest mouse tissue and homogenize using a Dounce homogenizer in lysis buffer. Mouse tissue can be stored and kept in −80° freezer for future use.
2. Centrifuge the homogenate at 10,000 × *g* for 10 min to obtain the supernatant.

3.1.2 Determination of Protein Concentration in Tissue Homogenate (See **Note 3**)

1. Using serial dilutions, prepare six standards with bovine serum albumin (BSA) concentrations ranging from 2 to 0.0625 mg/ml.
2. In a 96-well microplate, add in triplicates 25 μl of standards, blank (distilled water), and diluted tissue sample (1:20).
3. Prepare BCA reagent: Add BCA reagent A to reagent B in a ratio of 50 to 1.
4. Add 200 μl of the BCA reagent mixture to 25 μl of standard, sample, and blank in a microplate.

5. Mix well while avoiding bubbles and incubate the plate at 37 °C for 30 min and read the absorbance at 562 nm using a spectrophotometer plate reader. Using the standard curve constructed with the standards and blank, determine the protein concentrations of the tissue homogenates.

3.1.3 Caspase Activity Measurement

1. Transfer equal amounts of tissue homogenate protein (10–50 μg) to each well of a 96-well microplate containing 100 μl of caspase buffer with 50 μM of the fluorogenic substrate.
2. Incubate the reaction mixture in microplate for 1 h at 37 °C.
3. Determine the amount of liberated fluorescent group. To detect AMC use an excitation wavelength of 380 nm and an emission wavelength of 460 nm. AMC is used as a standard. Based on the standard curve made from a fluorescence reading with free AMC, the data for caspase activity are expressed as nanomoles of liberated AMC (*see* **Note 4**). If the caspase substrate coupled to the fluorescent group AFC is used for the assay, the same protocol is used as described for the AMC-conjugated substrate, except that the liberated AFC should be determined with an excitation wavelength of 405 nm and an emission wavelength of 500 nm. If the caspase substrate is conjugated to the chromophore pNA, the same protocol for the assay is used except that the liberated pNA should be quantified using a spectrophotometer microplate reader at an absorbance of 400 nm. The concentration of tetrapeptide with pNA is generally 0.2 mM to ensure estimation of activity within a linear range of substrate utilization.

3.2 Detection of Cleaved Caspases in Mouse Tissue Homogenates by Western Blot Analysis

Caspases are synthesized in the cytosol as inactive proenzymes containing a prodomain, a large subunit (~20 kDa), and a small subunit (~10 kDa). This method is based on the principal that procaspases are proteolytically processed to large and small subunits to become active enzymes and detecting these subunits by western blots. In response to an apoptotic stimulus, the initiator or apical caspases, caspase-8, -10, or -9, are activated and proteolytically process the effector caspases, caspase-3, -6, and -7 into large and small subunits that reassociate to become active heterodimers. In their latent inactive conformation, the effector caspases are dimeric whereas the initiator caspases are monomeric. The initiator caspases become dimers on activation by forming scaffolding oligomeric complexes via interaction with the adapter molecules. For example, caspase-8 is activated by forming DISC as the oligomeric complex and caspase-9 is activated by forming "apoptosome" as an oligomeric platform (*see* **Note 5**). The procaspase and the processed fragments can be detected by western blot analysis using specific antibodies that detect epitopes in procaspase and/or the cleaved fragments. The assay of cleaved (active) caspases has been previously performed in mouse tissues in cerebral ischemia [14], focal cerebral ischemia [16], lungs during renal ischemia [15], cardiac ischemia–reperfusion injury [17], and renal ischemia-reperfusion injury [18] (*see* **Notes 6** and 7).

3.2.1 Preparation of Tissue Homogenate

Follow protocol described in Subheading 3.1.1.

3.2.2 Determination of Protein Concentration in Tissue Homogenate

Follow protocol described in Subheading 3.1.2.

3.2.3 Sodium Dodecyl Sulfate–Polyacrylamide Gel Electrophoresis

1. Assemble a MiniProtean II (Bio-Rad) protein gel apparatus according to the manufacturer's instructions. Use clean glass plates. Rinse plates with distilled water and ethanol. After assembly, ensure that there is no leakage.
2. Prepare 12 % separating gel: Mix 8 ml of solution A with 6.7 ml of distilled water and 5 ml of solution B in a small flask and degas briefly. Add 200 μl of solution D, then 100 μl of solution E and 10 μl TEMED, and mix. This will start the polymerization. Quickly pour the solution into the gel assemblies and overlay each gel with 0.2 ml of a 50 % methanol/water mixture. Let the separating gel polymerize for 30 min.
3. Remove the overlay before pouring stacking gel. Prepare 4 % stacking gel: Mix 1.3 ml of solution A with 6.1 ml of distilled water and 2.5 ml of solution C and degas briefly. Add 100 μl of solution D, 50 μl of solution E, and 10 μl TEMED, mix, and pour on top of separating gel, insert comb, and let polymerize for 30 min.
4. Sample preparation and gel run: Mix 50 μg aliquots of tissue homogenate with 1 volume of 2× SDS-sample buffer and boil for 5 min at 95–100 °C. Spin briefly and load the samples on SDS gel. Run gels at 200 V (constant voltage) for 45 min.
5. Transfer proteins onto PVDF membrane (wet-transfer apparatus from Bio-Rad): Prepare a 1× transfer buffer containing 20 % methanol and precool at 4 °C before use. Assemble gel sandwich according to the manufacturer's instruction and run transfer at 100 V for 1 h.
6. Disassemble gel sandwich and wash the blot briefly in 1× PBS-T and mark the orientation of the gel. Block in 5 % dry milk in PBS-T at room temperature on a shaking platform for 1 h.
7. Incubate the blot with primary antibody diluted 1:1,000 in 1× PBS-T at 4 °C overnight.
8. Wash the blot with 1× PBS-T 4 × 5 min each.

9. Incubate with secondary antibody (e.g., goat anti-rabbit HRP-conjugated IgG; Santa Cruz cat# sc-2004) diluted 1:10,000 in 1× PBS-T at room temperature for 1 h.
10. Wash the blot with 1× PBS-T 4×5 min each.
11. Develop the blot with a chemiluminescence reagent and record the signal by exposing autoradiography films or record digitally.

3.2.4 Reprobing Western Blots for Loading Control

1. Wash membranes briefly in distilled water and incubate in 50 ml of stripping solution in a water bath shaker at 50 °C for 30 min.
2. Remove stripping solution by washing with several rinses of distilled water.
3. Equilibrate the blots in 1× PBS-T and perform blocking as described in Subheading 3.2.3 (**step 6**). Incubate with anti-GAPDH antibody diluted in 5 % dry milk in PBS-T at 1:500 at 4 °C, overnight.
4. Wash the blot with 1× PBS-T 4×5 min each.
5. Incubate with secondary antibody diluted in 1× PBS-T at room temperature for 1 h.
6. Wash the blot with 1× PBS-T 4×5 min each.
7. Develop the blot with a chemiluminescence reagent and record the signal by exposing autoradiography films or record digitally.

3.3 Immunodetection of Caspases in Tissue Sections with Antibodies Specific to Full-Length and Cleaved (Active) Caspases

Immunostaining can be used to detect both active and procaspases in tissue section using specific antibodies. Both fluorescent- and nonfluorescent-based methods for immunohistological detection can be used in tissue sections. In the fluorescent-based technique, the primary antibody to a specific antigen (active caspase) is first allowed to bind to the antigen in a tissue section and then the secondary antibody carrying the fluorophore is used to bind to the primary antibody. The fluorophore bound to the primary antibody is visualized using a fluorescent microscope. In the nonfluorescent based method, a histochemical technique known as immunoperoxidase labeling is commonly used. In this technique, the primary antibody first binds to corresponding antigen in the tissue section and then a secondary antibody conjugated with peroxidase enzyme is introduced that binds to the primary antibody. Treatment with a horseradish peroxidase (HRP) substrate solution (e.g., 3,3′-diaminobenzidine (DAB) tetrahydrochloride solution) results in brown stain which is visualized under a light microscope. Immunolocalization of active caspases is a commonly used technique to detect active caspases in a variety of mouse organs. For example, immunostaining with active caspase-3 in mouse tissue sections has been used to identify apoptotic cells [19–21]. The immunofluorescence technique has been used in brain tissue [22], liver injury [23], ischemic brain [24], a controlled cortical impact (CCI) model [25], and conjunctival epithelia in B6 mice [26]. The availability of specific antibodies to active caspase has helped in the detection of active caspase in paraffin-embedded, formalin-fixed mouse tissues.

3.3.1 Preparation of Tissue Sections

1. Cut mouse tissue appropriately from an organ of study and quickly immerse in a fixative buffer for 12 h to 2 days (*see* **Note 8**).
2. Paraffin-Embedding: Most histology labs use a tissue processor to initially process the fixed tissue for paraffin embedding. However, the procedure described in the following steps can be used.
3. Remove fixative buffer from the tissue and dehydrate the tissue specimen in a series of graded ethanol treatments starting with 70 %, proceeding through 80, 95–100 % ethanol.
4. Remove ethanol and add xylene for 15 min.
5. Following the removal of xylene, add paraffin that has been heated to about 57 °C, leave for 10–30 min at this temperature, and then remove paraffin as much as possible. Again treat the tissue specimen with new paraffin at about 57 °C for 10–30 min and remove the unbound paraffin. Repeat this step two more times.
6. Keep the paraffin block specimens for at least 1 h at room temperature and then store under refrigeration until used for making sections.
7. Cut sections at 5–10 μm thickness on a rotary microtome and then dry on a slide warmer prior to the staining process.

3.3.2 Deparaffinization and Antigen Retrieval

1. Immerse the sections in xylene for 5 min, repeat once.
2. Remove xylene and incubate in 100 % ethanol for 3 min, repeat one more time and then incubate in 95 % followed by 80 % ethanol for 1 min each.
3. After removing ethanol, rinse with double-distilled water.
4. Transfer slides into a glass jar filled with antigen retrieval buffer and microwave until solution starts boiling (e.g., heat three times for 20 s each) (*see* **Note 9**).
5. Let glass jar cool to room temperature (~20 min).
6. Wash the sections in PBS-T for 5 min, repeat two times.

3.3.3 Staining Procedure

3.3.3.1 Immunohistochemical Staining

1. Incubate the deparaffinized sections in 1 % H_2O_2 in PBS containing 0.1 % sodium azide for 10 min. The endogenous peroxidase activity is quenched by incubation with H_2O_2.
2. Incubate the sections in blocking buffer at room temperature for 30 min (*see* **Note 10**).

3. Incubate the sections with caspase antibody at appropriate dilution (at least 1:200) in antibody dilution buffer in a humidified chamber for 1 h at room temperature or overnight at 4 °C. To assess nonspecific staining or verify the binding specificities of primary antibodies, incubate the sections in antibody dilution buffer or equal amounts of nonspecific mouse, rabbit, or goat immunoglobulins without the primary antibody for negative controls.
4. Wash the sections with PBS-T for 5 min, repeat two times.
5. Incubate the sections with HRP-conjugated secondary antibody at appropriate dilution in antibody dilution buffer at room temperature in the dark for 1 h.
6. Wash the sections with PBS-T for 5 min, repeat two times.
7. Incubate the sections in dark with HRP substrate solution.
8. Wash slides with PBS-T for 5 min, repeat two times to remove excess chromogen.
9. Dehydrate the slides with ascending graded alcohols (for 20 s each in 35, 70, and 95 % ethanol, and 2 min in 100 % ethanol) and clear in xylene.
10. Mount the slides with mounting medium.

3.3.3.2 Immunofluorescence Staining

1. Incubate the sections in blocking buffer at room temperature for 30 min (*see* **Note 10**).
2. Incubate the sections with primary antibody at appropriate dilution (at least 1:200) in antibody dilution buffer in a humidified chamber for 1 h at room temperature or overnight at 4 °C.
3. Wash the sections with PBS-T for 5 min, repeat two times.
4. Incubate the sections with fluorescent secondary antibody at appropriate dilution in antibody dilution buffer at room temperature in the dark for 1 h (*see* **Note 11**).
5. Wash the sections with PBS-T for 5 min, repeat two times.
6. Counterstain nuclei with DAPI in mounting medium (e.g., VECTASHIELD).
7. Coverslip and cure slides in the dark at room temperature for 4 h or at 4 °C overnight.

*3.3.4 Imaging of Immunofluorescence (See **Note 12**)*

1. Record immunofluorescence using appropriate filter for secondary antibody and for DAPI and record bright-field or DIC images.
2. Compare with sections developed without primary antibody to assess background fluorescence of tissue.

3.4 Detection of Caspase-Specific Cleaved Products of PARP, Cytokeratin-18, and Lamin A in Tissue Homogenates and Tissue Sections as a Measure of Caspase Activation

At present many target substrates of caspases have been identified in vitro and in vivo [27]. Proteins cleaved by caspases can be detected by western blot analysis using specific antibodies to the cleaved products produced by caspases. For example, cleavage of PARP, cytokeratin-18, and lamin A is commonly used as an indicator of apoptosis and activation of caspases. PARP is a nuclear DNA-binding protein and is involved in DNA base excision repair. PARP is a well-known substrate for caspase-3 in vitro and in vivo. Caspase-7 also cleaves PARP in vivo. Caspase-mediated cleavage of PARP produces two fragments of 85 and 24 kDa [28, 29], which can be detected by western blot using specific antibodies. Cytokeratin-18 (CK18), an important component of intermediate filaments in epithelial cells, is a target substrate of all of the executioner caspases, caspase-3, -6, and -7 [30, 31]. In response to an apoptotic stimulus, active executioner caspases cleave CK18 and a caspase-generated neoepitope of CK18 at position Asp396 is selectively recognized by M30 monoclonal antibody (Roche Applied Sciences) by western blot or ELISA assay. Similarly, caspase activation results in a specific cleaved fragment of lamin A [32] that can be detected by western blot analysis. In response to an apoptotic stimulus, lamin A/C is specifically cleaved into a large (41–50 kDa) and a small (28 kDa) fragment particularly by active caspase-6 [33–35]. The caspase-6 recognition sequence VEID is present at amino acid residues 227–230 in lamin A/C protein.

1. Run western blots using specific antibodies to cleaved PARP, cytokeratin-18, or lamin A as outlined in Subheading 3.2.
2. Use the antibodies for immunohistochemistry as described in Subheading 3.3.

4 Notes

1. Caspase specificity for the synthetic short peptide substrates is highly promiscuous such that there is overlapping cleavage specificity for these short peptide synthetic substrates. It was reported that compared to other caspases, caspase-3 was more efficient to cleave most of the peptide substrates [36]. Also, DEVD, a commonly used caspase-3 substrate, is also cleaved by other caspases including caspase-2, -6, -7, -8, and -10 [27, 36].
2. In addition to the caspases, animal tissue homogenates also contain endogenous X chromosome-linked inhibitors of apoptosis (xIAPs) that can inhibit caspase-3, -7, and -9 activity [37].
3. Usually, 10–20 mg mouse tissue in 0.5 ml of lysis buffer will yield 8–10 mg/ml protein.
4. To ensure appropriate estimation of caspase activity, time-dependence and dose–response of the substrate concentration

should be examined. The time of incubation and the amount of substrate concentration should be used for the assay within the linear range.

5. Cleavage of initiator caspases is not necessary for their activation. The detection of proteolytic fragments due to internal proteolysis of these caspases may therefore, not represent true activation of the initiator caspases. Caspase-8 and -9 may be active without being cleaved or without altered mobility on western blots. Thus, additional methods including enzymatic activity assay should be performed.
6. Caspase-cleaved products in a mouse tissue may not remain stable and may become targets of other proteases for further proteolysis. Therefore, the level of cleaved product sometimes may not reflect true caspase activity.
7. The method relies on the source and availability of high quality antibodies for the pro-form and cleaved form of caspases.
8. Leaving samples longer than 3 days in fixative (over-fixation) should be avoided as it can lead to uneven and weak staining.
9. In case of incomplete antigen retrieval the duration of microwaving can be extended (timing needs to be adjusted by user as it depends on specimen preparation, thickness and fixation, and microwave power) or alternative methods like treatment of deparaffinized slides in a water bath or a steamer and enzymatic digestion can be applied. For a steamer or water bath, preheat glass jar filled with antigen retrieval buffer until boiling (temperature >95 °C), insert slides and continue heating for 20–40 min. Let glass jar cool to room temperature (~20 min). Wash sections twice in PBS-T for 5 min and continue with the staining procedure as described above. For enzymatic digestion of deparaffinized sections, transfer slides into a buffer containing 50 mM Tris–HCl, pH 8.0, 5 mM EDTA, and 10 μg/ml proteinase K and incubate at 37 °C for 5 min. Wash sections twice in PBS-T for 5 min and continue with the staining procedure as described above.
10. To reduce nonspecific staining, BSA in both the "Blocking buffer" and the "Antibody dilution buffer" can be substituted with serum from the species in which the secondary antibody was raised (e.g., goat serum instead of BSA in the specified buffers when using AlexaFluor-594 goat anti-rabbit antibody). The use of detergents in the blocking buffer can be varied to improve antigen detection (e.g., up to 0.1 % Triton X-100).
11. Primary antibodies directly conjugated to a fluorescent reporter, such as FITC, can be more efficient in detection of the antigen. FITC-conjugated rabbit polyclonal anti-active caspase-3 antibody is available with Cell Signaling Technology Inc. The secondary antibody is not required after incubation of tissue section with fluorochrome-conjugated primary antibodies.

12. Photobleaching is a common problem during immunofluorescence studies. This can be controlled by reducing the time-span or intensity of the light exposure. Photobleaching is generally less when efficient fluorophores such as Alexa Fluor are used.

Acknowledgement

This work was supported by NIH grant R01 DK081690 and VA Merit Award to G.P.K. and VA Merit Award to R.S.H.

References

1. Hyman BT, Yuan J (2012) Apoptotic and non-apoptotic roles of caspases in neuronal physiology and pathophysiology. Nat Rev Neurosci 13:395–406
2. Kuranaga E (2012) Beyond apoptosis: caspase regulatory mechanisms and functions in vivo. Genes Cells 17:83–97
3. Yi CH, Yuan J (2009) The Jekyll and Hyde functions of caspases. Dev Cell 16:21–34
4. Kerr JF, Wyllie AH, Currie AR (1972) Apoptosis: a basic biological phenomenon with wide-ranging implications in tissue kinetics. Br J Cancer 26:239–257
5. Coleman ML, Sahai EA, Yeo M et al (2001) Membrane blebbing during apoptosis results from caspase-mediated activation of ROCK I. Nat Cell Biol 3:339–345
6. Faleiro L, Lazebnik Y (2000) Caspases disrupt the nuclear-cytoplasmic barrier. J Cell Biol 151:951–959
7. Luthi AU, Martin SJ (2007) The CASBAH: a searchable database of caspase substrates. Cell Death Differ 14:641–650
8. Taylor RC, Cullen SP, Martin SJ (2008) Apoptosis: controlled demolition at the cellular level. Nat Rev Mol Cell Biol 9:231–241
9. Darzynkiewicz Z, Pozarowski P, Lee BW et al (2011) Fluorochrome-labeled inhibitors of caspases: convenient in vitro and in vivo markers of apoptotic cells for cytometric analysis. Methods Mol Biol 682:103–114
10. Tawa P, Tam J, Cassady R et al (2001) Quantitative analysis of fluorescent caspase substrate cleavage in intact cells and identification of novel inhibitors of apoptosis. Cell Death Differ 8:30–37
11. Thornberry NA, Rano TA, Peterson EP et al (1997) A combinatorial approach defines specificities of members of the caspase family and granzyme B. Functional relationships established for key mediators of apoptosis. J Biol Chem 272:17907–17911
12. Frances DE, Ingaramo PI, Mayoral R et al (2013) Cyclooxygenase-2 over-expression inhibits liver apoptosis induced by hyperglycemia. J Cell Biochem 114:669–680
13. Kumar S, Allen DA, Kieswich JE et al (2009) Dexamethasone ameliorates renal ischemia-reperfusion injury. J Am Soc Nephrol 20: 2412–2425
14. Namura S, Zhu J, Fink K et al (1998) Activation and cleavage of caspase-3 in apoptosis induced by experimental cerebral ischemia. J Neurosci 18:3659–3668
15. Lie ML, White LE, Santora RJ et al (2012) Lung T lymphocyte trafficking and activation during ischemic acute kidney injury. J Immunol 189:2843–2851
16. Yin XM, Luo Y, Cao G et al (2002) Bid-mediated mitochondrial pathway is critical to ischemic neuronal apoptosis and focal cerebral ischemia. J Biol Chem 277:42074–42081
17. Condorelli G, Roncarati R, Ross J Jr et al (2001) Heart-targeted overexpression of caspase3 in mice increases infarct size and depresses cardiac function. Proc Natl Acad Sci U S A 98: 9977–9982
18. Chen J, Wang W, Zhang Q et al (2013) Low molecular weight fucoidan against renal ischemia-reperfusion injury via inhibition of the MAPK signaling pathway. PLoS One 8: e56224
19. Duan WR, Garner DS, Williams SD et al (2003) Comparison of immunohistochemistry for activated caspase-3 and cleaved cytokeratin 18 with the TUNEL method for quantification of apoptosis in histological sections of PC-3 subcutaneous xenografts. J Pathol 199:221–228
20. Hughes J, Gobe G (2007) Identification and quantification of apoptosis in the kidney using morphology, biochemical and molecular markers. Nephrology (Carlton) 12:452–458
21. Jakob S, Corazza N, Diamantis E et al (2008) Detection of apoptosis in vivo using antibodies

against caspase-induced neo-epitopes. Methods 44:255–261
22. Ohsawa S, Hamada S, Yoshida H et al (2008) Caspase-mediated changes in histone H1 in early apoptosis: prolonged caspase activation in developing olfactory sensory neurons. Cell Death Differ 15:1429–1439
23. Canbay A, Feldstein A, Baskin-Bey E et al (2004) The caspase inhibitor IDN-6556 attenuates hepatic injury and fibrosis in the bile duct ligated mouse. J Pharmacol Exp Ther 308: 1191–1196
24. Li J, Li Y, Ogle M et al (2010) DL-3-n-butylphthalide prevents neuronal cell death after focal cerebral ischemia in mice via the JNK pathway. Brain Res 1359:216–226
25. You Z, Savitz SI, Yang J et al (2008) Necrostatin-1 reduces histopathology and improves functional outcome after controlled cortical impact in mice. J Cereb Blood Flow Metab 28:1564–1573
26. Zhang X, Chen W, De Paiva CS et al (2011) Interferon-gamma exacerbates dry eye-induced apoptosis in conjunctiva through dual apoptotic pathways. Invest Ophthalmol Vis Sci 52: 6279–6285
27. Timmer JC, Salvesen GS (2007) Caspase substrates. Cell Death Differ 14:66–72
28. Lazebnik YA, Kaufmann SH, Desnoyers S et al (1994) Cleavage of poly(ADP-ribose) polymerase by a proteinase with properties like ICE. Nature 371:346–347
29. Tewari M, Quan LT, O'Rourke K et al (1995) Yama/CPP32 beta, a mammalian homolog of CED-3, is a CrmA-inhibitable protease that cleaves the death substrate poly(ADP-ribose) polymerase. Cell 81:801–809
30. Caulin C, Salvesen GS, Oshima RG (1997) Caspase cleavage of keratin 18 and reorganization of intermediate filaments during epithelial cell apoptosis. J Cell Biol 138:1379–1394
31. Leers MP, Kolgen W, Bjorklund V et al (1999) Immunocytochemical detection and mapping of a cytokeratin 18 neo-epitope exposed during early apoptosis. J Pathol 187:567–572
32. Mintzer R, Ramaswamy S, Shah K et al (2012) A whole cell assay to measure caspase-6 activity by detecting cleavage of lamin A/C. PLoS One 7:e30376
33. Okinaga T, Kasai H, Tsujisawa T et al (2007) Role of caspases in cleavage of lamin A/C and PARP during apoptosis in macrophages infected with a periodontopathic bacterium. J Med Microbiol 56:1399–1404
34. Orth K, Chinnaiyan AM, Garg M et al (1996) The CED-3/ICE-like protease Mch2 is activated during apoptosis and cleaves the death substrate lamin A. J Biol Chem 271:16443–16446
35. Ruchaud S, Korfali N, Villa P et al (2002) Caspase-6 gene disruption reveals a requirement for lamin A cleavage in apoptotic chromatin condensation. EMBO J 21:1967–1977
36. McStay GP, Salvesen GS, Green DR (2008) Overlapping cleavage motif selectivity of caspases: implications for analysis of apoptotic pathways. Cell Death Differ 15:322–331
37. Altieri DC (2010) Survivin and IAP proteins in cell-death mechanisms. Biochem J 430:199–205

Chapter 10

Measurement of Caspase Activation in Mammalian Cell Cultures

Magnus Olsson and Boris Zhivotovsky

Abstract

The majority of caspases, cysteine-dependent aspartate-directed proteases, being in their activated state are involved in regulation of apoptosis by cleaving protein substrates harboring specific target motifs. Basically all biochemical and morphological changes in an apoptotic cell, including cell shrinkage, chromatin condensation, DNA fragmentation, and plasma membrane blebbing, are consequence of caspase-mediated proteolysis. Thus, uncovering activities of unique caspases are key determinants of the apoptotic process. This chapter describes a set of experimental protocols available for characterization, quantification and inhibition of caspase activities in mammalian cell cultures, including immunoblotting, usage of synthetic substrates, flow cytometry, and microscopic techniques.

Key words Apoptosis, Caspase, Caspase substrate, SDS-PAGE, Immunocytochemistry, FACS

Abbreviations

BSA	Bovine serum albumin
FACS	Fluorescence activated cell sorting
pAb	Polyclonal antibody
SDS	Sodium dodecyl sulfate
Tris	Tris(hydroxymethyl)aminomethane, mAb, monoclonal antibody

1 Introduction

1.1 Immunoblot Analysis of Procaspase Processing

In a simplified scheme, an individual caspase is synthesized as a zymogen, and proteolytical processing of the precursor is then required for the formation of a tetrameric active enzyme, composed of two repeating heterotypic subunits. The first processing event in a procaspase is separation of the large (p17) and small (p12) subunits, which is followed by deletion of the N-terminal pro-domain to form the catalytically mature protease [1]. Caspase zymogens are themselves substrates for caspases and

Peter V. Bozhkov and Guy Salvesen (eds.), *Caspases, Paracaspases, and Metacaspases: Methods and Protocols*, Methods in Molecular Biology, vol. 1133, DOI 10.1007/978-1-4939-0357-3_10,

sequential activations form the foundation for apoptotic cascades. The catalytic activity of an effector caspase is enhanced by several orders of magnitude by intrachain cleavage mediated by a specific initiator caspase. Hence, separation of denatured protein lysates isolated from apoptotic cells by sodium dodecyl sulfate polyacrylamide gel electrophoresis (SDS-PAGE) and taking advantage of specific antibodies in western blotting is one broadly used approach to discriminate processed caspases from their respective zymogens (Fig. 1). However, the activation process of initiator caspases has additional layers of complexity. A characteristic feature within this group of proteases is the presence of homotypic CARD (caspase-recruitment domain) or DED (death effector domain) sequence motifs in their N-terminal pro-domains, modules that are important for formation of relevant activity platform complexes in the cell [2].

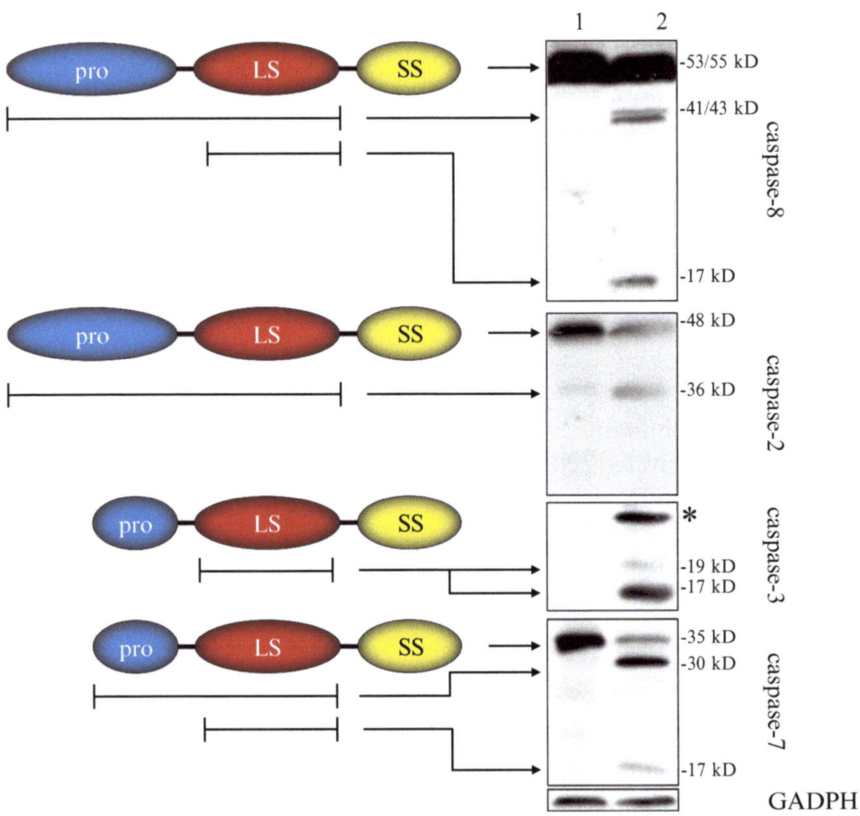

Fig. 1 Typical appearances of pro- and processed caspases in western blotting and a schematic illustration of their amino acid tertiary structure, connecting immunoblot signals from cleaved fragments to their corresponding subdomains. Western blotting was performed as described in the main text using protein lysates from control HCT116 cells (*lane 1*) and HCT116 cells incubated in the presence of 5-FU (5-fluorouracil, 768 μM) for 24 h (*lane 2*). Antibodies for immunodetection of caspase-8, caspase-2, cleaved caspase-3 and caspase-7 were used. Glyceraldehyde 3-phosphate dehydrogenase (G3PDH) was used as a control for equal loading of samples. The asterisk indicates a nonspecific band detected with the anti-cleaved caspase-3 pAb. Abbreviations: *pro*, N-terminal pro-domain, *LS* large subunit (p17), *SS* small subunit (p12)

Several theoretical models have been postulated aiming to clarify the molecular mechanism by which initiator caspases become activated [3]. Although these models still suffer from lack of supportive experimental evidences it is clear that initiator zymogens possess intrinsic enzymatic activity generated from clustering or dimerization per se, which in turn allows for autoprocessing of the first caspase in the apoptotic cascade. In summary, irrespectively of caspase type analyzed, detection of processed fragments by western blotting is undoubtedly a strong indication for enzymatic activity. On the other hand, the absence of these fragments may not necessarily mean that an initiator caspase is catalytically inactive.

1.2 Immunocytochemical Detection of Active Caspases

As an alternative to SDS-PAGE, active caspases can also be visualized by immunocytochemistry (Fig. 2). However, to date, only a limited number of antibodies suitable for immunostaining are available for specific detection of active caspases.

1.3 Immunoblot Analysis of Caspase Substrate Cleavage

Execution of apoptotic cell death is mediated by caspase cleavage of various vital proteins. To date, there are more than 500 caspase substrates presented in a continuously growing list [4, 5]. Several of these substrates have a characterized function in propagation of the

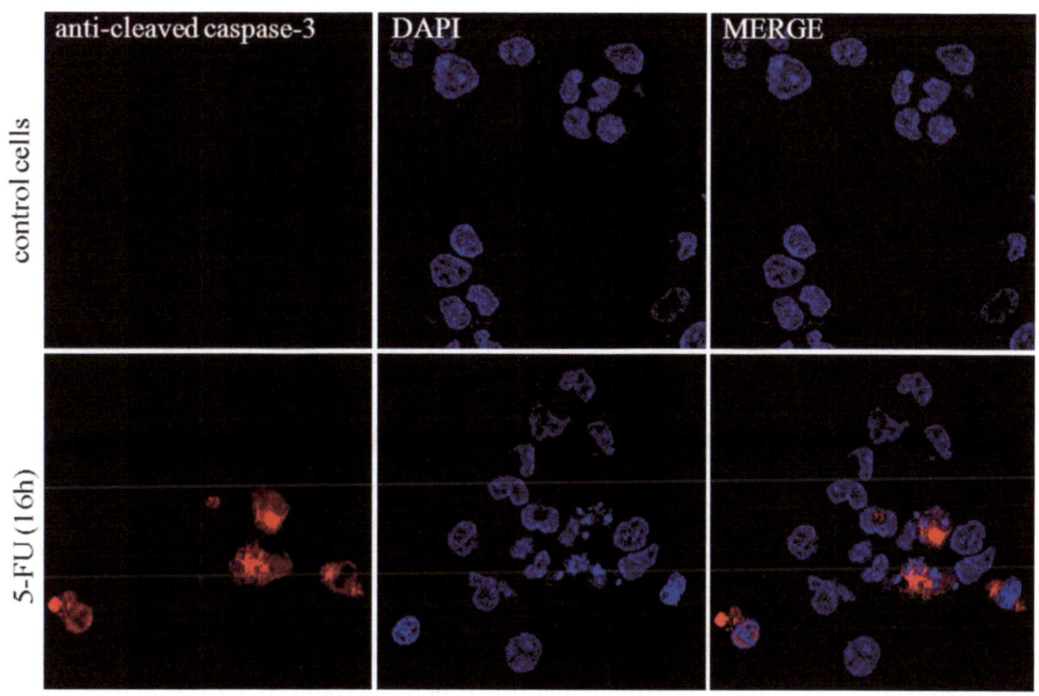

Fig. 2 Detection of processed caspase-3 in formaldehyde-fixed cells by immunocytochemistry. HCT116 cells were cultured on coverslips in the absence and in the presence of 5-FU (768 μM) for 24 h. Immunodetection of active caspase-3 was then performed as outlined in the main text (60 × magnification)

cell death process while others, since they happen to contain a caspase target motif in their sequence, may be cleaved as bystanders. The caspase substrates playing an active role in the progression of apoptosis can be divided in two groups. Firstly, there are apoptosis regulatory proteins which become activated by cleavage and thereby act as transducers or amplifiers that determine the apoptotic threshold and cell fate. Apart from caspases themselves, this can be exemplified by the DNase inhibitor ICAD/DFF45 (Inhibitor of Caspase Activated DNase/DNA fragmentation factor 45) which, upon cleavage by caspase-3, liberates the active CAD/DFF40 (Caspase Activated DNase/DNA fragmentation factor 40) nuclease that mediates apoptotic internucleosomal DNA fragmentation [6]. Correspondingly, a truncated version of the Bcl-2 protein family member Bid (BH3 interacting domain death agonist) results from death receptor signaling and caspase-8 proteolytic activity. In turn, tBid (truncated Bid) promotes release of mitochondrial pro-apoptotic factors, including cytochrome *c*, as a consequence of destabilized membrane potential [7]. The majority of substrates are, however, inactivated by caspases. This group consists of proteins implicated in a wide variety of cellular functions, including structural components, cell adhesion molecules, transcription factors, as well as factors involved in DNA repair and processing of RNA [8]. Detection of caspase substrate cleavage products by means of SDS-PAGE is often used to confirm apoptotic conditions in cultured cells. Depending on the experimental context, it may also be possible to verify a discrete apoptotic signaling route by this method. It must be noted, however, that substrates generally are cleaved by effector caspases and that overlapping cleaving activities between different caspases is common. Thus, conclusions regarding activity of a specific caspase deduced from substrate proteolysis should be supported by other methods, such as suppression of caspase expression by means of RNAi technology. One of the most widely used apoptotic markers in SDS-PAGE methodology is PARP-1 (Poly [ADP-ribose] polymerase 1) which is a target of caspase-3 in vivo but can be cleaved by many caspases in vitro. Disruption of PARP-1 activity by proteolytic cleavage results from a separation of the amino-terminal DNA binding domain (24 kDa) and the carboxy-terminal catalytic domain (89 kDa). Since PARP-1 is a nuclear poly (ADP-ribose) polymerase family member implicated in DNA repair and especially base excision repair a ceased activity of the protein will promote apoptosis by disruption of cellular homeostasis in general and genomic integrity in particular [9, 10].

1.4 Detection of Cytokeratin 18 Cleavage by Flow Cytometry

Cytokeratin 18 (CK18) is a member of the Type I intermediate filament family, that forms heterotetramers. This protein is widely expressed in normal single layer epithelial tissues and has also been found at high levels in some tumor tissues. During apoptosis, cleavage by caspase-3, -6, -7, and -9, exposes a neoepitope in CK18 molecule (M30) to which monoclonal antibodies have been raised [11]. Unconjugated primary antibody targeting the neoepitope can be used in western blotting but requires an additional secondary antibody conjugated with HRP (horseradish peroxidase). In flow cytometry applications, the use of a fluorochrome-conjugated anti-cleaved CK18 antibody creates a powerful tool facilitating quantification of apoptotic cells in a culture. However, the use of CK18 neoepitope as an marker for apoptosis is restricted to epithelium-derived cell lines. In non-epithelial cells, a similar protocol can be used to target the cleaved fragment of PARP-1 [12].

1.5 Caspase Activity Measurement Using the Peptide Cleavage Assay

The function of a particular caspase in apoptotic cascades can be determined by a strategy based on a series of peptide-based substrates and inhibitors developed from the favored substrates of each caspase. The caspase activity assay is a quantitative and sensitive method which utilizes synthetic tetrapeptide or pentapeptide sequences, conjugated with 7-amino-4-methylcoumarin (AMC), 7-amino-4-trifluoromethylcoumarin (AFC), or rhodamine-110 (Rh110). When the low molecular weight substrate is cleaved by a particular caspase, release of the fluorescent molecule is detected and the signal is proportional to the magnitude of the caspase proteolytic activity. Alternatively, spectrophotometric detection of the chromophore p-nitroanilide (p-NA) after cleavage from the labeled substrate can be used. The p-NA light emission can be quantified using microtiter plate reader at 405 nm. However, it is important to keep in mind that promiscuity for different cleavage motifs has been reported. For instance, by testing commercially available short peptide-based substrates in vitro and in cellulo, it was shown that caspase-3 was able to cleave several substrates more efficiently than those caspases to which the substrates were reported specific [13]. Accordingly, the use of synthetic substrates to define individual caspases and discrete apoptotic signaling routes is controversial.

1.6 Inhibition of Caspase Activity in Cell Cultures

In cells, the processing and activation of caspases is controlled by apoptosis regulatory factors such as FADD (FAS-associated death domain protein), APAF-1 (Apoptotic protease activating factor 1), Bcl-2 family members, FLIP (FLICE-like inhibitory protein), and IAPs (inhibitors of apoptosis proteins). Deregulation of these proteins by means of overexpression using cDNAs or suppression by RNAi methodology is in many experimental settings an efficient approach to manipulate apoptosis progression. Distinct from caspase regulators produced by the cell itself, active caspases can also be suppressed by artificial caspase inhibitors that have been developed both as research tools and as potential therapeutics to inhibit cell death in vivo [14]. The simplest and therefore also the most common method to accomplish inhibition of caspase activity in cells is to culture them in the presence of micromolar concentrations of cell-permeant and uncleavable substrate peptide motifs.

Z-VAD-FMK (carbobenzoxy-valyl-alanyl-aspartyl-fluoromethylketone) is a pan-caspase inhibitor that irreversibly binds to the catalytic site of caspases thereby preventing their activity, further processing and apoptosis. Similar to synthetic substrates used for the detection of caspase activities, tetrapeptides and pentapeptides designed to block the catalytic site of individual caspases are also commercially available. However, as previously discussed in the section describing the caspase activity assay, the use of synthetic amino acid target motifs to define individual caspases and isolated apoptotic signaling pathways is debated [13]. Since the peptides described bind to the catalytic site of active caspases they may also be used as markers. Exposure of live cells to fluorochrome-labeled inhibitors of caspases (FLICAs) results in uptake of these reagents by apoptotic cells. After fixation with formaldehyde, cells labeled with FLICAs can be examined by fluorescence microscopy, or quantified by flow cytometry [15].

2 Materials

2.1 Immunoblot Analysis of Procaspase Processing

All solutions should be prepared in ultrapure water (18.3 MΩ at 25 °C) and by using analytical grade chemicals. Unless indicated otherwise, all solutions can be prepared and stored at room temperature.

2.1.1 Sample Preparation

1. Phosphate-buffered saline (PBS), pH 7.4.
2. Pierce™ BCA Protein Assay Kit (Thermo Scientific) or similar.
3. 5× Sample (Laemmli) buffer: 625 μL 1 M Tris–HCl, pH 6.8, 1 mL glycerol, 2 mL 10 % SDS, 0.5 mL 0.5 % (w/v) bromophenol blue in water, 0.5 mL 2-mercaptoethanol. Add water to 10 mL, aliquot and store at −20 °C (*see* **Note 1**).

2.1.2 SDS-PAGE

1. Mini PROTEAN® 3 electrophoresis system or equivalent equipment.
2. Prot/Elec Tips (Bio-Rad) or similar tips.
3. Resolving gel buffer: 1.5 M Tris–HCl, pH 8.8. Weigh 181.7 g Tris and dissolve in 800 mL water. Titrate the solution of Tris with hydrochloric acid (HCl) until the correct pH is reached (*see* **Notes 2** and **3**). Make up to 1 L with water. Store at 4 °C.
4. Stacking gel buffer: 0.5 M Tris–HCl, pH 6.8. Weigh 60.6 g Tris and prepare a 1 L solution as in previous step. Store at 4 °C.
5. Protogel, 30 % (w/v) Acrylamide: 0.8 % (w/v) Bis-Acrylamide Stock solution (37.5:1).
6. *N*, *N*, *N'*,*N'*-tetramethylethylenediamine 99 % (TEMED).
7. Ammonium persulfate (APS): 10 % solution in water (*see* **Note 4**).

8. Electrode (running) buffer (10×): Dissolve 30 g of Tris base (247 mM), 144 g glycine (1.92 M), and 10 g sodium dodecyl sulfate (SDS) (0.1 % w/v) in 1 L of water, pH 8.3. Dilute to 1× running buffer in water prior to use. 1 L is sufficient for one electrophoretic run.
9. Prestained protein ladder ranging from 15 to 70 kDa (e.g., Fermentas PageRuler Prestained Protein Ladder, SM0671, Thermo Scientific, or equivalent).

2.1.3 Western Blot

1. Nitrocellulose membrane, 0.45 μm.
2. Western blot transfer buffer: 25 mM Tris base 192 mM glycine, and 20 % methanol (*see* **Notes 5** and **6**).
3. Mini-PROTEAN Tetra Cell and Mini Trans-Blot Module (Bio-Rad) or equivalent equipment.
4. Power supply (10–300 V, 4–400 mA).
5. 3MM Chr Blotting Paper (Whatman) or equivalent.

2.1.4 Immunodetection

1. Phosphate buffered saline (PBS; 10×): Dissolve 80 g of NaCl, 2 g of KCl, 14.4 g of Na_2HPO_4 a, and 2.4 g of KH_2PO_4 in 800 mL water. Make to 1 L with additional water. The pH in the final 1× PBS will become close to 7.4.
2. PBS containing 0.1–0.2 % Tween-20 (PBS-T).
3. Blocking solution: 5 % dry nonfat milk in PBS (*see* **Note 7**).
4. Primary antibody dilution buffer: 1 % bovine serum albumin (BSA), 0.1–0.2 % Tween-20 in PBS (*see* **Note 8**). Store at 4 °C.
5. Enhanced Chemiluminescence (ECL) solution (GE Healtcare) or equivalent.
6. Fuji medical X-ray film ^{super}RX (Fujifilm) or equivalent.
7. Primary antibodies (*see* **Note 9**):

   Anti-caspase-2 mAb, clone 35 (BD Biosciences).

   Anti-cleaved caspase-3 (Asp 175) pAb (Cell Signaling, Danvers).

   Anti-cleaved caspase-6 pAb (Millipore, Billerica).

   Anti-caspase-7 mAb, clone B94-1 (BD Biosciences).

   Anti-caspase-8 mAb, clone 1C12 (Cell Signaling).

   Anti-cleaved caspase-9 pAb (Cell Signaling).

   Anti-glyceraldehyde-3-phosphate dehydrogenase (G3PDH) pAb (e.g., from Trevigen).
8. Secondary antibodies:

   HRP-conjugated anti-mouse IgG and anti-rabbit IgG (e.g., from Thermo Scientific).
9. Stripping buffer: 12.5 mL of 1 M Tris–HCl, pH 6.8, 770 μL 2-mercaptoethanol, 20 mL 10 % SDS. Make up to 100 mL with water. Prepare immediately before use.

2.2 Immunocytochemical Detection of Active Caspases

2.2.1 Sample Preparation

1. Glass Coverslips: 12 mm circles (fit in 12 or 6-well culture plates), Thickness No. 1. Coverslips should be cleaned with ethanol and sterilized prior to use.
2. Phosphate-buffered saline (PBS), pH 7.4.
3. Fixative, 4 % Formaldehyde: Prepare formaldehyde using either paraformaldehyde powder solubilized in PBS or dilute 36.7 % formaldehyde in PBS. Methanol-free formaldehyde is recommended (*see* **Note 10**). The solution should be used fresh.

2.2.2 Immunostaining

1. Blocking/permeabilizing Buffer: 1–5 % bovine serum albumin (BSA) and 0.3 % Triton™ X-100 in PBS (*see* **Note 11**).
2. Antibody Dilution Buffer: 1 % BSA and 0.3 % Triton™ X-100 in PBS.
3. Primary antibodies:

   Anti-cleaved caspase-3 (Asp175) pAb (Cell Signaling).

   Anti-cleaved caspase-9 (Asp315) pAb (Thermo Scientific).

   Anti-cleaved caspase-6 pAb (Millipore).
4. Fluorochrome-conjugated secondary antibodies: Alexa Fluor® 488 (emission color-green) or Alexa Fluor® 594 (emission color-red) conjugated goat anti-rabbit IgG (H + L) (e.g., from Life Technologies).
5. Nuclear counterstaining: Hoechst 33342 or DAPI. Prepare a dilution of Hoechst 33342 or DAPI in PBS in a final concentration of 1 mg/mL.
6. VECTASHIELD® Mounting Media (Vector Laboratories) or equivalent.
7. Specimen glass slides.
8. Nail polish.

2.3 Immunoblot Analysis of Caspase Substrate Cleavage (See Note 12)

Apart from primary Abs listed below, the materials are identical to what is described in 2.1.1–2.1.4.

1. Anti-cleaved PARP (Asp214) Antibody (Cell Signaling).
2. Anti-Bid antibody (Cell Signaling) (*see* **Note 13**).
3. Anti-cleaved Lamin A (Asp 230) antibody (Abcam).

2.4 Detection of Cytokeratin 18 Cleavage by Flow Cytometry

2.4.1 Sample Preparation

1. Phosphate-buffered saline (PBS), pH 7.4.
2. Pure methanol.

2.4.2 Detection of the CK 18 Neoepitope

1. Anti-Cytokeratin 18 M30 CytoDEATH Monoclonal Antibody, FITC Conjugated (Roche Applied Science).
2. Antibody dilution buffer: PBS containing 1 % BSA.
3. FACS instrument.

2.5 Caspase Activity Measurement in Peptide Cleavage Assay

2.5.1 Sample Preparation

1. PBS, pH 7.4.
2. Pierce™ BCA Protein Assay Kit (Thermo Scientific) or similar.

2.5.2 Caspase Activity Measurement

1. Caspase activity buffer I, stock (analysis of caspase-1, caspase-3/-7, caspase-6, and caspase-8 activities): 100 mM HEPES (4-(2-hydroxyethyl)-1-piperazineethanesulfonic acid), pH 7.5, 10%(w/v)sucrose,0.1%(w/v)CHAPS(3-[(3-Cholamidopropyl) dimethylammonio]-1-propanesulfonate).
2. Caspase activity buffer II (analysis of caspase-2 and caspase-9 activities): 0.1 M MES (2-(N-Morpholino) ethanesulfonic acid), pH 6.5, 10 % (w/v) PEG (Polyethylene glycol), 0.1 % (w/v) CHAPS (*see* **Note 14**).
3. 1 M DTT (dithiothreitol): Dissolve 309 mg DTT in a final volume of 2 mL 0.01 M sodium acetate, pH 5.2. Sterilize by filtration, make aliquots, and store at −20 °C.
4. 0.1 % (v/v) NP-40 (Nonidet P-40): Add 1 μL of NP-40 to 1 mL water, vortex, and keep at 4 °C.
5. Caspase substrate stock solutions: Prepare stock solutions of each specific caspase substrate to be tested in DMSO as follows:

   Caspase-1 (Ac-YVAD-AMC) 100 μM.

   Caspase-2 (Ac-VDVAD-AMC) 100 μM.

   Caspase-3/-7 (Ac-DEVD-AMC) 40 μM.

   Caspase-6 (Ac-VEID-AMC) 100 μM.

   Caspase-8 (Ac-IETD-AMC) 100 μM.

   Caspase-9 (Ac-LEHD-AMC) 100 μM.

   Make aliquots and store at −20 °C.
6. Fluoroscan Ascent Microplate Fluorometer (Thermo Scientific) or similar.
7. Opaque 96-well assay plate (e.g., Corning Inc.)
8. Gen5 (Bio Tec) or similar software for acquisition and data processing.

2.6 Inhibition of Caspase Activity in Cell Cultures

Caspase inhibitor stock solutions: Prepare stock solutions of each specific caspase inhibitor (Z-VAD-FMK, Z-XXXD-FMK, or Z-VDVAD-FMK) to be tested by dissolving in DMSO at a concentration of 10–20 mM. Make aliquots and store at −20 °C (*see* **Note 15**).

3 Methods

3.1 Immunoblot Analysis of Procaspase Processing

All procedures are performed at room temperature unless otherwise stated.

3.1.1 Sample Preparation

1. Induce apoptosis in cells by desired way. Concurrently incubate a control culture without induction.
2. Collect 0.5–1 × 10^6 cells in a tube and centrifuge for 5 min at 150 × *g*, 4 °C.
3. Wash cells in PBS, centrifuge again, and discard supernatant.
4. Resuspend cells in cold 50–100 μL of cold PBS. Put the samples on ice (*see* **Note 16**).
5. Determine the total protein concentration in the samples by removal of 1–2 μL of homogenous cell suspension for further analysis using the BCA protein assay (*see* **Note 17**).
6. Add the appropriate volume of 5× sample buffer to a final dilution of 1×.
7. Boil samples for 5–10 min, preferentially in a high speed orbital shaker to eliminate the problem with high viscosity due to the release of chromosomal DNA (*see* **Note 18**). Cool down samples to room temperature before loading to the gel. Samples can be stored in −20 °C for several months (*see* **Note 19**).

3.1.2 SDS-PAGE

1. Assemble the glass plates for the Mini PROTEAN® 3 or equivalent electrophoresis system according to the manufacturer's instructions.
2. Prepare separating gel. Since the molecular masses of procaspases vary between 30 and 55 kDa, and their processed (active) subunits detected by most antibodies are around 18 kDa, a 15 % gel is recommended. To make one set (two gels) with 1.0-mm-thick spacers, combine the following components:

   4.8 mL H_2O.

   5 mL Resolving gel buffer, 1.5 M Tris–HCl, pH 8.8.

   200 μL 10 % SDS.

   10 mL Protogel.

   60 μL 10 % APS.

   33 μL TEMED.

3. Pour the separating gel in between the glass plates. Leave sufficient space for the stacking gel to be added later (the length of the comb teeth plus 0.5–1 cm). To provide a sharp interface, overlay the gel with water during polymerization.
4. Allow gel to polymerize for 30 min. Remove overlay water completely.
5. Prepare the stacking gel. For two gels (one set), combine the following components:

   6.05 mL H_2O.

   2.5 mL Stacking gel buffer, 0.5 M Tris–HCl, pH 6.8.

   100 μL 10 % SDS.

   1.35 mL Protogel.

   45 μL 10 % APS.

   25 μL TEMED.
6. Pour stacking gel solution on top of the polymerized separating gel. Immediately insert a clean Teflon comb into the stacking gel solution. Allow gel to polymerize for 30 min (*see* **Note 20**).
7. Remove the gel comb and assemble electrophoresis system according to the manufacturer's instructions. Fill it with 1× electrode buffer.
8. Load equal protein amounts from each sample (20–80 μg) into wells under the electrode buffer with Prot/Elec pipette tips. Load one well with 3 μL of the prestained SDS-PAGE protein ladder. In the well next to the last sample 1× sample buffer may be added to prevent horizontal spreading of sample proteins (*see* **Note 21**).
9. Attach the electrophoresis apparatus to an electric power supply. Run the gel at constant 80 V until the bromophenol blue front is reaching the separating gel. The remaining electrophoresis can be performed at constant 130 V.
10. When the bromophenol blue reaches the bottom of the resolving gel (approximately 2 h), turn off the power supply and disconnect the electrophoresis system. Remove the stacking gel and unused lanes. Submerge the remaining gel in transfer buffer.

3.1.3 Western Blotting

1. Cut nitrocellulose membranes and 3MM papers according to gel sizes. All the components, including supporting pads, should be soaked in transfer buffer in advance of assembly of the electroblot sandwich. Build the sandwich in the gel holder cassette as follows: foam pad, 3 3MM papers, the gel, nitrocellulose membrane, 3 3MM papers, and foam pad. Precautions should be taken to avoid air bubbles that would interfere with an accurate transfer (*see* **Note 22**).

2. Place the cassette into electrode assembly and fill the tank with cold transfer buffer.
3. Connect the electrophoresis unit to the power supply and make sure that the direction of transfer is correct.
4. Perform the transfer at a constant 100 V for 2 h (alternatively, 120 V for 1.5 h) at 4 °C. Transfer efficiency is indicated by transfer of prestained SDS-PAGE markers (*see* **Note 23**).

3.1.4 Immunodetection

1. Transfer the membrane into blocking solution. Incubate at least 1 h on a rocker.
2. Rinse the membrane several times in PBS.
3. Put the membrane into a small plastic container (*see* **Note 24**). Add the diluted primary antibody of interest (usually from 1:500 to 1:5,000 in diluent solution). Incubate overnight on a rocker at 4 °C (*see* **Note 25**).
4. Wash three times (10 min in PBS, 15 min in PBS-T, 10 min in PBS) on a rocker.
5. Add peroxidase-conjugated secondary antibody diluted in blocking buffer (usually 1:5,000 to 1:10,000). Incubate for 1 h on a rocker.
6. Wash as in **step 4**.
7. Develop membrane with ECL according to the manufacturer's instructions.
8. Remove excess PBS and wrap the membrane in plastic foil. Expose the membrane to X-ray film in a developing cassette. Bands corresponding to zymogen and processed caspases are normally detected within 1–10 min of exposure (*see* **Notes 26** and **27**).
9. Process the film in an automatic X-ray developer.
10. Before reprobing the membrane using other primary antibodies, incubate the membrane in stripping buffer for 30 min at 50 °C with agitation (*see* **Notes 28–30**).
11. Wash the membrane extensively with water and repeat **steps 1–9**.

3.2 Immunocytochemical Detection of Active Caspases

3.2.1 Sample Preparation

1. Place coverslips in 6-well or 12-well culture plates.
2. Seed cells and allow them to attach to the glass surface (*see* **Notes 31** and **32**).
3. Induce apoptosis in cells by desired way. Concurrently incubate a control culture without induction.
4. Aspirate cell culture medium and rinse cells twice in cold PBS.
5. After the last wash, cover cells with 4 % cold formaldehyde and incubate the samples at 4 °C for 15–30 min.
6. Wash the fixed cells 3× 5 min in PBS (*see* **Note 33**).

3.2.2 Immunostaining

All incubations should be carried out at room temperature unless otherwise noted, in a humid covered dish or plate to prevent drying and fluorochrome fading.

1. Aspirate PBS and incubate cells in blocking/permeabilizing buffer for 60 min.
2. While blocking, dilute the primary antibody of interest in antibody dilution buffer (*see* **Note 34**).
3. Using a thin forceps, pick up the coverslips and put them cell side up in a humid light-tight plate. Apply 70 μL of diluted primary antibody to each coverslip (*see* **Note 35**).
4. Incubate overnight at 4 °C.
5. Aspirate the primary antibody solution and rinse two to three times in 70–100 μL PBS for 5 min each.
6. Apply 70 μL fluorochrome-conjugated secondary antibody (usually 1:200) diluted in PBS and incubate for 1–2 h at room temperature in dark.
7. Wash specimens as in **step 5**.
8. Counterstain the nuclei for 10–15 min with 70 μL DAPI or Hoechst 33342 (both 1 μg/mL in PBS).
9. Wash once in PBS for 5–10 min.
10. Mount coverslips on specimen glass plates using 2–3 μL VECTASHIELD or equivalent and seal with nail polish.
11. For best results, allow mountant to cure overnight at room temperature. For long-term storage, store slides flat at 4 °C protected from light.
12. Analyze the slides in a confocal microscope.

3.3 Immunoblot Analysis of Caspase Substrate Cleavage

The method includes the same steps as described in Subheading 3.1.

3.4 Detection of Cytokeratin 18 Cleavage by Flow Cytometry

3.4.1 Sample Preparation

1. Induce apoptosis in cells by desired method. Concurrently incubate a control culture without induction.
2. Harvest and collect 1×10^6 cells in a tube. Centrifuge for 5 min at $150 \times g$, 4 °C.
3. Wash cells in PBS, centrifuge again, and discard supernatant.
4. Repeat **step 3**.
5. Fix the cells by resuspending in 0.5 mL ice-cold pure methanol at −20 °C for 30 min (*see* **Notes 36** and **37**).

3.4.2 Detection of the CK 18 Neoepitope

1. Wash the fixed cells twice in PBS.
2. Dilute the M30 Cytodeath-FITC antibody to 1 μg/mL in dilution buffer and incubate the cells with 100 μL of the working solution for 30–60 min at room temperature in the dark.

3. Wash the stained cells with PBS twice.
4. Add 0.3 mL PBS, protect samples from light, and measure the FITC fluorescence (FL1-H) with a flow cytometer, evaluating the percentage of cells positive for CK18 cleavage in control and apoptotic samples (*see* **Note 38**).

3.5 Caspase Activity Measurement in Peptide Cleavage Assay

3.5.1 Sample Preparation

1. Induce apoptosis in cells by desired method. For statistical analyzes, perform the experiment in triplicate wells. Concurrently incubate triplicate control cultures without induction.
2. Harvest $2.5–5 \times 10^5$ cells from each well and centrifuge for 5 min at $150 \times g$, 4 °C.
3. Wash cells in PBS and centrifuge for an additional 5 min at $150 \times g$, 4 °C.
4. Discard supernatant. Resuspend each pellet in 30 μL cold PBS and transfer into individual wells of an opaque 96-well assay plate on ice (*see* **Note 39**).
5. Pipette 5 μL of each sample for determination of protein concentration using the BCA (*see* **step 5** in Subheading 3.1.1).
6. Snap-freeze the samples in the plate by having it floating on liquid nitrogen or incubate the plate at −20 °C for 30 min (*see* **Note 40**).

3.5.2 Caspase Activity Measurement

1. Set up the computer control for microplate reader. Set temperature to 37 °C.
2. Prepare complete 1× caspase activity sample buffer (2.5 mL caspase activity stock buffer, 12.5 μL 1 M DTT, 2.5 μL 0.1 % NP40). This volume will be sufficient for analyzing 50 samples. Immediately before use, add 2.5 μL of the caspase substrate of interest.
3. Transfer the sample plate to room temperature and dispense 50 μL of complete sample buffer to each test well. Place the plate in the microplate reader.
4. Read samples. The maximum absorption for AMC is 354 nm; fluorometric detection for AMC cleaved from peptide is at excitation 380 nm and emission 460 nm.

3.6 Inhibition of Caspase Activity in Cell Cultures

1. The general working concentration of caspase inhibitors is 10–20 μM. Incubate cells for 1–2 h before inducing apoptosis by the method of choice. It is recommended to concurrently incubate control cultures without induction agent and inhibitor, containing only induction agent and containing only inhibitor (*see* **Note 41**).
2. Proceed to analysis (*see* **Note 42**).

4 Notes

1. To avoid SDS precipitation, the sample buffer needs to be heated prior to use.
2. Initially, concentrated HCl (12 N) can be used to drastically reduce the starting pH. As a near to desired pH is reached, HCl with lower ionic strengths should be used to avoid a sudden drop below the required pH.
3. Addition of HCl induces an exothermal reaction which in turn will affect the pH. Make sure that the final adjustments are performed on a room-tempered solution.
4. The solution can be stored for several weeks at 4 °C, or at −20 °C.
5. The buffer can be reused up to four times with maintained transfer capacity. Store at 4 °C.
6. A 10× transfer buffer stock solution with respect to Tris base (248 mM) and glycine (1,918 mM) can be prepared. Dilute in 700 mL water and add 200 mL of methanol, store at 4 °C until use.
7. The blocking solution does not contain any preservatives and the presence of lactic acid may influence the pH during prolonged storage. Therefore this buffer should be prepared fresh for each blotting procedure. Blocking buffers which contain casein (0.5–1 %) that is purified from milk by the Hammarsten method can be used as an alternative (Thermo Scientific).
8. Most diluted primary antibodies can be reused several times. For additional preservation capacity add 0.05 % sodium azide. Avoid this buffer when diluting horseradish peroxide (HRP) conjugated antibodies as the sodium azide is an inhibitor of HRP enzyme activity.
9. The itemized antibodies are targeting human caspases. Only caspases which have been described to be involved in apoptosis signaling are listed.
10. Heating to approximately 60 °C is required to dissolve paraformaldehyde powder in PBS. Once the paraformaldehyde is dissolved, the solution should be cooled and filtered. Formaldehyde is toxic, use only in fume hood.
11. Normal serum from the same species as the secondary antibody (e.g., normal goat serum, normal donkey serum) can be used as an alternative to BSA.
12. Only three examples of antibodies targeting common caspase substrates are provided here. For a more complete list of caspase substrates, *see* ref. [8] or the CASBAH/MerCASBA, a searchable databases of caspase substrates [4, 5].

13. Protein fragments produced as a result of caspase-mediated cleavage of specific substrates are not always possible to detect by SDS-PAGE using antibodies raised against the full-length protein. In this case, processing of target proteins in apoptotic samples may be indicated by a decrease of substrate quantity.
14. Both caspase activity buffers can be stored for several weeks at 4 °C.
15. In addition to classical FMK inhibitors, some suppliers also provide a new generation of caspase inhibitors. For e xample, Q-VD-OPh (Quinolyl-Valyl-*O*-methylaspartyl-[2,6-difluorophenoxy]-methyl ketone) is a cell-permeable, irreversible, broad spectrum caspase inhibitor that displays less toxicity and longer stability than the FMK inhibitors.
16. Normally, due to denaturing effects of sample buffer and boiling, the addition of protease inhibitors is unnecessary.
17. Cells will lyse in the BCA reaction buffer and release protein content.
18. Alternatively, shear the DNA either by sonication or by repeated passage through a 23-gauge needle.
19. Repeated loading of samples after storage in −20 °C requires 1–2 min heating at 95 °C in order to eliminate SDS precipitations.
20. Wrapped in damp paper towels, a polymerized SDS-PAGE gel can be stored at 4 °C for up to 1 week.
21. Since several factors, including quality of antibodies and efficiency of the ECL solution used will influence the chemiluminescence signal in SDS-PAGE immunodetection, the optimal protein amount to be loaded has to be determined experimentally.
22. Trapping of air bubbles is easily avoided if assembly of the sandwich is performed submerged in transfer buffer.
23. Transfer efficiency, quality of separation, and equal loading can be controlled by Ponceau S Staining. Using this method, microgram quantities of transferred protein can be detected with a clear background and red protein bands. This staining is reversible and allows further immunological detection.

    Ponceau S stain solution: Dissolve 0.033 g Ponceau S in 10 mL of water, add 0.3 mL of glacial acetic acid and make to 30 mL with water. Store the solution at room temp.

    Procedure for total protein detection:

    (a) Immerse the blotted membrane in Ponceau S staining solution and stain for 1–3 min.

    (b) After staining, rinse the membrane in water until a distinct protein pattern develops.

(c) Remove stain by repeated washing in water or PBS.

(d) Proceed to the blocking procedure.

24. To minimize evaporation and the required volume of primary antibody solution, choose a lidded container with a size similar to that of the membrane.

25. For high abundant protein targets, incubation for 1 h at room temperature is sufficient.

26. A number of factors, including quality of antibodies and the type of ECL used, may influence exposure times considerably.

27. In some experimental models, processed caspase fragments may be hard to reveal. This is, for some reason, especially true for caspase-2. One approach to overcome this problem is to firstly perform immunoprecipitation using agarose-conjugated VAD (agarose-valyl-alanyl-aspartyl), which bind the catalytic site of active caspases. Subsequently, precipitated material is separated in SDS-PAGE and immunoblotted with the antibody of interest [16].

28. By repeating the immunodetection step, it is possible to target other caspases in the same membrane Stripping is only required if the proteins targeted are overlapping in size or can be recognized by the same secondary Abs. If this is not the case, wash the membrane 3 × 10 min in PBS and proceed to immunodetection (**steps 1–9**) using a second primary antibody of interest.

29. Markers such as G3PDH or *b*-actin, which expression levels remain relatively constant across various cell culture conditions, should also be detected. Optical density values of target proteins bands are then correlated to the corresponding values of the marker protein. By this mean, comparisons in between individual samples become more accurate.

30. Long-term storage of nitrocellulose membranes is accomplished by drying. During storage, keep the membranes protected in between two 3MM papers. For repeated use, incubate the membranes in PBS for 30 min on a rocker and proceed to immunodetection.

31. For some cell types poly-L-lysine-coated coverslips may be required as culture substratum.

32. For cells growing in suspension, a cytospin protocol is required in order to attach cells to the coverslip.

33. Fixed and washed samples can be stored at 4 °C for up to 1 week.

34. When a primary antibody is used in immunocytochemistry for the first time, it should be titrated to determine the dilution which allows for a robust specific signal with a minimum of background (normally 1:200 to 1:1,000).

35. The surface tension will keep the liquid in place.

36. Prepared samples can be stored in methanol at −20 °C for several weeks.
37. To prevent clustering of cells which could interfere with flow cytometry analysis, ice cold methanol should be added drop wise to cells while wortexing.
38. For statistical calculations triplicate samples for each condition are recommended.
39. The results obtained have to be recalculated with respect to input of total protein amount, or the number of cells present in individual samples. Since this is critical in order to accomplish accurate assay readout, it is recommended to do both.
40. The freezing of samples is performed in order to permeabilize plasma membranes. If cells are lysed by a protocol which maintains protein function, cell lysates can equally well be used as templates for the caspase activity assay. Frozen samples can be stored for several days at −20 °C.
41. Caspase inhibitors are functionally stable in cell cultures at least 48 h.
42. Apoptosis in cells can be monitored by a phase-contrast microscope. Attached cell types are normally detached as a result of apoptosis. The dramatic change of morphology, due to plasma membrane blebbing and cell shrinkage, can be used to observe apoptotic conditions in suspension cells.

5 Conclusions

Since detection of caspase activities is one of the most common biochemical markers of apoptosis, several techniques have been developed to characterize this process. Although most of them are provided by various companies as method kits, comparable results can certainly be obtained by purchasing components and reagents separately and then, take advantage of shared protocols. Caspases are activated in a signaling cascade where a specific effector caspase is processed by one or several types of apical caspases. In turn, effector caspases may further activate apical caspases as part of an apoptosis amplifying loop. Thus, caspases are generally activated jointly. Once activated, caspases cleave target substrates with some degree of overlapping specificity. Consequently, all assays described in this chapter are appropriate for the detection of apoptosis per se. In order to make further conclusions regarding the role of individual caspases in a given system, it is highly recommended to also suppress the caspase of interest by additional techniques, such as. Another aspect that should be considered in advance of experimental design is timing. Initiation of cell death signaling may occur in minutes or days depending on the compound or culture condition used for apoptosis induction in a particular cell type. Moreover, once initiated, apoptosis is normally a fast process which in cell cultures is followed by secondary necrosis having morphological and biochemical unique signatures. Finally, it should also be noted that the function of caspases has been verified in processes separated from apoptosis and that a multitude of caspase-independent cell death modes exist [17]. To place caspase activity assay results in a biological context may therefore be a more complicated challenge, requiring a spectrum of additional cell and molecular techniques.

Acknowledgements

Work in the authors' laboratory was supported by grants from the Swedish and Stockholm Cancer Societies, the Swedish Research Council, and the Swedish Childhood Cancer Foundation.

References

1. Degterev A, Boyce M, Yuan J (2003) A decade of caspases. Oncogene 22:8543–8567
2. Fuentes-Prior P, Salvesen GS (2004) The protein structures that shape caspase activity, specificity, activation and inhibition. Biochem J 384: 201–232
3. Shi Y (2004) Caspase activation: revisiting the induced proximity model. Cell 117:855–858
4. Luthi AU, Martin SJ (2007) The CASBAH: a searchable database of caspase substrates. Cell Death Differ 14:641–650
5. Fridman A, Pak I, Butts BD, Hoek M, Nicholson DW, Mehmet H (2013) MerCASBA: an updated and refined database of caspase substrates. Apoptosis 18:369–371
6. Liu X, Zou H, Slaughter C, Wang X (1997) DFF, a heterodimeric protein that functions downstream of caspase-3 to trigger DNA fragmentation during apoptosis. Cell 89: 175–184
7. Li H, Zhu H, Xu CJ, Yuan J (1998) Cleavage of BID by caspase 8 mediates the mitochondrial damage in the Fas pathway of apoptosis. Cell 94:491–501
8. Fischer U, Janicke RU, Schulze-Osthoff K (2003) Many cuts to ruin: a comprehensive update of caspase substrates. Cell Death Differ 10:76–100
9. Oliver FJ, de la Rubia G, Rolli V, Ruiz-Ruiz MC, de Murcia G, Murcia JM (1998) Importance of poly(ADP-ribose) polymerase and its cleavage in apoptosis. Lesson from an uncleavable mutant. J Biol Chem 273: 33533–33539
10. Kim MY, Zhang T, Kraus WL (2005) Poly(ADP-ribosyl)ation by PARP-1: 'PAR-laying' NAD+into a nuclear signal. Genes Dev 19:1951–1967
11. Leers MP, Kolgen W, Bjorklund V, Bergman T, Tribbick G, Persson B et al (1999) Immunocytochemical detection and mapping of a cytokeratin 18 neo-epitope exposed during early apoptosis. J Pathol 187:567–572
12. Pozarowski P, Grabarek J, and Darzynkiewicz Z (2003) Flow cytometry of apoptosis. Curr Protoc Cytom Chapter 7, Unit 7 19
13. McStay GP, Salvesen GS, Green DR (2008) Overlapping cleavage motif selectivity of caspases: implications for analysis of apoptotic pathways. Cell Death Differ 15:322–331
14. Callus BA, Vaux DL (2007) Caspase inhibitors: viral, cellular and chemical. Cell Death Differ 14:73–78
15. Smolewski P, Bedner E, Du L, Hsieh TC, Wu JM, Phelps DJ et al (2001) Detection of caspases activation by fluorochrome-labeled inhibitors: multiparameter analysis by laser scanning cytometry. Cytometry 44:73–82
16. Tinel A, Tschopp J (2004) The PIDDosome, a protein complex implicated in activation of caspase-2 in response to genotoxic stress. Science 304:843–846
17. Galluzzi L, Vitale I, Abrams JM, Alnemri ES, Baehrecke EH, Blagosklonny MV et al (2012) Molecular definitions of cell death subroutines: recommendations of the Nomenclature Committee on Cell Death 2012. Cell Death Differ 19:107–120

Part II

Paracaspases and Metacaspases

Chapter 11

Detection and Measurement of Paracaspase MALT1 Activity

Stephan Hailfinger, Christiane Pelzer, and Margot Thome

Abstract

The paracaspase MALT1 is a Cys-dependent, Arg-specific protease that plays an essential role in the activation and proliferation of lymphocytes during the immune response. Oncogenic activation of MALT1 is associated with the development of specific forms of B-cell lymphomas. Through specific cleavage of its substrates, MALT1 controls various aspects of lymphocyte activation, including the activation of transcriptional pathways, the stabilization of mRNAs, and an increase in cellular adhesion. In lymphocytes, the activity of MALT1 is tightly controlled by its inducible monoubiquitination, which promotes the dimerization of MALT1. Here, we describe both in vitro and in vivo assays that have been developed to assess MALT1 activity.

Key words Immunomodulation, NF-kB, Lymphocyte activation, Lymphoma

1 Introduction

The gene encoding MALT1 was initially identified as a target of a chromosomal translocation occurring in B-cell lymphomas of the mucosa-associated lymphoid tissue (MALT lymphomas), which results in the formation of an oncogenic IAP2-MALT1 fusion protein [1]. Independently, MALT1 was described as a protein with sequence homology to caspases and metacaspases, and hence named paracaspase [2]. Subsequent genetic and biochemical studies have revealed a key function for MALT1 in the adaptive immune response, and in the development of diffuse large B-cell lymphomas (DLBCL) of the activated B-cell (ABC) subtype [3–5]. The most important function of MALT1 in lymphocytes seems to be its capacity to promote the activation of the transcription factor NF-kB [6, 7] via both its scaffold and its enzymatic function [8]. The activation of NF-kB then drives the expression of genes that are relevant for cellular proliferation and survival [9].

MALT1 contains a caspase-like domain that has high structural similarity to the proteolytic domain of caspases [10, 11] and that shares the presence of a conserved active site, the Cys-His dyad,

Peter V. Bozhkov and Guy Salvesen (eds.), *Caspases, Paracaspases, and Metacaspases: Methods and Protocols*, Methods in Molecular Biology, vol. 1133, DOI 10.1007/978-1-4939-0357-3_11, © Springer Science+Business Media New York 2014

with caspases and metacaspases [2]. However, MALT1 showed no caspase-like activity [2, 12]. Therefore, it was initially thought that MALT1 contributes to NF-kB activation mainly by its scaffold function, through physical recruitment of signaling proteins. This would in turn promote the activation of the IKK complex, which phosphorylates the NF-kB inhibitor IkB to induce its proteasomal degradation and to allow NF-kB to enter the nucleus and initiate transcription [9]. This concept is supported by the discovery that MALT1 binds the ubiquitin ligase TRAF6 [13] and, upon its TRAF6-mediated polyubiquitination, physically recruits the IKK complex [14]. Nevertheless, it was noticed in earlier studies that mutation of the conserved Cys residue in the caspase-like domain reduced the capacity of MALT1 or an oncogenic IAP2-MALT1 fusion protein to promote the activation of the NF-kB pathway [2, 15]. In 2008, two groups then independently reported the MALT1-dependent cleavage of Bcl10 and A20 directly C-terminal to an Arg residue, identifying MALT1 as an Arg-specific protease [16, 17]. The finding of an Arg-directed proteolytic activity of MALT1 is well compatible with prior studies reporting a similar Arg- or Lys-directed cleavage activity of metacaspases [18]. Indeed, both metacaspases and the MALT1 paracaspase contain conserved negatively charged residues in the substrate-binding S1 pocket that explain their preferential affinity for substrates with a positively charged Arg (or Lys) residue [18].

Over the last few years, several substrates of MALT1 have been identified, and the distinct roles of the cleavage of these individual substrates in the antigen-driven activation and proliferation of lymphocytes have been dissected. An important function of MALT1 in lymphocytes is to promote the activation of the transcription factor NF-kB [6, 7] via both its scaffold and its enzymatic function [8]. The scaffold function of MALT1 is required for IKK-mediated NF-kB activation [19, 20]. The fact that the protease activity of MALT1 is indeed similarly relevant for the NF-kB-dependent activation of lymphocytes was then established by the development of a MALT1 inhibitor [16] and by expression of catalytically inactive mutants of MALT1 in lymphocyte cell lines [17]. The protease activity of MALT1 controls NF-kB activation by cleaving the NF-kB family member RelB [20], which acts as negative regulator of T-cell activation [21]. RelB binds to the NF-kB subunits RelA- and c-Rel in the cytoplasm of lymphocytes [20, 22]. Therefore, MALT1-dependent cleavage of RelB and its subsequent proteasomal degradation are required to allow the DNA binding of RelA- or c-Rel-containing NF-kB complexes in the nucleus [20]. MALT1-dependent cleavage of the deubiquitinating enzyme A20 has been proposed as another way to promote NF-kB activation [17]. Since A20 can negatively regulate the activity of the IKK complex by deubiquitination of the IKK subunit NEMO [23], it was initially proposed that MALT1-dependent cleavage of A20 might serve to prolong the IKK-dependent NF-kB response [17]. However, pretreatment of cells with the MALT1 inhibitor does not affect IKK activity [19, 20], suggesting that A20 cleavage must contribute to lymphocyte activation in a distinct manner that remains to be identified.

MALT1-dependent cleavage of the deubiquitinating enzyme Cyld [24] and the adapter protein Bcl10 [16] promote the activation of the AP-1 transcriptional pathway and the adhesiveness of T cells, respectively, to regulate two other important aspects of lymphocyte activation. Recently, MALT1 was shown to promote T-cell activation in yet an additional manner, by cleavage of the RNAse Regnase-1 (also known as MCPIP1 or Zc3h12a) [25]. MALT1-dependent cleavage of MCPIP-1 leads to the stabilization of mRNAs of T-cell effector genes, such as growth-promoting cytokines [25]. Thus, MALT1 controls lymphocyte activation by transcription-dependent and -independent means.

The Ser/Thr kinase NIK has been identified as a specific substrate of the IAP2-MALT1 fusion protein that results from a chromosomal translocation found in aggressive forms of MALT lymphoma [26]. NIK is a protein with a short half-life that is normally rapidly turned over by proteasomal degradation. The cleavage of NIK by the IAP2-MALT1 fusion protein results in the generation of a stable C-terminal NIK fragment, containing the kinase activity [26]. The resulting active NIK fragment promotes the phosphorylation of the IKK subunit IKKα, which in turn drives the activation of the so-called alternative NF-kB pathway by the phosphorylation-dependent processing of the inactive NF-kB precursor p100 to form the biologically active p52 subunit. Thereby, the IAP2-MALT1 fusion seems to constitutively activate the alternative NF-kB pathway in MALT lymphomas [26].

In addition to its well-described role in the adaptive immune response, MALT1 is also required for NF-kB activation downstream of other immune receptors that activate NK cells, mast cells, dendritic cells or myeloid cells, and also downstream of certain G-protein coupled receptors or the receptor tyrosine kinase EGFR [3]. However, at present it remains largely unknown to which extent the protease activity of MALT1 contributes to these functions.

How can the enzymatic activity of MALT1 be monitored? One line of experimental evidence for MALT1 protease activity comes from the detection, by western blotting, of the presence of cleaved MALT1 substrates in cellular lysates of activated lymphocytes or lymphoma cells [16, 17, 20, 24, 26]. In addition, based on the cleavage site LRSR'G present in Bcl10, an in vitro assay was developed that uses the cleavage of a fluorogenic Ac-LRSR-AMC substrate by purified MALT1 or cellular lysates containing active MALT1 as a means to monitor MALT1 activity [16]. This in vitro cleavage assay has been further improved by the systematic testing of tetrapeptides in which the three amino acids (P2–P4) preceding the fixed C-terminal Arg residue were systematically varied [27]. Amongst the best tetrapeptide substrates identified with this approach were peptides containing a Ser residue in the P2 position, which is a feature that is common to 5 of the 6 presently identified substrates [16, 17, 20, 24, 26]. The P4 position showed a clear preference for Leu, a feature that is also present in the MALT1 substrates Bcl10, RelB and Regnase [16, 20, 25]. In order for purified recombinant MALT1 to be present in an active form, several approaches have been used that all favor dimerization or oligomerization of MALT1. In one study, recombinant GST-MALT1 protein was oligomerized by its binding to glutathione–sepharose/agarose beads [16]. Alternative approaches have been to generate fusion proteins of MALT1 with bacterial gyrase B or a leucine zipper domain [17, 28], or to perform assays in presence of a kosmotropic salt such as ammonium citrate, which is known to activate certain caspases by favoring their dimerization [29, 30]. Like caspases, MALT1 has a propensity to dimerize via its caspase-like domain, as evident from recently published crystallographic structures [10, 11]. The comparison of MALT1 crystals obtained in the absence or presence of a substrate analog, the irreversible tetrapeptide inhibitor z-VRPR-fmk, has revealed that while both conditions yielded MALT1 dimers, the dimer structure undergoes extensive conformational changes upon inhibitor binding, which lead to changes in the contact sites of the dimerization interface and a change in the orientation of the protease domain towards the C-terminal immunoglobulin (Ig)-like domain [10]. Thus, MALT1 dimerization or the adoption of a precise conformation of the dimer may be induced upon substrate binding. In vivo, MALT1 activation and dimerization requires an inducible monoubiquitination of MALT1 on a Lys residue (K644) that is localized within the C-terminal Ig-like domain [31]. A monoubiquitination-deficient K644R mutant of MALT1 was unable to sustain the activation of T cells and the growth of cell lines derived from ABC DLBCL [31] that critically depend on oncogenic MALT1 activity [32, 33]. In contrast, an in-frame fusion of MALT1 to a C-terminal ubiquitin moiety resulted in a hyperactive MALT1-Ub construct that constitutively dimerizes [31]. Thus, monoubiquitination of MALT1 most likely induces a conformational change that promotes the formation or stabilization of a constitutively active MALT1 dimer in vivo.

Below, we describe various methods that allow the detection of MALT1 activity, including the isolation of recombinant active MALT1 from bacteria, the detection of the cleavage of endogenous MALT1 substrates or of MALT1 monoubiquitination, and the quantification of MALT1 activity using fluorogenic substrate peptides or a FRET-based reporter construct.

2 Materials

Deionized water is used throughout. All reagents are prepared and stored at room temperature (unless indicated otherwise).

2.1 Purification of MALT1 from Bacteria

1. Plasmid coding for GST-MALT1: generated by cloning the MALT1 sequence into a vector allowing bacterial protein expression (such as pGEX from GE Healthcare).
2. LB agar dishes and LB medium with 100 μg/ml ampicillin.
3. Chemically competent BL21 bacteria (e.g., One Shot from Invitrogen).
4. Isopropyl-b-D-thiogalactopyranoside (IPTG).
5. PreScission protease (GE Healthcare).
6. Glutathione Sepharose 4B (GE Healthcare).
7. Poly-Prep columns (Bio-Rad).
8. Lysis buffer: 50 mM Hepes (pH 7.9), 300 mM NaCl, 1 mM EDTA, 0.1 % (v/v) NP-40, and 5 mM DTT. Always add DTT freshly before use, the other components can be mixed in advance and stored at 4 °C for a few weeks (*see* **Note 1**).
9. Wash buffer: 50 mM Tris–HCl (pH 7.6) and 150 mM NaCl. 5 mM DTT should be added before use.
10. SDS-PAGE reagents and Coomassie staining solution.
11. Bacteria incubator (37 and 18 °C).
12. Spectrophotometer (wavelength 600 nm).
13. Refrigerated centrifuge.
14. French press.
15. Shaker or rotation wheel.
16. SDS-PAGE components.

2.2 MALT1 In Vitro Cleavage Assay

1. Tetrapeptide substrate (Ac-LVSR-AMC or Ac-LRSR-AMC, Peptides International).
2. Cleavage assay buffer: 50 mM MES (pH 6.8), 150 mM NaCl, 0.1 % (w/v) CHAPS and 1 M ammonium citrate. 10 mM DTT has to be freshly added before use.
3. Black 96-well plates (e.g., OptiPlate-96 F, Perkin Elmer).
4. Microplate reader capable to excite at 350–380 nm and measure the emission at 460 nm (e.g., Synergy microplate reader, BioTek).

2.3 Reagents and Materials for the BCL10 and MALT1 Western Blot

1. Acrylamide (30 % w/v).
2. Bis-acrylamide (1 % w/v).
3. Ammonium persulfate (10 %).

4. TEMED.
5. Resolving buffer: 1.5 M Tris–HCl pH 8.8.

2.4 FRET-Based MALT1 Activity Assay

1. Reporter plasmid: constructed by linking eYFP and eCFP with the MALT1 cleavage recognition site LVSR, which originates from the MALT1 substrate RelB [20, 31]. A non-cleavable form with the linker sequence LVSG can be used as a control. Both constructs were cloned into a vector derived from pCR3 (Invitrogen).
2. Dulbecco's Modified Eagle Medium (DMEM) with 10 % fetal calf serum (FCS) and antibiotics.
3. 6-well cell culture plate.
4. To prepare the 2× HeBS buffer, dissolve 16.4 g NaCl, 11.9 g Hepes (acid form), and 0.2 g Na_2HPO_4 in 800 ml water. Titrate precisely to pH 7.05 with 5 M NaOH. Complete to 1 L and sterilize by filtration (0.22 μm).
5. Sterile 2.5 M $CaCl_2$ solution.
6. Flow cytometry buffer: 1 % FCS and 1 mM EDTA in PBS.
7. Flow cytometer equipped with a 405 nm laser (e.g., LSR II, BD Biosciences).

3 Methods

All incubation steps are at room temperature unless otherwise stated.

3.1 Purification of Active Recombinant MALT1 from Bacteria

1. To transform BL21 chemically competent E. coli with the GST-MALT1 (pGEX) plasmid by heat-shock, thaw the bacteria on ice, add 10–100 ng of the plasmid, mix by tapping gently (*see* **Note 2**) and incubate the vial on ice for 30 min.
2. Heat-shock the bacteria for 30 s at 42 °C in water bath without shaking.
3. Spread up to 200 μl from the transformation on a pre-warmed LB plate with 100 μg/ml ampicillin and incubate overnight at 37 °C.
4. Resuspend one single colony in a 25 ml LB liquid culture with ampicillin (100 μg/ml ampicillin) to produce a starter culture by incubation overnight at 37 °C with shaking.
5. Inoculate 500 ml LB media with antibiotic with 20 ml of the starter culture.
6. Incubate at 37 °C with shaking until OD_{600} reaches 0.6.
7. Place the bacteria culture on ice for 10 min.
8. Induce the expression of GST-MALT1 by adding 40 μM IPTG and incubate the bacteria overnight at 18 °C with shaking.

9. Pellet the bacteria by centrifugation at 3,700 × *g* for 20 min at 4 °C.
10. Resuspend the pelleted bacteria with 15 ml of lysis buffer, incubate for 20 min on ice and lyse the bacteria with a French Press at 4 °C.
11. Dilute the lysate with one volume of PBS.
12. Transfer the lysate to 2 ml vials and centrifuge for 15 min at 16,000 × *g* and 4 °C in a microcentrifuge.
13. Collect the supernatant into a falcon tube (*see* **Note 3**) and take 100 μl for western blot analysis of the lysate (*see* **Note 4**). For better storage of the lysate antimicrobial inhibitors can be added (*see* **Note 5**).
14. Wash 300 μl of Glutathione-(GSH)–Sepharose beads three times with PBS (*see* **Note 6**) and incubate with the lysis supernatants for 2.5 h at 4 °C on a shaker or a rotation wheel.
15. Transfer the beads with the supernatant into a column.
16. Wait until only beads are left in the column (do not let it run dry).
17. Wash the beads three times with 15 ml cold washing buffer at 4 °C.
18. Wait until the column stops dripping, close the column, and add 1.2 ml washing buffer with 80 U/ml PreScission protease (*see* **Note 7**).
19. Incubate for 2 h at 4 °C to remove MALT1 from the GSH-beads.
20. Collect the flow through (*see* **Note 8**) and load 10 μl together with the lysate and a BSA standard (0.5, 1, 2, 5, and 10 μg) on a SDS-PAGE and stain the gel with Coomassie Blue to estimate the concentration of MALT1 in the solution.
21. The flow through can now be used for further experiments, but also stored at −80 °C.

3.2 In Vitro MALT1 Cleavage Assay of Fluorogenic Peptides

1. Prepare cleavage assay buffer and add the tetrapeptide substrate (Ac-LRSR-AMC or Ac-LVSR-AMC, *see* **Note 9**) at a final concentration of 25 μM.
2. Add 95 μl of the cleavage assay buffer with the substrate and 2 μg of soluble MALT1 (in a volume of 5 μl) into the well of a black 96-well plate.
3. Cleavage assay buffer with substrate only—no MALT1 added—serves as a negative control; we also include the protease inactive mutant (MALT1 C464A) as a negative control (*see* **Note 10**).
4. Remove air bubbles in the wells since they might interfere with measurement.

5. Measure the fluorescence (excitation 350–380 nm, emission 460 nm) over time (measure every 10 min for 2–4 h) at 30 °C.
6. To get the MALT1 protease activity, calculate the slope of the fluorescence over time and normalize to the amount of MALT1 protein (*see* **Note 11**).

3.3 Detection of Endogenous BCL10 Cleavage or MALT1 Monoubiquitination by High-Resolution Gels and Western Blot

The detection of MALT1 substrate cleavage or MALT1 monoubiquitination can serve as a qualitative means to monitor MALT1 activity. The MALT1 mediated cleavage of A20, CYLD, RelB, and Regnase can easily be visualized by standard western blot [17, 20, 24, 25]. Only the detection of cleaved human BCL10 is challenging since MALT1 removes only five amino acids from its C-terminus. Cleaved BCL10 can therefore only be detected using high-resolution gels [16] or an antibody specific for cleaved BCL10 [32]. Moreover, the 8 kDa shift caused by the attachment of a single ubiquitin to MALT1 is more easily detectable using high-resolution gels. Here, we describe only the preparation of the high-resolution gels [34] for detection of cleaved BCL10 and MALT1 monoubiquitination since the further procedure does not vary from standard SDS-PAGE and western blot protocol.

1. To prepare a 500 ml stock solution of a high resolution 15 % acrylamide resolving gel mix to detect cleaved BCL10, add 250 ml acrylamide (30 % w/v), 43 ml bis-acrylamide (1 % w/v), 125 ml resolving buffer, and 82 ml water. For a 7,5 % high-resolution acrylamide gel mix, which is optimal to detect MALT1 monoubiquitination, add 125 ml acrylamide (30 % w/v), 97 ml bis-acrylamide (1 % w/v), 125 ml resolving buffer, and 150 ml water (*see* **Note 12**).
2. Add 50 μl of 10 % ammonium persulfate and 5 μl of TEMED per 10 ml of mix. Cast the gels within minutes and gently overlay with isopropanol.
3. Proceed with standard SDS-PAGE and western blot protocol.

3.4 FRET-Based Assay of Protease Activity

To quantify MALT1 protease activity in intact cells we generated an eYFP–Leu-Val-Ser-Arg–eCFP expression construct as well as the respective non-cleavable negative control, eYFP–Leu-Val-Ser-Gly–eCFP [31]. Here we describe the use of the FRET-based cleavage assay in HEK293T cells, but it might be adapted to other cells.

1. Plate approximately 150.000 HEK293T cells in DMEM (with 10 % FCS and antibiotics) in a 6-well cell culture plate.
2. After 24 h, transfect the sub-confluent HEK293T cells with 0.1 μg of the reporter construct together with 1 μg MALT1 and 0.2 μg BCL-10 expression constructs using the calcium phosphate transfection method. Mix 125 μl water with 12.5 μl $CaCl_2$ (2.5 M), add the respective amount of plasmids and add 125 μl 2× HeBS dropwise while gently shaking on a vortex.

3. Incubate the mix for 10 min at room temperature. Distribute the transfection mix carefully and equally on the cells.
4. Change the media 6–14 h after transfection with fresh completed DMEM.
5. 24 h after onset of transfection, add 1 ml of PBS to the wells and detach the cells with a 1 ml pipette by washing the cells off (by physical force).
6. Spin the cell suspension for 1 min at 1,000 ×*g* in a table-top centrifuge and after removing the PBS, add flow cytometry buffer.
7. Filter the cells into suitable FACS tubes through a 70 μm mesh to remove cell clumps.
8. Analyze the cells by flow cytometry. To measure the eCFP and FRET signals, the transfected cells are excited by a 405-nm laser and their emission is recorded from the standard 450/50 filter for eCFP fluorescence, and a 585/42 filter for FRET fluorescence. For an example of the FACS analysis please see ref. 31.
9. The expression levels of all constructs and the proportion of reporter cleavage can be controlled by western blot (*see* **Note 13**).

4 Notes

1. DTT prevents the oxidation of the cysteine, which is located in the active site of MALT1.
2. Do not mix the bacteria by pipetting.
3. At this step the lysate can be stored at −20 °C for several months.
4. To confirm the induction of protein expression, the lysate can be loaded immediately on a SDS-PAGE gel followed by a Coomassie Blue staining. GST-MALT1 migrates at around 115 kDa.
5. To protect the supernatants from microbial growth add sodium azide (5 % w/v stock solution, use 1:1,000).
6. To wash the beads, add at least 1 ml PBS to the beads and centrifuge at 1,000 ×*g* for 1 min in a microcentrifuge to remove the ethanol. Repeat two times.
7. To increase the yield of MALT1 protein, the beads can be washed several times with 1 ml washing buffer. The respective collected fractions will still contain MALT1 protease even if the activity is lower than the first fraction.
8. MALT1 activity can be induced by concentrating it on beads or by the use of kosmotropic buffers, which favor MALT1 dimerization. Therefore, MALT1 protease activity is also present

when GST-MALT1 is bound to glutathione-coupled beads. This approach might be considered if the planned experiment is not compatible with the presence of kosmotropic salts.

9. The sequence of the Ac-LRSR-AMC peptide derives from the substrate BCL-10 [16], whereas Ac-LVSR-AMC originates from RelB [20]. MALT1 protease activity for the RelB-mimicking substrate is roughly four times higher than for Ac-LRSR-AMC [20]. Both peptides can be stored frozen at a concentration of 10 mM dissolved in DMSO.
10. In case protease inactive recombinant MALT1 is not available, the addition of 1 μM of the MALT1 inhibitor Z-VRPR-FMK (Bachem) might be useful as additional negative control.
11. Only consider the linear part of the fluorescence increase for the calculation of the MALT1 activity. During the first 20 min there might be variations in the fluorescence data due to mixing effects.
12. The stock solutions can be stored at 4 °C for a few weeks. Please note that the acrylamide–bis-acrylamide ratio is much higher than in standard commercial pre-mixes.
13. Using an anti-GFP antibody (e.g., ALX 210-199; Enzo LifeSciences) the full-length eYFP-LVSR-eCFP (around 48 kDa) but also the cleaved fragments (around 24 kDa) are detectable since both YFP and CFP are recognized by the antibody [31].

Acknowledgments

We thank Katrin Cabalzar, Maike Jaworski, and Chantal Decaillet for critical reading of the manuscript. Work in the Thome laboratory is supported by the Swiss National Science Foundation, the Swiss Cancer League, the foundations Leenaards and Helmut Horten, the Novartis Foundation for Medical-Biological Research, and a collaboration agreement with Ono Pharmaceuticals.

References

1. Du MQ (2011) MALT lymphoma: many roads lead to nuclear factor-kappaB activation. Histopathology 58(1):26–38
2. Uren AG, O'Rourke K, Aravind LA, Pisabarro MT, Seshagiri S, Koonin EV, Dixit VM (2000) Identification of paracaspases and metacaspases: two ancient families of caspase-like proteins, one of which plays a key role in MALT lymphoma. Mol Cell 6(4):961–967
3. Rosebeck S, Rehman AO, Lucas PC, McAllister-Lucas LM (2011) From MALT lymphoma to the CBM signalosome: three decades of discovery. Cell Cycle 10(15):2485–2496
4. Thome M, Charton JE, Pelzer C, Hailfinger S (2010) Antigen receptor signaling to NF-kappaB via CARMA1, BCL10, and MALT1. Cold Spring Harb Perspect Biol 2(9):a003004
5. Staudt LM (2010) Oncogenic activation of NF-kappaB. Cold Spring Harb Perspect Biol 2(6):a000109
6. Ruland J, Duncan GS, Wakeham A, Mak TW (2003) Differential requirement for Malt1 in T

and B cell antigen receptor signaling. Immunity 19(5):749–758

7. Ruefli-Brasse AA, French DM, Dixit VM (2003) Regulation of NF-kappaB-dependent lymphocyte activation and development by paracaspase. Science 302(5650):1581–1584
8. Thome M (2008) Multifunctional roles for MALT1 in T-cell activation. Nat Rev Immunol 8:495–500
9. Li Q, Verma IM (2002) NF-kappaB regulation in the immune system. Nat Rev Immunol 2(10):725–734
10. Wiesmann C, Leder L, Blank J, Bernardi A, Melkko S, Decock A, D'Arcy A, Villard F, Erbel P, Hughes N, Freuler F, Nikolay R, Alves J, Bornancin F, Renatus M (2012) Structural determinants of MALT1 protease activity. J Mol Biol 419(1–2):4–21
11. Yu JW, Jeffrey PD, Ha JY, Yang X, Shi Y (2011) Crystal structure of the mucosa-associated lymphoid tissue lymphoma translocation 1 (MALT1) paracaspase region. Proc Natl Acad Sci U S A 108(52):21004–21009
12. Snipas SJ, Wildfang E, Nazif T, Christensen L, Boatright KM, Bogyo M, Stennicke HR, Salvesen GS (2004) Characteristics of the caspase-like catalytic domain of human paracaspase. Biol Chem 385(11):1093–1098
13. Sun L, Deng L, Ea CK, Xia ZP, Chen ZJ (2004) The TRAF6 ubiquitin ligase and TAK1 kinase mediate IKK activation by BCL10 and MALT1 in T lymphocytes. Mol Cell 14(3): 289–301
14. Oeckinghaus A, Wegener E, Welteke V, Ferch U, Arslan SC, Ruland J, Scheidereit C, Krappmann D (2007) Malt1 ubiquitination triggers NF-kappaB signaling upon T-cell activation. EMBO J 26(22):4634–4645
15. Lucas PC, Yonezumi M, Inohara N, McAllister-Lucas LM, Abazeed ME, Chen FF, Yamaoka S, Seto M, Nunez G (2001) Bcl10 and MALT1, independent targets of chromosomal translocation in malt lymphoma, cooperate in a novel NF-kappa B signaling pathway. J Biol Chem 276(22):19012–19019
16. Rebeaud F, Hailfinger S, Posevitz-Fejfar A, Tapernoux M, Moser R, Rueda D, Gaide O, Guzzardi M, Iancu EM, Rufer N, Fasel N, Thome M (2008) The proteolytic activity of the paracaspase MALT1 is key in T cell activation. Nat Immunol 9:272–281
17. Coornaert B, Baens M, Heyninck K, Bekaert T, Haegman M, Staal J, Sun L, Chen ZJ, Marynen P, Beyaert R (2008) T cell antigen receptor stimulation induces MALT1 paracaspase-mediated cleavage of the NF-kappaB inhibitor A20. Nat Immunol 9(3):263–271
18. Vercammen D, Declercq W, Vandenabeele P, Van Breusegem F (2007) Are metacaspases caspases? J Cell Biol 179(3):375–380
19. Duwel M, Welteke V, Oeckinghaus A, Baens M, Kloo B, Ferch U, Darnay BG, Ruland J, Marynen P, Krappmann D (2009) A20 negatively regulates T cell receptor signaling to NF-kappaB by cleaving Malt1 ubiquitin chains. J Immunol 182(12):7718–7728
20. Hailfinger S, Nogai H, Pelzer C, Jaworski M, Cabalzar K, Charton JE, Guzzardi M, Decaillet C, Grau M, Dorken B, Lenz P, Lenz G, Thome M (2011) Malt1-dependent RelB cleavage promotes canonical NF-kappaB activation in lymphocytes and lymphoma cell lines. Proc Natl Acad Sci U S A 108(35):14596–14601
21. Weih F, Durham SK, Barton DS, Sha WC, Baltimore D, Bravo R (1996) Both multiorgan inflammation and myeloid hyperplasia in RelB-deficient mice are T cell dependent. J Immunol 157(9):3974–3979
22. Marienfeld R, May MJ, Berberich I, Serfling E, Ghosh S, Neumann M (2003) RelB forms transcriptionally inactive complexes with RelA/p65. J Biol Chem 278(22):19852–19860
23. Mauro C, Pacifico F, Lavorgna A, Mellone S, Iannetti A, Acquaviva R, Formisano S, Vito P, Leonardi A (2006) ABIN-1 binds to NEMO/IKKgamma and co-operates with A20 in inhibiting NF-kappaB. J Biol Chem 281(27): 18482–18488
24. Staal J, Driege Y, Bekaert T, Demeyer A, Muyllaert D, Van Damme P, Gevaert K, Beyaert R (2011) T-cell receptor-induced JNK activation requires proteolytic inactivation of CYLD by MALT1. EMBO J 30(9):1742–1752
25. Uehata T, Iwasaki H, Vandenbon A, Matsushita K, Hernandez-Cuellar E, Kuniyoshi K, Satoh T, Mino T, Suzuki Y, Standley DM, Tsujimura T, Rakugi H, Isaka Y, Takeuchi O, Akira S (2013) Malt1-induced cleavage of regnase-1 in CD4 helper T cells regulates immune activation. Cell 153(5):1036–1049
26. Rosebeck S, Madden L, Jin X, Gu S, Apel IJ, Appert A, Hamoudi RA, Noels H, Sagaert X, Van Loo P, Baens M, Du MQ, Lucas PC, McAllister-Lucas LM (2011) Cleavage of NIK by the API2-MALT1 fusion oncoprotein leads to noncanonical NF-kappaB activation. Science 331(6016):468–472
27. Hachmann J, Snipas SJ, van Raam BJ, Cancino EM, Houlihan EJ, Poreba M, Kasperkiewicz P, Drag M, Salvesen GS (2012) Mechanism and specificity of the human paracaspase MALT1. Biochem J 443(1):287–295
28. Fontan L, Yang C, Kabaleeswaran V, Volpon L, Osborne MJ, Beltran E, Garcia M, Cerchietti L,

Shaknovich R, Yang SN, Fang F, Gascoyne RD, Martinez-Climent JA, Glickman JF, Borden K, Wu H, Melnick A (2012) MALT1 small molecule inhibitors specifically suppress ABC-DLBCL in vitro and in vivo. Cancer Cell 22(6):812–824

29. Boatright KM, Renatus M, Scott FL, Sperandio S, Shin H, Pedersen IM, Ricci JE, Edris WA, Sutherlin DP, Green DR, Salvesen GS (2003) A unified model for apical caspase activation. Mol Cell 11(2):529–541
30. Roschitzki-Voser H, Schroeder T, Lenherr ED, Frolich F, Schweizer A, Donepudi M, Ganesan R, Mittl PR, Baici A, Grutter MG (2012) Human caspases in vitro: expression, purification and kinetic characterization. Protein Expr Purif 84(2):236–246
31. Pelzer C, Cabalzar K, Wolf A, Gonzalez M, Lenz G, Thome M (2013) MALT1 protease activity is controlled by monoubiquitination. Nat Immunol 14:337–345
32. Hailfinger S, Lenz G, Ngo V, Posvitz-Fejfar A, Rebeaud F, Guzzardi M, Penas EM, Dierlamm J, Chan WC, Staudt LM, Thome M (2009) Essential role of MALT1 protease activity in activated B cell-like diffuse large B-cell lymphoma. Proc Natl Acad Sci U S A 106(47): 19946–19951
33. Ferch U, Kloo B, Gewies A, Pfander V, Duwel M, Peschel C, Krappmann D, Ruland J (2009) Inhibition of MALT1 protease activity is selectively toxic for activated B cell-like diffuse large B cell lymphoma cells. J Exp Med 206(11):2313–2320
34. Anderson NL, Anderson NG (1978) Analytical techniques for cell fractions. XXII. Two-dimensional analysis of serum and tissue proteins: multiple gradient-slab gel electrophoresis. Anal Biochem 85(2):341–354

Chapter 12

Leishmania Metacaspase: An Arginine-Specific Peptidase

Ricardo Martin, Iveth Gonzalez, and Nicolas Fasel

Abstract

The purpose of this chapter is to give insights into metacaspase of *Leishmania* protozoan parasites as arginine-specific cysteine peptidase. The physiological role of metacaspase in *Leishmania* is still a matter of debate, whereas its peptidase enzymatic activity has been well characterized. Among the different possible expression systems, metacaspase-deficient yeast cells (*Δyca1*) have been instrumental in studying the activity of *Leishmania major* metacaspase (LmjMCA). Here, we describe techniques for purification of LmjMCA and its activity measurement, providing a platform for further identification of LmjMCA substrates.

Key words *Leishmania*, Cysteine peptidase, Arginine-specific peptidase, Metacaspase, Enzymatic assay, Protease inhibitors

1 Introduction

In 2000, Uren et al. described a group of cysteine proteases orthologous to caspases but absent in mammals, which was named metacaspases [1]. Metacaspases belong to the C14 family of CD clan of cysteine proteases [2] and possess caspase-like domain with a highly conserved catalytic dyad of histidine and cysteine. They are divided into two structurally different types: type I metacaspases with an additional N-terminal extension that is similar to initiator/inflammatory caspases and type-II metacaspases with an insertion of around 200 amino acids between two caspase-like subunits. Interestingly, genomic analysis reveals the presence of a large pool of metacaspases in unicellular and filamentous cyanobacteria that are still poorly studied [3], whereas metacaspases from plants, *Saccharomyces cerevisiae*, and protozoan parasites have rather been extensively investigated.

Depending on the species, *Leishmania* protozoan parasites induce different forms of diseases ranging from cutaneous, mucocutaneous or visceral leishmaniasis. All the *Leishmania* species express a unique type of metacaspase harboring a central catalytic

Peter V. Bozhkov and Guy Salvesen (eds.), *Caspases, Paracaspases, and Metacaspases: Methods and Protocols*, Methods in Molecular Biology, vol. 1133, DOI 10.1007/978-1-4939-0357-3_12, © Springer Science+Business Media New York 2014

domain (containing the conserved catalytic dyad histidine and cysteine) flanked by an N-terminal domain containing a mitochondrial localization signal and a less conserved proline-rich C-terminal domain (61.4–100 % homology), which probably plays a role in protein–protein interactions. Interestingly, although the N-terminal mitochondrial localization signal is functional, most of *Leishmania major* metacaspase (LmjMCA) is detected in the cytoplasm either in a full length or in a processed form corresponding to the central catalytic domain lacking the N- and the C-terminal domains [4].

Due to the mitochondrial localization signal and the proline-rich sequences, LmjMCA N- and C-terminal domains could preclude expression and activity measurement of metacaspase. Therefore, it is necessary to limit expression and activity measurement of LmjMCA to the 251 amino acids (amino-acid residues 63–314 of LmjF35.1580) predictive of the catalytic domain (cd-LmjMCA). To do so, the DNA sequence encoding the catalytic domain was amplified and the PCR product was inserted into the pESC-His vector (Stratagene) using appropriate cloning sites [5]. This vector contains a galactose inducible promoter and N-terminally 6× His and C-terminal FLAG epitope encoding sequences respectively allowing purification with Ni-NTA resin or with murine monoclonal antibodies against the Penta-His-epitope (α-His5; Qiagen) or the FLAG epitope (α-FLAG; Stratagene). A single step was sufficient to enrich for enough material for specific enzymatic activity tests (Subheading 3.6; Fig. 1).

In contrast to caspases that have strict substrate specificity towards aspartic acid, metacaspases rather cleave arginines or lysines at the substrate P1 position [5–8].

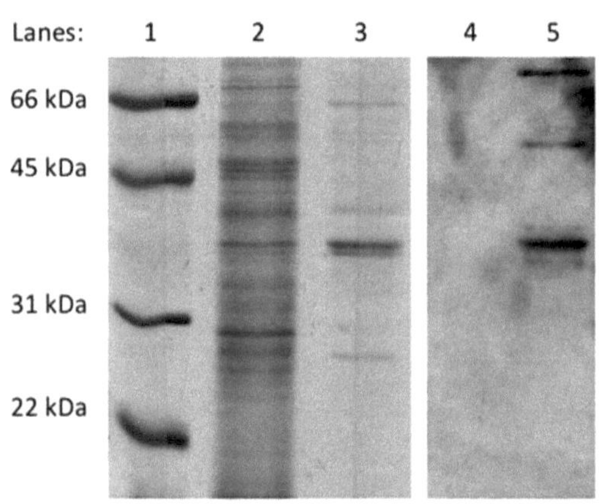

Fig. 1 cd-LmjMCA was purified from yeast expressing cells on an Ni-NTA resin and analyzed by 12 % SDS-PAGE and staining with Coomassie or by immunoblotting using the α-5His antibody. *Lanes 1–3*, Coomassie staining. *Lane 1*, molecular mass markers; *Lane 2*, whole cell lysate; *Lane 3*, cd-LmjMCA purified on Ni-NTA column. *Lanes 4* and *5*, immunoblotting with anti-5His antibody. *Lane 4*, whole cell lysate expressing cd-LmjMCA; *Lane 5*, cd-LmjMCA purified on Ni-NTA column

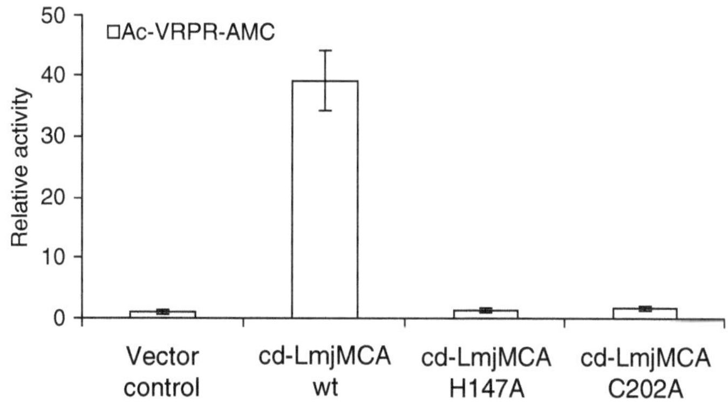

Fig. 2 Enzymatic activity of cd-LmjMCA with the peptidyl substrate Ac-VRPR-AMC. Protein extracts from Δ*yca1* yeast cells transformed with the pESC-His vector alone (vector control) and expressing the catalytic domain of LmjMCA (cd-LmjMCA) wild type (wt) and its respective H147A and C202A mutants, were evaluated for their activity towards Ac-VRPR-AMC substrate. The AMC release was measured every 15 min for 2 h to determine the activity as the slope of the resulting linear regression. Relative activity is expressed as the fold-increase relative to the activity of the vector control. Data show mean ± standard deviation

LmjMCA has been found to be an arginine-specific cysteine protease able to complement the yeast metacaspase (YCA1). In the evaluation of specific recognition of the *A. thaliana* metacaspase AtMC9 using a peptide library, amino acids valine, arginine, proline, and arginine were found to be important in positions P4, P3, P2, and P1, respectively, allowing the design of the optimized tetrapeptide substrate VRPR [9]. To examine the specificity of LmjMCA for this peptide, the catalytic domain of LmjMCA (cd-LmjMCA) can be expressed in *Δyca1* yeast cells and tested with the fluorogenic substrate (Subheading 3.9; Fig. 2).

Enzymatic activity of cd-LmjMCA can be tested in whole yeast cell lysate providing that specific substrates and inhibitors are available. Total protein extracts of *Δyca1* yeast cells expressing cd-LmjMCA were tested for their enzymatic activity with Boc-GRR-AMC, z-GGR-AMC, and Ac-VRPR-AMC substrates in the presence of different inhibitors such as a broad caspase inhibitor z-VAD-fmk, the cysteine protease inhibitor E64, and the serine protease inhibitors PMSF, leupeptin, and aprotinin (Subheading 3.5; Fig. 3). The caspase inhibitor z-VAD-fmk produced a low but significant inhibition of cd-LmjMCA activity with both Boc-GRR-AMC (p value = 0.0008) and z-GGR-AMC (p value < 0.0001) but not with the Ac-VRPR-AMC substrate. The cysteine protease inhibitor E64 had no significant effect on cd-LmjMCA activity with the three substrates. The serine protease inhibitors PMSF and aprotinin had no effect on cd-LmjMCA activity with both Boc-GRR-AMC

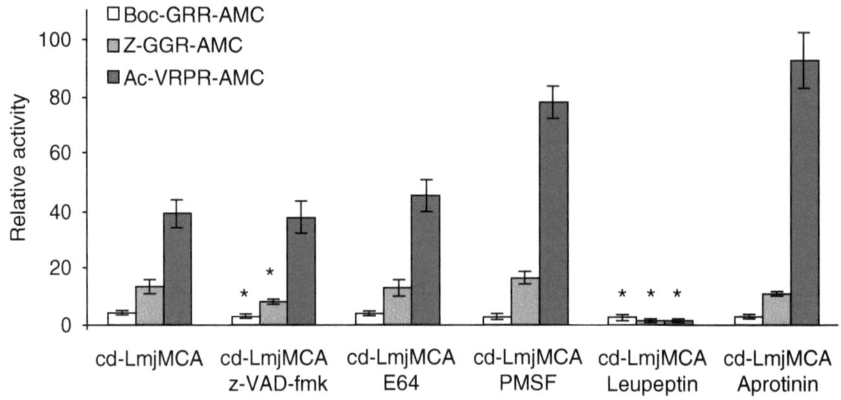

Fig. 3 Effect of protease inhibitors on cd-LmjMCA enzymatic activity. Protein extracts from Δ*yca1* yeast cells transformed with the pESC-His vector expressing the catalytic domain of LmjMCA (cd-LmjMCA) were tested for enzymatic activity with the Boc-GRR-AMC, Z-GGR-AMC, and Ac-VRPR-AMC substrates in absence or presence of 100 μM z-VAD-fmk, 100 μM E64, 10 mM PMSF, 1 mM leupeptin, and 100 μM aprotinin. The AMC release was measured every 15 min for 2 h to determine the activity as the slope of the resulting linear regression. Relative activity was calculated as the fold increase relative to the activity of the vector control (with and without protease inhibitors). Data show mean ± standard deviation. **P* < 0.05

and z-GGR-AMC substrates, however, cd-LmjMCA activity with the Ac-VRPR-AMC substrate was increased when these two latter inhibitors were added. However, this increase was not always observed. The increase of activity of cd-LmjMCA with the Ac-VRPR-AMC substrate in the presence of PMSF and aprotinin could be due to a protective effect over cd-LmjMCA by inhibition of its degradation by other proteases. Since these experiments were done with total protein extracts, the influence of other yeast proteases cannot be excluded. Interestingly, leupeptin, a serine protease inhibitor, which can also inhibit some cysteine proteases such as calpains and cathepsins, completely abrogated cd-LmjMCA activity with all three substrates (Boc-GRR-AMC p-value < 0.0002; z-GGR-AMC p-value < 0.001; and Ac-VRPR-AMC p-value < 0.0001). Although the structural similarity of cd-LmjMCA with caspases could explain the slight inhibition found with z-VAD-fmk for Boc-GRR-AMC and z-GGR-AMC, this inhibitor was not able to affect the activity of cd-LmjMCA towards Ac-VRPR-AMC, the most preferred substrate of this metacaspase (Fig. 3).

2 Materials

All chemicals used are of Molecular Biology grade unless specified and solutions are prepared with deionized water. When not specified, incubations are performed at room temperature.

2.1 *Leishmania* Metacaspase Gene

1. *L. major* metacaspase gene: LmjF.35.1580 (Gene ID: 3684453).

2.2 YCA1 Disrupted Yeast Cells Expressing cd-LmjMCA

1. Metacaspase disrupted yeast cells: Euroscarf YCA1 disrupted strain (*yca1Δ* cells) Accession Number Y02453 (BY4741; MAT a; his3Δ1; leu2Δ0; met15Δ0; ura3Δ0; YOR197w::kanMX4) transformed with the pESC-His vector (Stratagene, La Jolla) or with the pESC-His expressing LmjMCA or its inactive forms cd-LmjMCA H147A and cd-LmjMCA C202A [5].
2. Inactive forms of cd-LmjMCA obtained by using the QuikChange®multi site-directed mutagenesis kit (Stratagene).

2.3 Yeast Media and Transformation

1. YPD medium: 20 g/l Difco peptone, 10 g/l Yeast extract, H_2O to 950 ml, pH 6.5. Autoclave, let cool down to 55 °C, and add 50 ml of 40 % glucose filtered through a 0.22 μm size filter (Final concentration of glucose: 2 %).
2. YPD plates: same as YPD medium but supplemented with 20 g/l agar.
3. 10× Dropout amino acid solution (without histidine when using the pESC-His vector): 200 mg/ml L-adenine hemisulfate salt, 200 mg/ml L-arginine HCl, 200 mg/ml L-histidine, 300 mg/ml L-isoleucine, 1,000 mg/ml L-leucine, 300 mg/ml L-lysine HCl, 200 mg/ml L-methionine, 500 mg/ml L-phenylalanine, 2,000 mg/ml L-threonine, 200 mg/ml L-tryptophan, 300 mg/ml L-tyrosine, 200 mg/ml L-uracil, 1,500 mg/ml L-valine. Pass the solution through a 0.22 μm size filter and aliquot in 50-ml tubes. Store at 4 °C.
4. SD/DO/Glucose medium: for 1 L weigh 6.7 g of Yeast nitrogen base without amino acids and add H_2O to 850 ml. Control the pH (pH 5.8). Autoclave and let cool down to 55 °C. Add 100 ml of the 10× DO solution (without histidine) and then 50 ml of 40 % glucose or galactose (filtered; final concentration 2 %).
5. SD/DO/Glucose plates: same as SD/DO medium (without histidine) supplemented with 20 g/l agar.
6. 10× TE: 0.1 M Tris–HCl, 10 mM EDTA, pH 7.5. Pass the solution through a 0.22 μm filter and store at −20 °C.
7. 10× LiAc: 1 M LiAc, pH 7.5. Pass the solution through a 0.22 μm filter and store at −20 °C.
8. 1× TE/1× LiAc solution: 500 μl 10× TE, 500 μl 10× LiAc, adjust to 4 ml with H_2O. Use freshly prepared solution.
9. PEG 1,000/Tris/LiAc solution: 4 ml of 50 % PEG 1,000 solution, 500 μl 10× TE, and 500 μl 10× LiAc.
10. 87 % glycerol.

2.4 Yeast Lysis (TCA Protocol) and Protein Extraction (Glass Beads) for SDS-PAGE Analysis

1. Solution B: 3.67 ml H_2O, 925 μl 10 M NaOH, 370 μl β-Mercaptoethanol, 50 μl 100 mM PMSF in isopropanol, 50 μl 0.5 M EDTA/KOH, pH 7.0.
2. Trichloroacetic acid (TCA).
3. Acetone.
4. Lysis buffer: 0.5 % NP40, 20 mM HEPES, pH 8.0, 84 mM KCl, 10 mM $MgCl_2$, 0.2 mM EDTA, 0.2 mM EGTA, 1 mM DTT, 5 μg/ml Aprotinin, 5 μg/ml Leupeptin, 1 μg/ml Pepstatin, 1 mM PMSF.
5. Pierce BCA protein assay kit (Thermo Scientific) with BSA as standard.

2.5 Enzymatic Activity Test in Whole or Purified Cell Lysates

1. Lysis buffer for whole cell lysate activity test: 50 mM KH_2PO_4, pH 7.5, 500 mM NaCl, 1 mM EDTA, 5 mM DTT, 1 % CHAPS, 5 μg/ml Aprotinin.
2. Lysis buffer for purified cell lysate activity test: 50 mM NaH_2PO_4, pH 8.0, 300 mM NaCl, 10 mM imidazole, 1 % Triton X-100.
3. Washing Buffer for purified cell lysate activity test: 50 mM NaH_2PO_4, pH 8.0, 300 mM NaCl, 20 mM imidazole.
4. Elution Buffer for purified cell lysate activity test: 50 mM NaH_2PO_4, pH 8.0, 300 mM NaCl, 250 mM imidazole).
5. Amicon Ultra-4 centrifugal device (Millipore).
6. Base buffer: 150 mM NaCl, 25 mM HEPES, 10 % glycerol, pH 7.5. Store at 4 °C.
7. 10 % CHAPS solution: 1 g CHAPS in 10 ml H_2O. Store at 4 °C.
8. Activity buffer for whole cell lysate activity test: 3.92 ml Base buffer, 40 μl 10 % CHAPS, 20 μl 2 M DTT.
9. Activity buffer for purified cell lysate activity test: 3.52 ml Base buffer, 40 μl 10 % CHAPS, 20 μl 2 M DTT, 400 μl 1 M $CaCl_2$.
10. Fluorogenic substrates: Boc-Gly-Arg-Arg-7-amino-4-methylcoumarin (Boc-GRR-AMC), Z-Gly-Gly-Arg-7-amino-4-methylcoumarin (Z-GGR-AMC) (both from Bachem AG, Switzerland), and *N*-acetyl-Val-Arg-Pro-Arg-7-amino-4-methylcoumarin (Ac-VRPR-AMC) (Sigma).
11. Protease inhibitors: Benzyloxycarbonyl-Val-Ala-Asp (*1010*) fluoromethylketone (z-VAD-fmk), E64, PMSF, leupeptin, and aprotinin.
12. Trypsin powder (Boehringer Manheim GmbH, Germany).
13. 96-well black plates (Optiplate-96 F, PerkinElmer).
14. Spectrophotometer.

2.6 Sodium Dodecyl Sulfate–Polyacrylamide Gel Electrophoresis (SDS-PAGE)

1. The Bio-Rad Power Pac 3000 system or similar.
2. Separating gel (12 %): mix 1.625 ml H_2O, 2 ml 30 % acrylamide–Bis-acrylamide solution (29.2:0.8 acrylamide–Bis-acrylamide), 1.3 ml 1.5 M Tris/HCl buffer, pH 8.8, 25 μl 20 % SDS; 50 μl 10 % ammonium persulfate (APS) and 2 μl tetramethylethylenediamine (TEMED).
3. Stacking gel (3.9 %): mix 2.64 ml H_2O, 0.67 ml 30 % acrylamide–Bis-acrylamide solution, 0.5 ml 1 M Tris/HCl buffer, pH 6.8, 20 μl 20 % SDS; 40 μl 10 % APS, 8 μl TEMED, and 20 μl bromophenol blue.
4. 4× Tris/HCl/SDS solution: 91 g Tris base, 2 g SDS, 500 ml H_2O, pH 8.8.
5. 2× SDS sample loading buffer (100 ml final volume): 25 ml 4× Tris/HCl/SDS solution, 20 ml glycerol, 4 g SDS, 2 ml β-Mercaptoethanol (or 3.1 g DTT), 1 mg bromophenol blue, adjust to 100 ml with H_2O.
6. 10× SDS electrophoresis running buffer: 250 mM Tris base, 1.92 M glycine, 1 % SDS. Adjust pH to pH 8.3 when diluting to 1×.
7. Gel staining solution: 50 % methanol, 0.05 % Coomassie Brilliant Blue R-250, 10 % acetic acid.
8. Gel destaining solution: 5 % methanol, 7 % acetic acid.

2.7 Western Blotting

1. The Bio-Rad Power Pac 3000 system or similar.
2. 10× gel transfer buffer: 250 mM Tris base, 1.92 M glycine.
3. 1× gel transfer buffer solution: 10 ml 10× gel transfer buffer, 20 ml methanol, pH 8.3–8.4, adjust to 1 L with H_2O.
4. Nitrocellulose membrane (Whatman, GE Healthcare Life sciences or similar).
5. Filter paper, cut to the size of the gel.
6. Ponceau S Solution: 0.5 g Ponceau S, 1 ml glacial acetic acid, adjust to 100 ml with H_2O.
7. 1× Tris Buffered Saline supplemented with tween 20 (TBST): 25 mM Tris, 150 mM NaCl, 3 mM KCl, pH 7.5, 0.1 % Tween 20.
8. Blocking buffer: 1× TBST, 5 % nonfat dry milk (NFDM).
9. Primary antibody: 1/200 anti-Flag antibody or 1/1,000 anti-Histidine antibody in TBST with 1 % NFDM.
10. Secondary antibody: 1/2,500 horseradish peroxidase-conjugated antibody in TBST with 1 % NFDM.
11. Glass plate.
12. ECL Western Blotting Detection Reagent (GE Healthcare).

13. Filter papers.
14. Plastic wrap.
15. Cassette and X-Ray film.

3 Methods

3.1 Yeast Transformation

1. Plate *Δyca1* cells (*see* **Note 1**) from frozen stock onto YPD plates using a platinum loop, which has been previously sterilized by flaming and then cooled quickly on the plate.
2. Incubate at 30 °C for 4 days and then inoculate 1 ml of YPD medium with 1 colony and vortex for 2 min.
3. Transfer to 49 ml of YPD medium (total volume 50 ml) and place on a shaker at 30 °C overnight.
4. The next day, dilute the overnight culture to OD_{600} 0.2–0.3 in 300 ml (*see* **Note 2**) and further incubate at 30 °C with shaking for 2 h or until OD_{600} reaches 0.4–0.6.
5. Centrifuge at 1,000 × *g* for 5 min in 50 ml tubes, dilute and pool pellets in 50 ml H_2O, centrifuge at 1,000 × *g* for 5 min at room temperature.
6. Resuspend pellet in 1.5 ml of 1× TE/1× LiAc fresh solution.
7. Add 10 μl of 10 mg/ml herring sperm carrier DNA in a 1.5-ml vial, heat at 95 °C for 5 min and quick chill on ice.
8. Leave on ice and add 1 μg of cd-LmjMCA plasmid and mix.
9. Add 100 μl of yeast cell suspension and vortex.
10. Add 600 μl of PEG1000/Tris/LiAc fresh solution and vortex for 10 s.
11. Incubate at 30 °C with shaking for 30 min.
12. Add 70 μl of DMSO from stock solution and mix by inversion at 42 °C for 15 min (heat shock).
13. Leave on ice for 2 min, then microfuge at 10,000 × *g* for 5 s.
14. Resuspend the pellet in 500 μl of 1× TE.
15. Dilute with 1× TE and plate 100 μl of dilutions 1:1, 1:10, 1:100, and 1:1,000 on YPD plates and incubate at 30 °C for 3 days to obtain colonies.
16. Verify that the transformation was efficient and that your cells have the desired plasmid by using standard minilysate protocol.
17. Grow overnight culture: inoculate one transformed colony into 1 ml of SD/DO/Glucose medium, vortex, transfer to 9 ml of SD/DO/Glucose medium, and incubate at 30 °C with continous shaking overnight.
18. Prepare frozen stock of transformed yeast cells: mix 700 μl of the overnight culture and 300 μl of 87 % glycerol, mix and store at −70 °C.

3.2 Induction of the cd-LmjMCA Expression in Transformed Yeast Cells

1. Inoculate one transformed colony into 1 ml of SD/DO/Glucose medium and vortex.
2. Transfer to 9 ml of SD/DO/Glucose medium and incubate at 30 °C with continous shaking overnight.
3. Dilute overnight culture to OD_{600} 0.05–0.1 in 10 ml (*see* **Note 3**) in a 100-ml Erlenmeyer flask (ten times culture volume) and incubate at 30 °C with shaking for 6 h or until OD_{600} reaches 0.4–0.6.
4. Centrifuge 1 ml of culture at 10,000 × *g* for 1 min and store pellet at −70 °C (non induced control).
5. For the galactose induction, centrifuge the culture at 1,000 × *g* for 5 min and dilute the pellet with 10 ml of SD/DO/Galactose medium and then incubate at 30 °C with shaking overnight.
6. Measure OD_{600} after at least 16 h of induction.
7. Centrifuge the culture at 1,000 × *g* for 5 min and store pellet at −70 °C (galactose induced culture) until use.
8. The pellets are ready for lysis and analysis.

3.3 Yeast Lysis (TCA Protocol) for SDS-PAGE Analysis

1. Dilute frozen pellet of the 10 ml cultures (non-induced and galactose induced) with 500 μl of 1× TE and centrifuge at 10,000 × *g* for 1 min at 4 °C.
2. Resuspend the pellet with 500 μl of H_2O and add 75 μl of Solution B.
3. Shake 10 min at 4 °C and add 280 μl of 72 % TCA.
4. Put on ice for 5 min and then centrifuge at 10,000 × *g* for 10 min at 4 °C.
5. Add 700 μl of acetone to the pellet and centrifuge at 10,000 × *g* for 10 min at 4 °C.
6. Repeat the wash with acetone.
7. Let dry the pellet and then resuspend in 50 μl of 1× PBS (*see* **Note 4**).
8. Store at −70 °C until use.

3.4 Yeast Protein Extraction (Glass Beads)

1. Dilute frozen pellet with 50 μl of lysis buffer, transfer to a 1.5-ml vial with 0.08 g glass beads.
2. Vortex ten times, 1 min each, and collect supernatant.
3. Wash beads with 50 μl of lysis buffer and collect supernatant.
4. Pool supernatants and centrifuge at 10,000 × *g* for 1 h at 4 °C.
5. Collect and store supernatant at −70 °C in lysis buffer containing protease inhibitors.
6. Protein concentration in supernatant can be measured using a BCA protein assay reagent with BSA as standard.

3.5 cd-LmjMCA Enzymatic Activity in Whole Yeast Cell Lysate

1. Harvest transformed yeast cells from a 50 ml culture following 24 h of induction. The pellet can be kept frozen at –70 °C.
2. Resuspend the frozen pellet in 100 µl of lysis buffer, transfer to a 1.5-ml vial and add 0.08 g of glass beads.
3. Vortex ten times, 1 min each.
4. Collect and save supernatant.
5. Wash the beads with 50 µl of lysis buffer, collect and save supernatant.
6. Pool supernatants and centrifuge at 10,000 × *g* for 1 h at 4 °C. Collect and store supernatant at –70 °C in lysis buffer containing protease inhibitors.
7. Measure protein concentration in the supernatant using a BCA protein assay reagent with BSA as standard.
8. For one black-plate well, add 196 µl of Activity buffer and 4 µl of 10 µg/µl total protein extract (40 µg total protein per well). Prepare duplicate or triplicate wells.
9. Dilute 50 mM of substrate-AMC to 5 mM with Activity buffer and add 2 µl of diluted substrate per well (final concentration 50 µM). Read fluorescence each 15 min for 2 h at 24 °C with 360 nm excitation and 460 nm emission wavelengths.
10. As a positive control use 10 ng of trypsin per well in the 200 µl reaction volume. As negative controls, use protein extracts from yeast cells transformed with the pESC-His vector or expressing cd-LmjMCA (H147A) and cd-LmjMCA (C202A).
11. Determine enzymatic activity by calculating the slope of the linear regression. Express results in arbitrary milli-fluorescence units per minute per µg of protein (mFU/min/µg), or as the fold increase relative to the activity of the vector control (*see* **Note 5**).
12. To test the effect of different protease inhibitors on the enzymatic activity, supplement activity reactions with the following concentrations of inhibitors: 100 µM z-VAD-fmk, 100 µM E64, 10 mM PMSF, 1 mM leupeptin, and 100 µM aprotinin.

3.6 Purification of Leishmania Metacaspase Catalytic Domain (cd-LmjMCA) from Yeast on Ni-NTA Resin

1. Resuspend frozen pellet from a 500 ml culture after induction with galactose for 18 h in 2.5 ml of lysis buffer.
2. Add 2.5 g of glass beads (0.25–0.5 mm) and vortex ten times, 1 min each (*see* **Note 6**).
3. Collect and save supernatant.
4. Wash the glass beads with 2.5 ml of lysis buffer, collect and save supernatant.
5. Pool the supernatants, centrifuge at 10,000 × *g* for 1 h at 4 °C, and save supernatant (contains soluble proteins).

6. Wash 1 ml 50 % Ni-NTA resin with 2 ml of lysis buffer and add the supernatant (soluble proteins) to the washed resin.
7. Incubate overnight at 4 °C on a wheel.
8. Centrifuge at 1,000 × *g* for 5 min at 4 °C, wash the resin twice with 500 μl of Washing Buffer.
9. Elute protein by adding three aliquots of 500 μl of Elution Buffer and then pool the eluates.
10. Centrifuge the three pooled elutions at 10,000 × *g* for 1 min at 4 °C.
11. Pool the supernatants and concentrate eluted proteins in 1× PBS with an Amicon Ultra-4 centrifugal device prior to protein concentration measurement.
12. Store at −80 °C until use for the activity test.

3.7 Sodium Dodecyl Sulfate–Polyacrylamide Gel Electrophoresis (SDS-PAGE)

1. Wash gel glass plates and mount the electrophoresis system according to manufacturer's protocol.
2. Prepare separating gel, fill to the three quarters the glass plate, add some isopropanol on the gel to obtain a flat surface and wait for the gel to polymerize.
3. Prepare stacking gel, fill the gel glass plate up to the edge, insert the comb and wait for the gel to polymerize.
4. Mix each sample (20 μg of total protein from yeast lysates) with 2× SDS sample loading buffer in a ratio 1:1 (v/v), boil samples for 5 min at 95 °C, spin in microfuge and load on the gel.
5. Run gel for 20 min at 80 V and then for 45 min at 180 V with chamber on ice.
6. Stain the gel with Coomassie Blue (*see* **Note 7**): Soak the gel in a staining solution and incubate with shaking at room temperature for 1 h to overnight.
7. Soak the gel in a destaining solution and incubate with shaking at room temperature for 30 min. Repeat until background disappears. Store the gel in water or dry (*see* **Notes 8** and **9**).

3.8 Western Blotting

1. Equilibrate the gel, four filter papers, and sponges in 1× transfer buffer.
2. Mount a sandwich in the following way: white sponge, two filter papers, nitrocellulose membrane, gel, two filter papers, green sponge (white sponge oriented to the cathode—red face).
3. Remove bubbles by rolling a 15-ml tube over the sandwich.
4. Run in 1× transfer buffer for 1 h at 100 V with chamber on ice.
5. After protein transfer, incubate the nitrocellulose membrane on a shaker at room temperature for 5 min in Ponceau S Solution.

6. Incubate on the shaker at room temperature for 2 min in water to remove excess of Ponceau S red.
7. Take a picture and mark molecular weights with a pencil.
8. Incubate on a shaker at room temperature for 10 min in water to complete destaining.
9. Incubate the membrane on the shaker for 1 h at room temperature or overnight at 4 °C in the blocking buffer.
10. Incubate the membrane on a shaker overnight at 4 °C with the first antibody.
11. Wash four times for 15 min each with TBST.
12. Incubate the membrane on a shaker for 1 h at room temperature with the secondary antibody.
13. Wash four times for 15 min each with TBST (*see* **Note 10**).
14. For membrane development, deposit the membrane on a clean glass plate.
15. Dry quickly with a filter paper.
16. Overlay 1.5 ml of a developing solution (1:1 of ECL solutions A:B for a 0.125 ml/cm^2 membrane) and wait for 2 min.
17. Dry with a filter paper and cover the membrane with a plastic wrap.
18. Insert the membrane in a cassette and expose to an X-Ray film for different times (e.g. 2 s, 10 s, 2 min, 10 min); develop the film (*see* **Note 11**).

3.9 cd-LmjMCA Activity Measurement with the Ac-VRPR-AMC Substrate

1. Use 1 μg of purified protein in a total 200 μl volume per well of a 96-well black plate.
2. Add 196 μl of activity buffer and 1–4 μl of Ni-NTA purified cd-LmjMCA per well. Prepare duplicate or triplicate wells.
3. Dilute Ac-VRPR-AMC in Activity buffer to the final concentration 5 mM and add 2 μl of diluted substrate per well (final concentration 50 μM). Read fluorescence each 15 min for 2 h at 24 °C with 360 nm excitation and 460 nm emission wavelengths.
4. As a positive control use 10 ng of trypsin per well in the 200 μl reaction volume.
5. Determine enzymatic activity by calculating the slope of the linear regression. Express results in arbitrary milli-fluorescence units per minute per μg of protein (mFU/min/μg), or as the fold increase relative to the activity of the vector control (*see* **Note 5**).

4 Notes

1. *Δyca1* cells can be obtained from Euroscarf Accession Number Y02453.
2. ~10 ml overnight culture in 290 ml of YPD medium.
3. ~400 μl overnight culture in 9.6 ml of SD/DO/Glucose medium
4. Protein concentration can be measured using the BCA protein assay reagent (Pierce Biotechnology, Inc., Rockford, IL) with BSA as standard.
5. Enzymatic activity tests must be performed at least three times and means and standard deviations must be calculated. The Student *t* test is used in statistical analysis and significance is considered when $p < 0.05$.
6. Use 1 g of beads per 1 ml of lysis buffer.
7. Detection limit is 0.3–1 μg/protein band.
8. First dilute Coomassie Blue in methanol.
9. If you want to keep your gel (after staining), we suggest to put a plastic sheet (candy wrap plastic) over a filter paper; Put the gel over the plastic sheet and soak with water; Put another plastic sheet over the gel; Perforate with a needle around the gel border; Put it on the desiccator at 70 °C under vacuum for 1 h.
10. For competitive blot, first incubate with the peptide at 10 μg/ml, then add antibody and incubate on wheel 60 for min at room temperature.
11. Stripping membranes: 15 min shaking in 0.1 M of glycine–HCl pH 2–3; rinse with 1 M of NaCl in 1× PBS; wash 2 × 5 min in 1× TBS–0.1 % Tween-20; rinse with H_2O; expose film for 10 min to detect former signal. If there is no signal the membrane is ready to be blocked and exposed to a new primary antibody.

Acknowledgements

The authors are grateful to the members of the Fasel's group and to Dr. Frank Madeo who provided the pFM21 construct, which served as a control. This work was funded by the grants FNRS N° 3100A0-116665/1 and IZ70Z0-131421 to NF.

References

1. Uren AG, O'Rourke K, Aravind LA, Pisabarro MT, Seshagiri S, Koonin EV, Dixit VM (2000) Identification of paracaspases and metacaspases: two ancient families of caspase-like proteins, one of which plays a key role in MALT lymphoma. Mol Cell 6(4):961–967
2. Mottram JC, Helms MJ, Coombs GH, Sajid M (2003) Clan CD cysteine peptidases of parasitic protozoa. Trends Parasitol 19(4):182–187
3. Jiang Q, Qin S, Wu QY (2010) Genome-wide comparative analysis of metacaspases in unicellular and filamentous cyanobacteria. BMC Genomics 11:198. doi:10.1186/1471-2164-11-198
4. Zalila H, Gonzalez IJ, El-Fadili AK, Delgado MB, Desponds C, Schaff C, Fasel N (2011) Processing of metacaspase into a cytoplasmic catalytic domain mediating cell death in Leishmania major. Mol Microbiol 79(1):222–239. doi:10.1111/j.1365-2958.2010.07443.x
5. Gonzalez IJ, Desponds C, Schaff C, Mottram JC, Fasel N (2007) Leishmania major metacaspase can replace yeast metacaspase in programmed cell death and has arginine-specific cysteine peptidase activity. Int J Parasitol 37(2):161–172. doi:10.1016/j.ijpara.2006.10.004
6. Vercammen D, van de Cotte B, De Jaeger G, Eeckhout D, Casteels P, Vandepoele K, Vandenberghe I, Van Beeumen J, Inze D, Van Breusegem F (2004) Type II metacaspases Atmc4 and Atmc9 of Arabidopsis thaliana cleave substrates after arginine and lysine. J Biol Chem 279(44):45329–45336. doi:10.1074/jbc.M406329200
7. Watanabe N, Lam E (2005) Two Arabidopsis metacaspases AtMCP1b and AtMCP2b are arginine/lysine-specific cysteine proteases and activate apoptosis-like cell death in yeast. J Biol Chem 280(15):14691–14699. doi:10.1074/jbc.M413527200
8. Lee N, Gannavaram S, Selvapandiyan A, Debrabant A (2007) Characterization of metacaspases with trypsin-like activity and their putative role in programmed cell death in the protozoan parasite Leishmania. Eukaryot Cell 6(10):1745–1757. doi:10.1128/EC.00123-07
9. Vercammen D, Belenghi B, van de Cotte B, Beunens T, Gavigan JA, De Rycke R, Brackenier A, Inze D, Harris JL, Van Breusegem F (2006) Serpin1 of *Arabidopsis thaliana* is a suicide inhibitor for metacaspase 9. J Mol Biol 364(4):625–636. doi:10.1016/j.jmb.2006.09.010

Chapter 13

Purification, Characterization, and Crystallization of *Trypanosoma* Metacaspases

Karen McLuskey, Catherine X. Moss, and Jeremy C. Mottram

Abstract

Metacaspases are cysteine peptidases found in trypanosomes but absent in mammals, and despite being distantly related to the mammalian caspases they show significant disparity in their cellular and enzymatic functions. The genome of the parasitic protozoa *Trypanosoma brucei* (the causative agent of African sleeping sickness) encodes five metacaspases: TbMCA1-TbMCA5. Of these TbMCA2, TbMCA3, and TbMCA5 are active cysteine peptidases expressed in the bloodstream form of the parasite. To investigate the structure–function relationship of the trypanosome metacaspases and the structural basis for their divergence from the caspases, paracaspases, and other Clan CD cysteine peptidases (or *vice versa*), we purified and characterized TbMCA2 and determined the three-dimensional structure of an inactive mutant using X-ray crystallography. The methods presented in this chapter describe the recombinant expression of active TbMCA2 and inactive TbMCA2^{C213A}. The protocols produce large amounts of recombinant protein for use in structural, biochemical, and kinetic studies and include detailed information on how to produce diffraction quality crystals of TbMCA2^{C213A}.

Key words Trypanosome, Metacaspase, TbMCA2, Parasite, X-ray crystallography, Protein purification, Enzyme assay, Cysteine protease

1 Introduction

Metacaspases are cysteine peptidases, which are distantly related to the mammalian caspases. They share a conserved histidine/cysteine (His/Cys) catalytic dyad and a basic caspase hemoglobinase fold, but metacaspases are only found in phylogenetic kingdoms lacking caspases such as plants, yeast, and lower eukaryotes including trypanosomes. A number of plant metacaspases have been shown to play a role in cell death pathways [1], suggesting a link with the cell death control exhibited by the caspases, which are key players in mammalian programmed cell death. However, trypanosomatid parasitic protozoa such as *Trypanosoma brucei* and *Leishmania*, the causative agents of Human African Trypanosomiasis and Leishmaniasis, respectively, appear to lack such regulated cell death [2].

Peter V. Bozhkov and Guy Salvesen (eds.), *Caspases, Paracaspases, and Metacaspases: Methods and Protocols*, Methods in Molecular Biology, vol. 1133, DOI 10.1007/978-1-4939-0357-3_13, © Springer Science+Business Media New York 2014

Indeed, trypanosomatid metacaspases seemed to have evolved alternative functions, e.g., as a negative regulator of amastigote proliferation in *Leishmania mexicana* [3] and as a pseudopeptidase virulence factor in *T. brucei* (TbMCA4) [4]. In addition, metacaspases from other species have also been implicated in a variety of cellular functions including clearance of protein aggregates [5], endoplasmic reticulum stress and cell proliferation (reviewed in [6]), suggesting a significant divergence from the cellular processes defined for the caspases.

Along with their divergent roles in host cells, biochemical classification of metacaspases and caspases has revealed that they form two functionally distinct groups of cysteine peptidases, differing in both their substrate specificities and modes of activation. Caspases are homodimeric aspartic acid-specific peptidases for which dimerization is critical, and proteolytic processing is often necessary for activation [7–9]. Conversely, metacaspases are arginine/lysine-specific fully functional monomers that do not always require processing for activation but are activated by calcium [10–13]. Interestingly, the genome of *T. brucei* encodes five metacaspases (denoted TbMCA1–TbMCA5) which share between 40 and 89 % primary sequence identity [14] and can have distinct in vivo functions [4, 15]. Metacaspases TbMCA2, TbMCA3, and TbMCA5 all encode proteins with the canonical His/Cys dyad and as a group are thought to be essential for the bloodstream form of the parasite [15]. TbMCA2 and TbMCA3 share 89 % sequence identity (differing only in their N-terminal regions), whereas TbMCA5 contains a large (~16 kDa) C-terminal extension. Interestingly, TbMCA1 and TbMCA4 contain amino acid substitutions in their catalytic residues (His to Tyr and Cys to Ser, respectively). In TbMCA4, even though the substitution results in a catalytically inactive pseudopeptidase, it still functions as a membrane-linked virulence factor, which is processed by TbMCA3 [4]. TbMCA4 provides an illustration of the functional diversity of the trypanosome metacaspases and the challenges faced in deciphering individual functions.

As part of a *T. brucei* drug discovery program we undertook to further understand the structural and functional relationships of trypanosome metacaspases, and to allow an accurate structural comparison with the caspases, paracaspases, and other Clan CD [16] cysteine peptidases. A large part of this work included determining the three dimensional crystal structure of a metacaspase from *T. brucei* and to this end we successfully obtained the X-ray crystal structure of an inactive mutant of TbMCA2 (TbMCA2^{C213A}) to around 1.5 Å resolution ([17]; PDB ID: 4AFR). The structure revealed a core caspase fold with an eight-stranded β-sheet that stabilized the enzyme as a monomer and a well-ordered N-terminus, which wrapped around the molecule covering the catalytic dyad. In addition, biochemical and kinetic assays along with samarium binding studies were used to identify the calcium binding site, study potential TbMCA2 substrates, and identify Z-VRPR-FMK as a TbMCA2 inhibitor [17]. These results collectively gave new insights into the mode of activation and substrate specificity of metacaspases and in particular of TbMCA2.

This chapter presents a detailed overview of the methods used in our laboratory to overexpress, purify, and crystallize recombinant $TbMCA2^{C213A}$. It includes details on crystal improvement using seeding techniques and on obtaining a heavy atom derivative using samarium in order to investigate the structure of the calcium-binding site. In addition, it describes the use of fluorogenic substrate enzyme assays to calculate the activity of TbMCA2 and/or the K_m of a chosen substrate. It also describes how to perform inhibitor-binding assays and calculate the IC_{50} for inhibitors. Many of the methods presented here will be useful when working with metacaspases from other species and for Clan CD cysteine peptidases in general.

2 Materials

Expression plasmids for this work were originally produced using the TbMCA2 coding sequence, amplified from *T. brucei* genomic DNA, inserted into the NdeI and XhoI sites of pET28a+ (Novagen). This produced an enzyme with an N-terminal His-tag and thrombin cleavage site, the latter of which was subsequently mutated to remove the cleavage site arginine residue [10] (*see* **Note 1**). All mutations to TbMCA2 were constructed from this plasmid (pGL1573) using the QuickChange Site-Directed Mutagenesis Kit (Stratagene) and the appropriate oligonucleotide primers [17].

All solutions should be prepared with ultrapure water and stored at room temperature unless otherwise stated. In general, all chemicals should be of ACS reagent grade of 99 % purity, prepared in double distilled water (ddH_2O), and sterile filtered before use.

2.1 Transformation of Cells and Protein Expression

1. Expression vector for protein expression containing the metacaspase genes and an N-terminal His-tag for protein purification (*see* **Note 2**).
2. Competent cells, *E. coli* BL21 (DE3) or Rosetta (DE3) (e.g., One Shot BL21 (DE3) chemically competent *E. coli*, Invitrogen; Rosetta 2 (DE3) Singles, Millipore) (*see* **Note 3**).
3. Magnetic stirrer and stirring bar.
4. Sterile bacterial cell spreaders.
5. Water bath set at 42 °C and a shaking incubator set at 37 °C.
6. Kanamycin stock (1,000×, 25 mg/mL) prepared in ddH_2O, sterile filtered, and stored at −20 °C (*see* **Note 4**).

7. Prewarmed LB Agar plates: Dissolve 1 % (w/v) tryptone, 0.5 % (w/v) yeast extract, 1 % (w/v) NaCl, and 1.5 % (w/v) microagar in ddH_2O and sterilize by autoclaving. Allow the mixture to cool to around 50 °C before adding 25 μg/mL kanamycin. Gently mix by swirling and pour a thin layer (~10 mL and 5 mm thick) into a sterile petri dish. Allow the plates to cool completely before storing upside down at 4 °C and prewarm to room temperature before use.
8. Prewarmed SOC medium: 2 % (w/v) tryptone, 0.5 % (w/v) yeast extract, 10 mM NaCl, 2.5 mM KCl, 10 mM $MgSO_4$, 10 mM $MgCl_2$, 20 mM glucose prepared in ddH_2O and filter-sterilized. Store at 4 °C and prewarm to room temperature before use.
9. Auto induction medium: Add 1 L (*see* **Note 5**) of ddH_2O and 10 mL of glycerol to 60 g Overnight Express Instant TB medium (Novagen) and stir using a magnetic stirrer until the medium is dissolved. Microwave to sterilize and allow cooling to at least 50 °C before adding kanamycin to 25 μg/mL. This medium can be stored at 4 °C and should be prewarmed to room temperature before use.

2.2 SDS-PAGE Analysis

The Novex NuPAGE SDS-PAGE Gel System (Invitrogen) is described here and while any SDS-PAGE system would suffice the percentage of the gels and the chosen running buffer are important for different parts of the analysis.

1. Mini gel system for SDS-PAGE analysis (XCell Surelock, Invitrogen).
2. SDS-PAGE gels (10 % and 4–12 %) (NuPAGE Bis-Tris Gels, Invitrogen).
3. MOPS and MES SDS running buffers (NuPage 20× solution, Invitrogen) diluted to a 1× working solution using ddH_2O (MOPS and MES Buffers, respectively).
4. Coomassie protein strain for SDS-PAGE gels (SimplyBlue SafeStain, Invitrogen).
5. Sample buffer (NuPAGE LDS Sample Buffer (4×), Invitrogen).
6. Sample reducing agent (NuPAGE Reducing Agent (10×), Invitrogen).
7. General reducing agent (NuPAGE antioxidant, Invitrogen).
8. Protein molecular weight markers (SeeBlue Plus2, Invitrogen).
9. Microcentrifuge tubes.

2.3 Protein Purification

All purification buffers should be filtered (0.2 μm) before use and additionally buffers for size exclusion chromatography (gel filtration) should be degassed for around 20 min.

1. ÄKTA liquid chromatography system (GE Healthcare) or similar protein purification system.
2. Bottle top vacuum filter (e.g., Nalgene reusable bottle top filters).
3. Cellulose nitrate membrane filters (e.g., Whatman, 0.2 μm).
4. Vacuum pump.
5. 10 mL syringes and syringe-top filters (e.g., Millipore, 0.2 μm).
6. Mechanical cell disrupter such as a French pressure cell or a One-Shot cell disruptor (Constant Systems).
7. 2 mL and 20 mL centrifugal concentrators with a molecular weight cutoff of 30 kDa (e.g., Vivaspin, Sartorius; Amicon Ultra, Millipore).
8. 5 mL HisTrap Ni^{2+} Sepharose (GE Healthcare) or equivalent IMAC (immobilized metal affinity chromatography) column.
9. HiLoad 16/60 Superdex 200 gel filtration column (S200 column, GE Healthcare).
10. Lysis Buffer: 20 mM sodium phosphate pH 7.5, 500 mM NaCl, 0.1 mg/mL DNase, and 5 mM $MgCl_2$.
11. IMAC buffer A (Buffer A): 20 mM sodium phosphate pH 7.5 and 500 mM NaCl.
12. IMAC buffer B (Buffer B): 20 mM sodium phosphate pH 7.5, 500 mM NaCl, and 500 mM imidazole.
13. Gel filtration (GF) buffer: 10 mM Tris–HCl pH 7.0 and 5 mM DTT (filtered and degassed).

2.4 Buffer Exchange

A variety of materials can be used to exchange protein buffers. Below is just a selection of the most commonly used.

1. Desalting columns (e.g., PD-10, GE Healthcare; Zeba Spin columns, Pierce).
2. Dialysis tubing with a MWCO of 10–30 kDa (e.g., SnakeSkin dialysis tubing, Pierce).
3. Centrifugal concentrators with MWCO of 30 kDa (*see* **Note 6**).
4. Size exclusion columns (e.g., S200 column, GE Healthcare).

2.5 Crystallization, Seeding, Cryoprotection, and Heavy Atom Soaks

1. Spectrophotometer capable of absorbance measurements at 280 nm (A_{280}).
2. Benchtop vortex mixer and microcentrifuge.
3. Stereomicroscope for crystallography.
4. Microcentrifuge tubes (0.5 mL and 1.8 mL).
5. Liquid nitrogen.
6. Crystallization incubator or temperature controlled room (4 °C and ~21 °C).

7. 24-well sitting drop crystallization plates (e.g., Cryschem, Hampton).
8. 96-well sitting drop crystallization plates (optional) (e.g., MRC crystallization plate, Molecular Dimensions).
9. Glass coverslip or microscope slide with depression.
10. Sealant for crystallization trays (e.g., ClearVue Sheets, Molecular Dimensions).
11. Adhesive PCR foil for sealing trays prior to crystallization (optional) (Fischer Scientific).
12. Cryogenic loops (cryoloop), caps, and vials (e.g., LithoLoops, CryoCaps, and CryoVials, Molecular dimensions).
13. Seed beads (e.g., Seed bead kit, Hampton Research; MicroSeed beads, Molecular Dimensions).
14. Crystallization Solution: 50 mM Hepes pH 7.0, 0.1 % (w/v) tryptone, and 20 % (w/v) PEG3350 (*see* **Note 7**).
15. Sparse matrix crystallization screening kits (commercially available from Hampton Research, Sigma, or homemade) (*see* **Note 8**).
16. Cryoprotectant: 2-Methy-2-4-pentandiol (MPD).
17. Samarium acetate (SmAc).

2.6 Metacaspase Activity Assays

1. Spectrophotometer capable of absorbance measurements at 280 nm (A_{280}).
2. Fluorescence multi-well plate reader (e.g., Envision 2101, PerkinElmer), equipped with excitation and emission filters of 355 nm and 460 nm respectively (*see* **Note 9**).
3. Black, flat-bottomed, 96-well plates.
4. 1.8 mL microcentrifuge tubes.
5. Data analysis software (e.g., GraFit, Erithacus Software or Prism, GraphPad software).
6. Exchange Buffer (for buffer exchange): 50 mM Tris pH 7.5, 100 mM NaCl.
7. 1× Assay Buffer: 50 mM Tris–HCl pH 7.5, 100 mM NaCl, 5 mM DTT, 1 mM $CaCl_2$.
8. 2× Assay Buffer: 100 mM Tris pH 7.5, 200 mM NaCl, 10 mM DTT, 2 mM $CaCl_2$ (allow for 100 μL per well).
9. 7-Amino-4-methylcoumarin (AMC) reference standard (e.g., Sigma) made up to 20 mM in DMSO.
10. Commercial fluorogenic substrates made up to 100 mM in 100 % DMSO, e.g., Z-GGR-AMC, Z-RR-AMC, Bz-R-AMC (all Bachem) Z-FPR-AMC (MP Biomedicals), Z-VRPR-AMC [18].

2.7 Inhibitor Studies

All inhibitors for testing (e.g., Z-VRPR-FMK) should be made up in 100 % DMSO or ddH_2O at a concentration of 10–100 mM, depending on individual inhibitor solubility.

2.8 Auto-processing and Inhibitor Binding Gel-Shift Assays

1. Temperature controlled shaking incubator set at 37 °C.
2. 1.8 mL microcentrifuge tubes.
3. Working Buffer (for buffer exchange): 10 mM Tris pH 7.5, 50 mM NaCl, 5 mM DTT.
4. 10× Working Buffer: 100 mM Tris pH 7.5, 500 mM NaCl, 50 mM DTT.
5. 10 mM EGTA stock solution, prepared in ddH_2O, pH-adjusted to around 8.0 with NaOH and stored at room temperature.
6. 10 mM $CaCl_2$, prepared in ddH_2O and stored at room temperature.
7. Metacaspase peptide inhibitors, e.g., Z-VRPR-FMK (Enzo Life Sciences) prepared to ~500 μM.
8. SDS-Sample Buffer (90 μL NuPAGE LDS Sample Buffer supplemented with 10 μL NuPAGE Reducing Agent).

3 Methods

3.1 Expression

Sterile techniques and equipment should be used throughout cell transformation and protein expression.

3.1.1 Transformation

1. Thaw 50 μL of the chemically competent *E. coli* on ice.
2. Add 5–10 ng of the expression plasmid DNA (*see* **Note 10**) (typically 1–5 μL) to the cells and mix by tapping gently.
3. Place the cells on ice and incubate for 30 min.
4. Transform the cells by heat-shock, by placing the vial at 42 °C for 30 s.
5. Return the tube to ice for 2 min to allow the cells to recover.
6. Add 250 μL of prewarmed SOC medium (*see* **Note 11**) to the vial and incubate at 37 °C with shaking (~225 rpm) for 1 h.
7. Pipette 20–200 μL (*see* **Note 12**) onto the prepared and prewarmed LB Agar plates and spread the samples.
8. Invert the plates and incubate overnight at 37 °C.

3.1.2 Metacaspase Expression

1. Aliquot 2×10 mL of the prepared auto induction medium (containing kanamycin) into sterile tubes, to be used for starter cultures, before dividing the remaining medium into four 1 L baffled flasks (~250 mL per flask).
2. Pick a single colony from an LB Agar plate and inoculate the two starter cultures.

3. Incubate the starter cultures for at least 8 h at 37 °C with constant shaking at ~225 rpm (the cultures can be left to grow overnight if required).
4. Add 5 mL of a starter culture to each 1 L culture flask and incubate at 37 °C for 16 h with shaking at ~225 rpm.
5. Harvest the cells by centrifugation at 5,000 × *g* for 20 min at 4 °C; discard the supernatant and store each 250 mL cell culture separately at −20 °C, until required.

3.2 Metacaspase Purification

A single-step IMAC purification often suffices for metacaspase assays, but the protein must be exchanged into a suitable assay buffer prior to the experiment. For crystallization experiments the protein should be further purified by size-exclusion chromatography.

1. Thaw a 250 mL cell culture (*see* **Note 13**) at room temperature and resuspend in Lysis Buffer using around 5 mL of buffer per 1 g of cultured cells.
2. Lyse the cells by three passages through a French pressure cell at 14,000 psi (internal pressure) or equivalent.
3. Remove cell debris by centrifugation at 48,000 × *g* for 20 min at 4 °C.
4. Collect the clarified supernatant and filter-sterilize (0.22 μm).
5. Apply the supernatant to a 5 mL HisTrap (IMAC) column equilibrated in Buffer A with 25 mM imidazole (5 % Buffer B) and wash for 10 column volumes (CV) in the same buffer.
6. Wash the column with Buffer A containing 50 mM imidazole (5 CV, 10 % Buffer B) and elute the protein with 250 mM imidazole (5 CV, 50 % Buffer B) (*see* **Note 14**), collecting 1 mL fractions.
7. Detect the fractions containing protein using the A_{280} trace, from the protein purification system, and identify which fractions contain the target metacaspase using SDS-PAGE analysis with a 4–12 % gel and MES Buffer.
8. Pool the fractions containing the purified protein. NB: If the protein is to be used solely in activity assays move on to Subheading 3.5.
9. Concentrate the protein from the IMAC column to around 3 mL using a centrifugal concentrator with a 30 kDa molecular weight cutoff.
10. Sterile filter the sample before loading onto an S200 column equilibrated with 2 CV of GF buffer and run for 1.5 CV in the same buffer.
11. Pool all fractions containing the protein.

3.3 SDS-PAGE Analysis

1. Prepare a 20 μL reduced sample by adding 5 μL sample buffer, 2 μL sample reducing agent, and 13 μL purified protein (or a 13 μL sample containing purified protein plus ddH_2O) in a microcentrifuge tube.
2. Heat samples for 10 min at 70 °C or for 3 min at 90 °C.
3. Assemble mini-gel equipment and add 200 mL of MOPS or MES Buffer to the upper chamber and 600 mL to the lower chamber.
4. Add 200 μL of general reducing agent to the upper chamber.
5. Load ~8 μL of the protein molecular weight marker and all of the denatured protein samples (20 μL) on to the gel.
6. Run at 200 V for around 40 (MES Buffer) to 60 (MOPS Buffer) min.
7. Remove gel from tank and wash three times (5 min) in water with shaking.
8. Add Coomassie stain to the gel and leave shaking for ~1 h.
9. Dispose of the staining solution and destain the gel with ddH_2O or appropriate destaining solution for ~1 h or until the gel background is clear (*see* **Note 15**).

3.4 Crystallization and Crystal Handling

Crystallization of active recombinant TbMCA2 has so far proved unsuccessful. This may be a consequence of the mixed products resulting from the enzyme exhibiting auto-processing during over-expression in *E. coli*. To combat this, the catalytically inactive mutant TbMCA2^{C213A} was used in crystallization experiments and was found to be stable and routinely produce crystals. Unfortunately, many of the initial crystals were composed of multiple plates that were unsuitable for diffraction experiments. However, optimizing them in 24-well sitting drop plates using microseeding techniques proved very useful in obtaining diffraction-quality crystals (Fig. 1) and is highly recommended if the initial crystals grow in clusters.

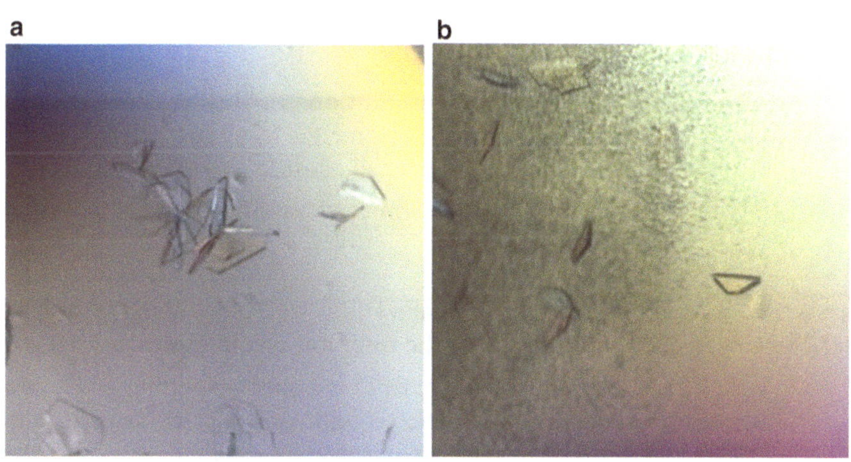

Fig. 1 TbMCA2^{C213A} crystallization experiments. (**a**) Initial clusters of crystals were unsuitable for diffraction experiments. (**b**) Improved singular crystals were obtained using microseeding techniques

3.4.1 Crystallization of Inactive TbMCA2^{C213A}

1. Measure protein absorbance at 280 nm (A_{280}) using a spectrophotometer and use this and the extinction coefficient (M^{-1} cm^{-1}) (*see* **Note 16**) to calculate the approximate protein concentration.
2. Concentrate the protein to around 7 mg/mL using a centrifugal concentrator. The protein can be stored at 4 °C but for best results it should be used immediately after purification (*see* **Note 17**).
3. Prior to crystallization, centrifuge the sample at 12,000 × *g* for 5 min in a benchtop centrifuge to remove any precipitates.
4. Pipette 500 μL of the Crystallization Solution into the reservoir of a 24-well sitting drop plate.
5. Without crystallization seeds: Pipette 3 μL of the purified protein into the crystallization well, followed by 3 μL of the reservoir solution.
6. With crystallization seeds: Pipette 3 μL of the purified protein into the crystallization well, followed by 1 μL of the seed stock and 2 μL of the reservoir solution.
7. Seal the tray and place at 4 °C for around 3 weeks after which time they can be moved to ~21 °C for storage and crystal manipulation.
8. Check trays weekly for crystals.

3.4.2 Preparation of Seed Stocks

1. Pipette 50 μL of the Crystallization Solution into a microcentrifuge tube (*see* **Note 18**).
2. Collect crystals from a crystallization drop using an appropriately sized cryoloop.
3. Place the cryoloop into the solution and shake it to displace the crystals into the solution.
4. After crystal collection, transfer the entire solution into a microcentrifuge tube containing a seed bead and vortex for 90 s.
5. Add a further 450 μL of the Crystallization Solution and vortex for a further 90 s.
6. Prepare five serial dilutions of this stock solution using the Crystallization Solution e.g., 1:2, 1:10, 1:100, 1:500, 1:1,000 and store seeds in small aliquots at −80 °C or −20 °C.
7. Set up crystallization trays varying the seed stocks and monitor the results.

3.4.3 Cryoprotection of TbMCA2^{C213A} Crystals

1. Open the crystallization well containing the crystals and pipette 4 μL of the reservoir solution onto a glass coverslip.
2. Add 1 μL of 100 % MPD and mix by pipetting gently up and down (this results in the cryo-solution, which contains 20 % MPD)
3. Collect a single crystal from the crystal tray using an appropriately sized cryoloop and transfer to the cryo-solution.

4. After a few seconds collect the crystal from the cryo-solution and immediately flash freeze by plunging into liquid nitrogen.
5. Cryo-cooled crystals can be stored at 100 K (−173 °C) indefinitely, until required for diffraction studies and data collection.

3.4.4 Sm^{3+} Heavy Atom Soak

The structure of $TbMCA2^{C213A}$ was the first crystal structure of a metacaspase available in the Protein Data Bank (http://www.pdb.org [19]) and consequently the incorporation of a heavy atom (HA) into the crystals was required to determine the structure (various attempts at Molecular Replacement using caspase structures failed). However, searching for a HA to achieve this proved to be time-consuming and laborious with all attempts at co-crystallizing the enzyme with a heavy atom (or calcium) failing. Crystal soaking with many of the more traditional HA compounds, for example compounds containing mercury (e.g., pcmb, $HgCl_2$), platinum (e.g., $KPtCl_6$, $KPtCl_4$), lead ($PbNO_3$), and gold (AuCN), proved unsuccessful typically resulting in crystal damage (and unusable diffraction) as soon as the HA was incorporated. However, lanthanides are often used to define calcium-binding sites in proteins [20] and consequently a lanthanide, in the form of samarium acetate (SmAc), was tested as a HA derivative for the structure solution. This resulted in a fully occupied Sm^{3+} binding site on the surface of the molecule (Fig. 2), which allowed the structure to be determined and also defined the Ca^{2+} binding site of TbMCA2 [17].

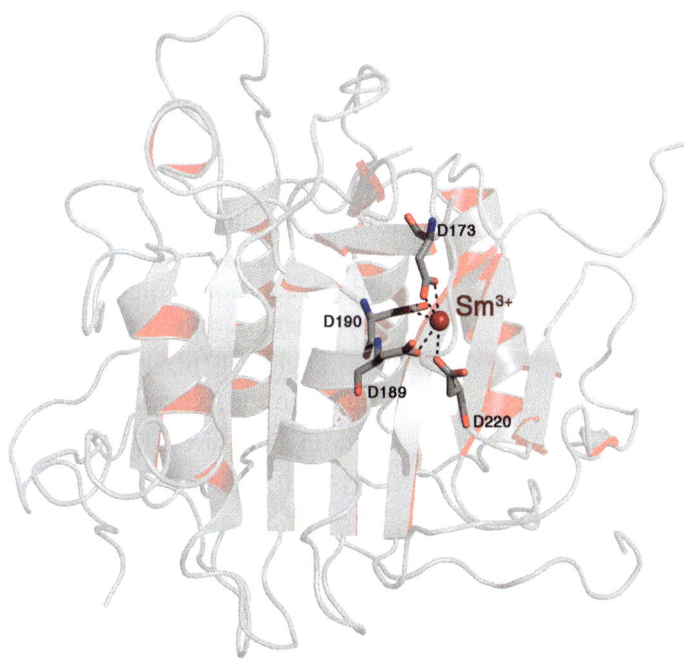

Fig. 2 A samarium ion (Sm^{3+}) binds tightly on the surface of $TbMCA2^{C213A}$. Sm^{3+} is coordinated by four aspartic acid residues (D173, D189, D190, and D220) and defines the calcium-binding site in the enzyme

Consequently, the methods presented below may be useful for the structure solution of other Ca^{2+} binding proteins where the more traditional HA methods fail.

1. Prepare a solution containing 60 mM SmAc in the crystallization solution.
2. Pipette 2–3 μL of this solution onto clean well of a 24-well crystallization tray.
3. Open a crystallization well containing the crystals and collect a single crystal using a cryoloop and transfer to the heavy atom condition.
4. Cover the well for 2 min before transferring the crystal briefly to the cryo-solution and freezing at 100 K (*see* above).

3.5 Activity Assays

Fluorogenic substrate assays are used to assess the activity of active TbMCA2 and associated mutants, by measuring the release of a fluorophore (e.g., AMC) from the hydrolysis of a fluorescent peptide by the enzyme. All assays should be run in technical triplicate and include a negative control (containing no protein). In addition, all replicates and controls should be included on the same 96-well plate. The amount of each reagent is given for a standard assay, but the volume of protein should be optimized as required.

3.5.1 Fluorogenic Substrate Activity Assays

1. Exchange the protein into the Exchange Buffer using a PD-10 desalting column.
2. Concentrate the protein using a centrifugal concentrator, or dilute as necessary, to around 0.5–1 mg/mL. The protein can be stored at 4 °C but for best results it should be used as soon as possible after purification (*see* **Note 17**).
3. Dilute the fluorogenic substrates to a 100 μM working stock in ddH_2O. Prepare enough for 20 μL per well.
4. Immediately prior to running the assay, dilute protein to between 0.2 and 0.4 mg/mL in 1× Assay buffer. This is a 40× working solution requiring 5 μL per well and giving a final assay concentration of 5–10 μg/mL.
5. Place a 96-well black bottomed plate on ice and add the reagents in the following order:
   (a) ddH_2O to give a final volume of 200 μL
   (b) 100 μL 2× Assay Buffer
   (c) 5 μL protein (or buffer for a negative control). This volume can be adjusted as necessary.
   (d) 20 μL substrate
6. Remove the plate from the ice, place in the fluorescence plate reader, and equilibrate to ambient temperature (20–25 °C) for 5 min.

7. Read the fluorescence using an excitation wavelength of 355 nm and emission wavelength of 460 nm. Fluorescence measurements should be taken approximately every 5 s for around 10 min, but the interval time and total read time can be varied, depending on the reaction velocity.
8. Plot the relative fluorescence units (FU) versus time (T) (in min).
9. Select a time interval over which the increase in fluorescence is linear and calculate the slope of the graph over this time frame. This gives the reaction rate ($\Delta FU/T$), which is a relative measure of protein activity. This can generally be used to compare the activities of different enzymes within the same experiment (on the same 96-well plate).

3.5.2 AMC Standard Curve

In order to calculate the specific activity of the protein an AMC standard curve should be generated.

1. Make a 10 μM solution of AMC in ddH_2O (AMC 1) in a microcentrifuge tube.
2. Make a 1/2 (2-fold) dilution of AMC 1 with ddH_2O (e.g., 200 μL AMC + 200 μL ddH_2O) and place in tube 2.
3. Repeat this 2-fold serial dilution in to further nine tubes.
4. Add 200 μL of each dilution to a single row (wells 1–11) of a 96-well plate and add 200 μL ddH_2O to well 12.
5. Remove the plate from the ice, place in the fluorescence plate reader, and equilibrate to ambient temperature (20–25 °C) for 5 min.
6. Read the fluorescence using an excitation wavelength of 355 nm and emission wavelength of 460 nm (*see* **Note 9**).
7. Plot the relative fluorescence units (FU) versus the total amount of AMC (μmoles per well; [AMC]/5,000 to give total moles AMC in 200 μL well).
8. Fit a linear regression line to this data with the formula $y = mx + b$, where y is FU and x is μmoles of AMC per reaction.

3.5.3 Converting Relative Activity to Specific Activity (See **Note 19**)

1. Using the AMC standard curve, calculate the amount of AMC released (increase in AMC; ΔAMC), using the ΔFU value obtained earlier from the plot of FU versus time. Ensure the amount of AMC is in μmoles.
2. Divide ΔAMC by T (in min) to give the number of enzyme units (U) in your 200 μL well (U = amount of substrate released in μmol/min).
3. Divide the number of enzyme units (U) by the amount of protein (in mg) in the reaction to give the specific activity (U/mg).

3.5.4 Calculation of K_m

1. Choose a working protein concentration of around 5–10 μg/mL or an optimized concentration from an earlier activity assay.
2. Dilute the protein into 2× Assay Buffer to a final concentration of 2× the chosen working concentration (~10–20 μg/mL).
3. Place a 96-well black-bottomed plate on ice.
4. Add 100 μL ddH_2O to 11 wells on single row (wells 1–11) of the plate.
5. In well 12 add 200 μL of substrate, diluted to 200 μM in ddH_2O.
6. Using a pipette, remove 100 μL of the solution in well 12 and place into the adjacent well (well 11). Mix by gently pipetting.
7. Repeat this 2-fold (1/2) dilution (transferring 100 μL from well 11 to well 10 and so on) for all remaining wells.
8. Discard 100 μL of the diluted substrate in well 1, leaving 100 μL substrate in each well.
9. Keeping the plate on ice, add 100 μL of the 2× protein–buffer mixture to each well.
10. Remove the plate from the ice, place in the fluorescence plate reader and equilibrate to ambient temperature (20–25 °C) for 5 min.
11. Read the fluorescence using an excitation wavelength of 355 nm and emission wavelength of 460 nm, using time scales and intervals described above.
12. Plot the FU versus *T* for each substrate concentration.
13. Calculate the reaction rate (FU/*T*) for each substrate concentration, as above.
14. Plot substrate concentration versus reaction rate. This should give a hyperbolic plot, where the *y* value approaches *V*max. If the plot is not hyperbolic, the substrate range used in the assay may need to be altered.
15. Plot 1/[substrate] versus 1/reaction rate (Lineweaver–Burk plot). The slope of the fitted line is equal to $1/K_m$. Alternatively, a program such as GraFit or Prism can be used to automatically calculate K_m.

3.5.5 Calculation of IC_{50}s for Inhibitors

1. Prepare protein by concentrating or diluting to between 200 and 400 μg/mL in 1× Assay buffer. This is a 40× working solution, giving a final assay concentration of 5–10 μg/mL.
2. Dilute the inhibitor to 100 μM in ddH_2O (stock inhibitor) and place a 96-well black-bottomed plate on ice.
3. Dilute a fluorogenic substrate to 100 μM in ddH_2O, preparing enough for 20 μL per well.

4. Add 100 μL ddH_2O to 11 individual wells on single row (wells 1–11) of the 96-well plate.
5. In well 12 add 150 μL of the stock inhibitor.
6. Using a pipette, remove 75 μL of the solution in well 12 and place into the adjacent well (well 11) and mix by gently pipetting.
7. Repeat this 2-fold dilution (transferring 75 μL from well 11 to well 10 etc.) for the remaining wells.
8. Discard 75 μL of the diluted inhibitor from well 1.
9. To the inhibitor solutions add the following reagents in order:
   (a) 100 μL 2× Assay Buffer.
   (b) 5 μL protein (40× working solution).
   (c) 20 μL substrate.
10. Remove the plate from the ice, place in the fluorescence plate reader, and equilibrate to ambient temperature (20–25 °C) for 5 min.
11. Read the fluorescence using an excitation wavelength of 355 nm and emission wavelength of 460 nm, using time scales and intervals described above.
12. Plot FU versus *T* for each concentration of inhibitor.
13. Calculate the reaction rate for each inhibitor concentration following the method set out in Subheading 3.5.1.
14. Use Grafit or Prism to plot [inhibitor] versus reaction rate and calculate the IC_{50}.

3.5.6 In Vitro Auto-processing of Active Metacaspases and Mutant Variants

Many metacaspases exhibit in vitro auto-processing. This is typically enhanced in the presence of 1 mM $CaCl_2$, inhibited by 1 mM EGTA and can be visualized using SDS-PAGE.

1. Exchange the protein into the Working Buffer and concentrate to around 10 μM.
2. For each sample for analysis add:
   (a) 10 μL of protein.
   (b) 2 μL of EGTA or $CaCl_2$ (optional).
   (c) 2 μL 10× assay buffer.
   (d) ddH_2O to 20 μL.
3. Incubate the samples for around 30 min at 37 °C with or without shaking (*see* **Note 20**).
4. Stop the reaction by adding 11 μL SDS-Sample buffer.
5. Heat samples for 10 min at 70 °C or for 3 min at 90 °C.
6. Load each sample onto a 4–12 % SDS-PAGE gel and run for 40 min in MES buffer.

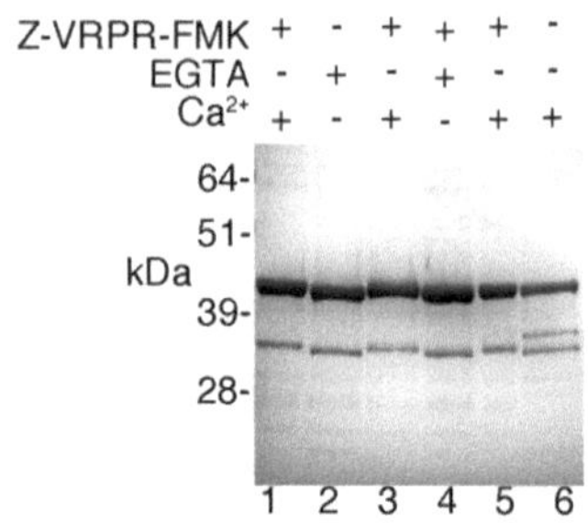

Fig. 3 TbMCA2 gel-shift assay. Inhibitor Z-VRPR-FMK binds to active TbMCA2 on the addition of Ca^{2+} (as indicated by a small increase in apparent molecular weight in *Lanes 1*, *3* and *5*) but not in the presence of EGTA (*Lane 4*). Z-VRPR-FMK also protects active TbMCA2 from autoproteolysis in the presence of Ca^{2+} (autoproteolysis shown in *Lane 6*)

3.5.7 Inhibitor Binding Gel-Shift Assays

1. Exchange the protein into the Working Buffer and concentrate to around 10 μM.
2. Set up four 20 μL reactions in 1.8 mL microcentrifuge tubes to contain 5 μM protein (10 μL), 10× Working Buffer (2 μL) and the following reagents:
   (a) 1 mM EGTA (2 μL of 10 mM stock).
   (b) 1 mM $CaCl_2$ (2 μL of 10 mM stock).
   (c) 1 mM EGTA (2 μL)+25 μM inhibitor (1μL of 1 mM stock).
   (d) 1 mM $CaCl_2$ (2 μL)+25 μM inhibitor (1 μL).
3. Add ddH_2O to a total volume of 20 μL.
4. Incubate the reactions for 20 min at 37 °C with or without shaking.
5. Stop the reaction by adding 11 μL SDS-Sample Buffer.
6. Heat samples for 10 min at 70 °C or for 3 min at 90 °C.
7. Load each sample onto a 10 %, 10-lane, SDS-PAGE gel and run for 50 min in MOPS buffer.
8. Analyze the results (Fig. 3).

4 Notes

1. An arginine residue in the thrombin cleavage site of the pET28a+vector was mutated to glycine for the expression of recombinant and active TbMCA2. This mutant was made after it was realized that the protein linker in this region was being cleaved by the enzyme and adversely removing the His-tag and may be something to consider when working with arginine-specific peptidases in general.

2. The expression vector pET28a+(Novagen) was used for the protein expression of metacaspases from *T. brucei*. However, other expression vectors may be used provided potential cleavage sites in the linker are considered (*see* **Note 1**).
3. Most of the *T. brucei* metacaspases express well in both BL21 (DE3) and Rosetta (DE3) cell strains although TbMCA4 generally produces a much higher yield in Rosetta (DE3).
4. Varying cell lines and vectors will require different antibiotics and in general a 1,000× stock solution is prepared in ddH_2O and stored at −20 °C, until required.
5. *T. brucei* metacaspases generally express very well giving a purified yield (after size exclusion chromatography) of around 5 mg from a 250 mL cell culture and these amounts can be scaled up/down as required.
6. Centrifugal concentrators can be used to exchange the buffer for any protein that can tolerate the procedure and *T. brucei* metacaspases are generally amenable to it. The protein sample should be concentrated to a minimum volume and diluted back to the starting volume in the new buffer, this should be repeated two/three times to ensure buffer exchange.
7. It might be useful to vary the PEG concentration in the crystallization experiments but in general seeding was the most effective method for crystal improvement.
8. These may be useful for other *T. brucei* metacaspases, but for the inactive $TbMCA2^{C213A}$ mutant the stated crystallization condition should be most successful.
9. The excitation and emission maxima for the fluorescent molecule (AMC) are 345 nm and 445 nm, but wavelengths at either side of these values can be used, albeit with decreased fluorescence (*see* http://www.invitrogen.com/site/us/en/home/Products-and-Services/Applications/Cell-Analysis/Labeling-Chemistry/Fluorescence-SpectraViewer.html).
10. Expression plasmids for the proteins described in this chapter, can be identified by their plasmid numbers (pGL): Active TbMCA2, pGL1573; $TbMCA2^{C213A}$, pGL2128; TbMCA3, pGL1593; TbMCA4, pGL1967.
11. LB (Luria–Bertani) medium can also be used for recovery and a 30 min incubation may also suffice.
12. It is advisable to plate out two different volumes on to the LB-Agar plate and keep excess cells overnight at 4 °C in order to plate out an optimized volume the next day, if required.
13. Using auto-induction media results in a large amount of proteins in the cell lysate. In order to recover a reasonable amount of *T. brucei* metacaspases from the IMAC column it is better to limit the amount of cell culture used to a maximum of 500 mL.

14. All fractions should be checked by SDS-PAGE to identify the protein, as the elution profile may change depending on the temperature and pH of the buffers. This protocol is good for inactive TbMCA2^{C213A}, but other forms of this enzyme and other *T. brucei* metacaspases may require additional purification steps.
15. SimplyBlue SafeStain (Invitrogen) can be disposed of safely by flushing down the sink and the gels destained using ddH_2O. Other types of Coomassie stain may require specific disposal methods and destaining protocols, and if chosen these should be investigated.
16. The extinction coefficient can be calculated from the primary sequence using tools such as ProtParam ([21], http://web.expasy.org/protparam/). The extinction coefficient of TbMCA2^{C213A}, including the linker and His-tag is 29,910 M^{-1} cm^{-1} and the Beer–Lambert Law ($A = Ecl$) can then be used to calculate the relative protein concentration where $A = A_{280}$, E = extinction coefficient, c = concentration, and l = path length of the cell. While this method may not give an exact measure of protein concentration (relying on the absorbance from the Tyr and Trp residues) it is sufficiently accurate and reproducible for preparation of samples for crystallization trials.
17. Purified *T. brucei* metacaspases should be stored at 4 °C rather than at −20 °C, as freeze–thaw cycles are known to reduce peptidase activity and affect the crystallizability of TbMCA2^{C213A}. In addition, TbMCA2^{C213A} should be used in crystallization trials as soon as possible after purification as any delay in this will affect the success rate of reproducing crystals.
18. In the case of TbMCA2^{C213A}, the crystallization solution can be used as an appropriate stabilizing solution for seeding.
19. One enzyme unit (U) is the amount of enzyme that catalyzes the reaction of 1 μmol of substrate.
20. The incubation time is enzyme specific and will have to be optimized for different proteins and possibly different batches of the same protein. The given times are based on assays using active recombinant TbMCA2.

Acknowledgements

This work was supported by Wellcome Trust Grant 091790 and Medical Research Council Grant 0700127. The Wellcome Trust Centre for Molecular Parasitology is supported by core funding from Wellcome Trust Grant 085349.

References

1. Coll NS, Vercammen D, Smidler A, Clover C, Van Breusegem F, Dangl JL, Epple P (2010) Arabidopsis type I metacaspases control cell death. Science 330:1393–1397
2. Proto WR, Coombs GH, Mottram JC (2012) Cell death in parasitic protozoa: regulated or incidental? Nat Rev Microbiol 11:58–66
3. Castanys-Muñoz E, Brown E, Coombs GH, Mottram JC (2012) *Leishmania mexicana* metacaspase is a negative regulator of amastigote proliferation in mammalian cells. Cell Death Dis 3:e385
4. Proto WR, Castanys-Munoz E, Black A, Tetley L, Moss CX, Juliano L, Coombs GH, Mottram JC (2011) *Trypanosoma brucei* metacaspase 4 is a pseudopeptidase and a virulence factor. J Biol Chem 286:39914–39925
5. Lee REC, Brunette S, Puente LG, Megeney LA (2010) Metacaspase Yca1 is required for clearance of insoluble protein aggregates. Proc Natl Acad Sci USA 107:13348–13353
6. Tsiatsiani L, Van Breusegem F, Gallois P, Zavialov A, Lam E, Bozhkov PV (2011) Metacaspases. Cell Death Differ 18: 1279–1288
7. MacKenzie SH, Clark AC (2012) Death by caspase dimerization. Adv Exp Med Biol 747:55–73
8. Fuentes-Prior P, Salvesen GS (2004) The protein structures that shape caspase activity, specificity, activation and inhibition. Biochem J 384:201–232
9. Pop C, Salvesen GS (2009) Human caspases: activation, specificity, and regulation. J Biol Chem 284:21777–21781
10. Moss CX, Westrop GD, Juliano L, Coombs GH, Mottram JC (2007) Metacaspase 2 of *Trypanosoma brucei* is a calcium-dependent cysteine peptidase active without processing. FEBS Lett 581:5635–5639
11. Watanabe N, Lam E (2011) Calcium-dependent activation and autolysis of Arabidopsis metacaspase 2d. J Biol Chem 286(12):10027–10040
12. Watanabe N, Lam E (2005) Two Arabidopsis metacaspases AtMCP1b and AtMCP2b are arginine/lysine-specific cysteine proteases and activate apoptosis-like cell death in yeast. J Biol Chem 280:14691–14699
13. Vercammen D, van de Cotte B, De Jaeger G, Eeckhout D, Casteels P, Vandepoele K, Vandenberghe I, Van Beeumen J, Inzé D, Van Breusegem F (2004) Type II metacaspases Atmc4 and Atmc9 of *Arabidopsis thaliana* cleave substrates after arginine and lysine. J Biol Chem 279:45329–45336
14. Mottram JC, Helms MJ, Coombs GH, Sajid M (2003) Clan CD cysteine peptidases of parasitic protozoa. Trends Parasitol 19:182–187
15. Helms MJ, Ambit A, Appleton P, Tetley L, Coombs GH, Mottram JC (2006) Bloodstream form *Trypanosoma brucei* depend upon multiple metacaspases associated with RAB11-positive endosomes. J Cell Sci 119: 1105–1117
16. Rawlings ND, Barrett AJ, Bateman A (2012) MEROPS: the database of proteolytic enzymes, their substrates and inhibitors. Nucleic Acids Res 40:D343–D350
17. McLuskey K, Rudolf J, Proto WR, Isaacs NW, Coombs GH, Moss CX, Mottram JC (2012) Crystal structure of a *Trypanosoma brucei* metacaspase. Proc Natl Acad Sci USA 109:7469–7474
18. Vercammen D, Belenghi B, van de Cotte B, Beunens T, Gavigan J-A, De Rycke R, Brackenier A, Inzé D, Harris JL, Van Breusegem F (2006) Serpin1 of *Arabidopsis thaliana* is a suicide inhibitor for metacaspase 9. J Mol Biol 364:625–636
19. Berman HM, Westbrook J, Feng Z, Gilliland G, Bhat TN, Weissig H, Shindyalov IN, Bourne PE (2000) The protein data bank. Nucleic Acids Res 28:235–242
20. Martin B (1979) Lanthanides as probes for calcium in biological systems. Q Rev Biophys 12(2):181–209
21. Wilkins MR, Gasteiger E, Bairoch A, Sanchez JC, Williams KL, Appel RD, Hochstrasser DF (1999) Protein identification and analysis tools in the ExPASy server. Methods Mol Biol 112: 531–552

Chapter 14

Monitoring the Proteostasis Function of the *Saccharomyces cerevisiae* Metacaspase Yca1

Amit Shrestha, Robin E.C. Lee, and Lynn A. Megeney

Abstract

The functional versatility of metacaspase proteases has been established by reports of their involvement in non-apoptotic cellular processes, in addition to their canonical role in apoptosis/programmed cell death. While the budding yeast metacaspase Yca1 has been well characterized for its role in cell death regulation, more recent examinations suggest that the protease may be involved in key processes that increase survival and fitness. More specifically, examinations suggest that Yca1 is central to maintaining cellular proteostasis as it interacts with major components involved in protein biosynthesis and functions to limit aggregate deposition. Here, we describe the methods utilized to analyze the role Yca1 in proteostasis.

Key words Aggregates, Filtration, Heat stress, Immunoprecipitation, Metacaspase, Proteostasis, Protocols, Vacuole morphology

1 Introduction

Caspase enzymes, and related proteases, are broadly conserved across the phyla of extant organisms. Structural investigations have revealed that proteases within the clan CD family C14 share similar quaternary architecture and an invariant cysteine–histidine catalytic dyad [1, 2]. In fungi, plants, and protozoa, such proteases are termed metacaspases [3]. Metacaspases can be further subdivided based on the presence or absence of a proline-rich N-terminal prodomain, referred to as "type I" and "type II" classes respectively. When compared across organisms, the number and type of metacaspases expressed is highly variable [4]. To date, only a single type I metacaspase has been identified in *Saccharomyces cerevisiae* [5]. This protease, termed Yca1 (also referred to as Mca1), harbors a unique poly Q/N motif in its prodomain, upstream of its caspase-like domain. The functional importance of this motif remains largely unknown; however, it is suggested from metazoan initiator caspases that the motifs within the prodomains of type I metacaspases

Peter V. Bozhkov and Guy Salvesen (eds.), *Caspases, Paracaspases, and Metacaspases: Methods and Protocols*, Methods in Molecular Biology, vol. 1133, DOI 10.1007/978-1-4939-0357-3_14, © Springer Science+Business Media New York 2014

facilitate protein–protein interactions and contribute to the regulation of proteolytic activity [6, 7].

Metacaspase function in *S. cerevisiae* has been extensively explored within a death-centric perspective, owing to the preexisting view of caspase proteases as de facto mediators of programmed cell death. Under such circumstances, various studies have shown Yca1 involvement as an essential feature that progresses the apoptotic phenotype induced by various insults [4, 8]. However, observations of cell-death events independent of Yca1, such as the inability of the proapoptotic hBax protein to engage Yca1 and accelerate toxicity in yeast [9], argue against a death-only function for Yca1. Importantly, the evolutionary retention of death-specific machinery in single celled organisms, such as *S. cerevisiae*, supports the premise that this protease may have evolved or coevolved non-death functions [10].

Investigation of metacaspase biology in *S. cerevisiae* has since elucidated "non-death" functions for Yca1, which may be shared by other counterparts in trypanosoma and other yeast species [11]. Initially, a loss of function model demonstrated that Yca1 coordinated processes that affect cell cycle timing and cellular fitness; *Δyca1* cells displayed a delay in the G1/S transition and disruption of microtubule assembly, via nocodazole exposure, did not halt progression of *Δyca1* cells through the G2/M checkpoint [7]. Subsequently, attempts to identify a molecular role for Yca1 under these pro-survival parameters suggested that the protease functions to ensure fidelity of protein turnover. Immunoprecipitation [12] using the tandem affinity purification (TAP) method led to identification of Cdc48, Hsp40, and Hsp70 as high-affinity interaction partners with Yca1 [13]. These Yca1-interacting proteins are central contributors to cellular proteostasis; Cdc48 is involved in ER-associated degradation and its level within the insoluble proteome depends on the presence of Yca1 [14]; and Hsp40/70 together with Hsp100 constitute the "bi-chaperone" system that actively re-solubilize misfolded proteins under stress conditions [15, 16]. Furthermore, fluorescent microscopy studies showed that Yca1 and Hsp104, a protein involved in aggregate dissolution that serves as a marker for aggregated material, can co-localize in cells collected from heat-stressed and aged cultures [13].

Additional comparisons between wild-type and *Δyca1* cells implicated *Δyca1* cells with a protein turnover phenotype. Sedimentation of cellular protein [17–19] to separate the insoluble protein fraction followed by membrane filtration showed that *Δyca1* cells accumulate a greater amount of insoluble material [13]. In addition, analysis of vacuolar morphology via fluorescence microscopy and lipophilic styryl dyes [20] demonstrated that the loss of Yca1 resulted in increased vacuole formation, possibly the consequence of an increased protein aggregate load. Together, these findings support the hypothesis that Yca1 increases cellular fitness as a bona fide player in cellular proteostasis.

The discovery of a well-entrenched non-death function of Yca1 has established metacaspases as versatile biological entities. As such, the ability of Yca1 to cycle between death and non-death activities demands stringent forms of analysis to accurately differentiate the mechanisms that contribute to divergent cell fates. For example, screening for known hallmark features of death such as the integrity of the nucleus via DNA stains [13], may aid in delineating processes that contribute to cell fate, yet these measures do not distinguish the physiological roles of Yca1. This chapter consists of the detailed protocols, which are instrumental in defining and assessing the non-death, proteostasis function of the budding yeast metacaspase Yca1.

2 Materials

2.1 Yeast Growth Conditions

1. YPD medium: 1 % (w/v) yeast extract, 2 % (w/v) Bacto Peptone, 2 % (v/v) dextrose, pH 3.5 (*see* **Note 1**).
2. Orbital incubator (a range of 25–42 °C).

2.2 Protein Extraction

1. Buffer A: 50 mM Tris–HCl, 1 mM EDTA, 150 mM NaCl, 1 % (v/v) glycerol, 1 % (v/v) NP-40, pH 7.4. Store at 4 °C.
2. Buffer B: 50 mM Tris–HCl, 1 mM EDTA, 1 % (v/v) glycerol, 0.1 % (v/v) NP-40, pH 7.4. Store at 4 °C.
3. Protease inhibitors (Calbiochem Cocktail Set IV or similar).
4. Acid washed glass beads, 0.4–0.6 mm in diameter (Sigma).
5. 27 G × ½ (0.4 mm × 13 mm) needles (Becton Dickinson).
6. 15 mL tubes (Sarstedt).

2.3 Sedimentation Assay

1. Buffer B: 50 mM Tris–HCl, 1 mM EDTA, 1 % (v/v) glycerol, 0.1 % (v/v) NP-40, pH 7.4. Store at 4 °C.
2. Protease inhibitors (Calbiochem Cocktail Set IV or similar).
3. Wash buffer: 98 % (v/v) buffer B, 2 % (v/v) NP40.

2.4 Filter Trap Assay

1. Bio-Dot SF microfiltration apparatus (Bio-Rad Labs).
2. Bio-Dot SF Filter paper (Bio-Rad Labs).
3. 0.45 μm PVDF membrane (e.g., Immobilon).
4. Buffer B: 50 mM Tris–HCl, 1 mM EDTA, 1 % (v/v) glycerol, 0.1 % (v/v) NP-40, pH 7.4. Store at 4 °C.
5. Protease inhibitors (Calbiochem Cocktail Set IV or similar).
6. Wetting buffer: 80 % (v/v) buffer B, 20 % (v/v) methanol.
7. Coomassie Blue solution: 0.1 % (w/v) Coomassie Blue R-250, 45 % (v/v) methanol, 10 % (v/v) acetic acid, 45 % (v/v) water.
8. Destain solution: 45 % (v/v) methanol, 10 % (v/v) acetic acid, 45 % (v/v) water.

9. Vacuum source.
10. Scotch Tape.

2.5 Vacuole Staining

1. YPD medium: 1 % (w/v) yeast extract, 2 % (w/v) Bacto Peptone, 2 % (v/v) dextrose, pH 3.5.
2. FM lipophilic styryl dye (e.g., Molecular Probes).
3. Phosphate buffered saline (PBS), pH 7.4.
4. Aluminum foil.
5. 37 % (v/v) Formaldehyde.
6. Coverslips: 24 × 50 mm (Fisher Scientific).
7. Poly-L-Lysine coated glass slides (Sigma).

2.6 Immuno-precipitation

1. Dynabeads (Dynal Biotech).
2. Magnetic stand (Dynal Biotech).
3. Buffer C: 20 mM HEPES, 0.1 % (v/v) Tween 20, 2 mM $MgCl_2$, 300 mM NaCl, pH 7.4. Store at 4 °C.
4. Protease Inhibitors (Calbiochem Cocktail Set IV or similar).
5. Buffer D:50 mM Tris–HCl, 2 % (w/v) SDS, 0.1 % (w/v) bromophenol blue, 10 % (v/v) glycerol, 150 mM NaCl, pH 6.8.
6. 0.1 M sodium phosphate buffer: 19 mM monosodium phosphate, 81 mM disodium phosphate, pH 7.4.
7. Ammonium sulfate buffer: 3 M ammonium sulfate, 19 mM monosodium phosphate, 81 mM disodium phosphate, pH 7.4.
8. Phosphate buffered saline (PBS), pH 7.4.
9. 0.1 M Citric acid, pH 3.1.
10. Dimethylformamide (DMF).
11. Triton-X 100.
12. Antibody: nonspecific Rabbit IgG (Chemicon, *see* **Note 2**).

2.7 DNA Staining

1. 70 % (v/v) Ethanol.
2. 50 mM Sodium citrate.
3. 0.2 mg/mL RNase A.
4. Phosphate buffered saline (PBS), pH 7.4.
5. SYTOX Green (Molecular Probes).

3 Methods

3.1 Normal and Heat Stress Growth Conditions

1. Inoculate a 3 mL aliquot of medium with a single yeast colony. This will be the starter culture.
2. Grow for 12–16 h at 30 °C, 225 rpm until OD_{660} is at least between 0.8 and 1.0.

3. Transfer between 5 and 10 μL of culture to fresh 50 mL of media.
4. Incubate freshly inoculated 50 mL culture at 30 °C, 225 rpm. Incubation time depends on doubling time of yeast strain used.
5. At OD_{660} 0.6–0.8, collect cells via centrifugation at 4,600 × *g* for 5 min at room temperature (*see* **Note 3**).
6. Discard supernatant and resuspend the cells in sterile autoclaved water.
7. Centrifuge at 4,600 × *g* for 5 min at room temperature to collect the cells.
8. Discard supernatant and store cell pellet at −80 °C. Cells may be stored up to 3 months at −80 °C.
9. For heat stress and recovery, prepare four additional flasks with 50 mL media and inoculate as stated in **steps 3** and **4**. Use the same starter culture for all samples.
10. At OD_{660} 0.6–0.8, transfer culture to 42 °C and incubate for 1 h, 225 rpm.
11. Collect heat-stressed cells as described in **steps 5–8** (*see* **Note 4**).
12. For recovery, after 42 °C treatment, transfer cultures to 30 °C and incubate further at 225 rpm.
13. Collect cells as described in **steps 5–8** every 30 min.

3.2 Protein Extraction

1. Thaw frozen cells on ice.
2. Add 300 μL of ice-cold buffer A or buffer B supplemented with 0.5 % (v/v) protease inhibitors (*see* **Note 5**).
3. Transfer cell suspension to a chilled microcentrifuge containing 0.7 g of acid-washed glass beads.
4. Lyse cells by placing the microcentrifuge tube containing the cell suspension onto the disruptor vortex for 1 min at 4 °C.
5. After vortex, let suspension stand at 4 °C for 1 min.
6. Repeat **steps 4** and **5** for five more cycles.
7. After lysis, use needle to pierce the bottom of the microcentrifuge tube.
8. Remove needle and place into chilled 15 cm culture tube on ice (*see* **Note 6**).
9. Elute mixture from microcentrifuge tube via centrifugation at 260 × *g* for 1 min at 4 °C.
10. Transfer entire extract from culture tube to fresh microcentrifuge tube on ice.
11. Clear cell debris via centrifugation at 350 × *g* for 1 min at 4 °C.
12. Transfer supernatant to a new microcentrifuge tube on ice.

13. Clear additional cell debris via centrifugation at 800 × *g* for 1 min at 4 °C.
14. Transfer supernatant to chilled new microcentrifuge tube. This is the total cellular protein extract. The extract may be split into aliquots and stored in −80 °C.

3.3 Sedimentation Assay

1. Quantify total protein extract and take equal amount of protein from all samples.
2. Normalize volume with buffer B containing 0.5 % (v/v) protease inhibitors.
3. Sediment protein by centrifugation at 15,000 × *g* for 15 min at 4 °C.
4. Transfer supernatant to separate microcentrifuge tube. This fraction contains soluble proteins. Store on ice for the remainder of the procedure. The remaining pellet fraction contains insoluble proteins.
5. Add 400 μL of wash buffer to the tube containing the protein pellet.
6. Vortex at setting 3 for 10 s.
7. Centrifuge at 15,000 × *g* for 15 min to collect the protein pellet.
8. Discard supernatant and repeat **steps 5**–**7**.
9. Add original amount of buffer B containing 0.5 % (v/v) protease inhibitors to the protein pellet.
10. Vortex sample at 4 °C to completely dissolve the pellet.
11. The resulting fractions may be analyzed via SDS-PAGE or Filter Trap Assay.

3.4 Filter Trap Assay

1. Prepare protein fractions as described under Subheading 3.3 starting with 2,000 μg of total protein lysate. Adjust volume to set concentration to 5 μg/μL.
2. Serially dilute both of the fractions for each experimental sample: Start with 75 μL of the sample plus 125 μL buffer B containing 0.5 % (v/v) protease inhibitors. Mix well by gentle vortex (setting 3) for 10 s. Label this mixture as D1.
3. For dilution 2 (D2), take 100 μL of D1 and add that to 100 μL of buffer B containing 0.5 % (v/v) protease inhibitors. Mix by gentle vortex for 10 s.
4. For dilution 3 (D3), take 100 μL of D2 and add that to 100 μL of buffer B containing 0.5 % (v/v) protease inhibitors. Mix by gentle vortex for 10 s.
5. For dilution 4 (D4), take 100 μL of D3 and add that to 100 μL of buffer B containing 0.5 % (v/v) protease inhibitors. Mix by gentle vortex for 10 s. Discard 100 μL.

6. Normalize volume of all samples and place on ice. A minimum volume of 200 μL is recommended.
7. Assemble Bio-Dot SF apparatus as described in the product manual.
8. Soak 3× Bio-Dot SF filter paper in wetting buffer for 10 min.
9. Cut a 9 × 12 cm sized piece of 45 μm PVDF membrane and place in 100 % (v/v) methanol for 2 min.
10. After activation, discard methanol and rinse membrane with water. Transfer the membrane to wetting buffer.
11. Place the filter papers above the support plate followed by the activated membrane on top of the filter paper.
12. Place the sample template on top of the membrane and tighten the screws on the sample template.
13. Attach the apparatus to the vacuum source. Set the valve in the apparatus to atmosphere setting. Ensure that the vacuum is not turned on.
14. To equilibrate the wells, add 100 μL of buffer B to all wells and turn on vacuum for 30 s to filter the buffer through. Ensure that the membrane does not dry out during filtration (*see* **Note 7**).
15. Turn off vacuum and discard any leftover liquid in the wells.
16. Load the samples on the wells starting with the most dilute for each set (in the order of D4, D3, D2, and then D1). Fill all unused wells with buffer (same volume as sample) and seal them off using Scotch Tape.
17. Let sample stand in the wells for 10 min at room temperature to allow for protein binding to the membrane.
18. After incubation, filter the sample through by turning on the vacuum source (*see* **Note 8**).
19. Seal off wells with Scotch Tape as soon as the sample has filtered through to prevent drying (*see* **Note 9**).
20. Add 200 μL of buffer B containing 0.5 % (v/v) protease inhibitors to each sample well and let stand for 2 min at room temperature.
21. Repeat the filtration process as described in **steps 18** and **19**.
22. After filtration, leave the vacuum on and unscrew the sample template.
23. Turn off the vacuum, remove the membrane and let it air-dry completely.
24. Activate the membrane in methanol as described in **steps 9** and **10**.
25. Place the membrane into Coomassie Blue R-250 solution and stain for 15 min on a rocking platform.

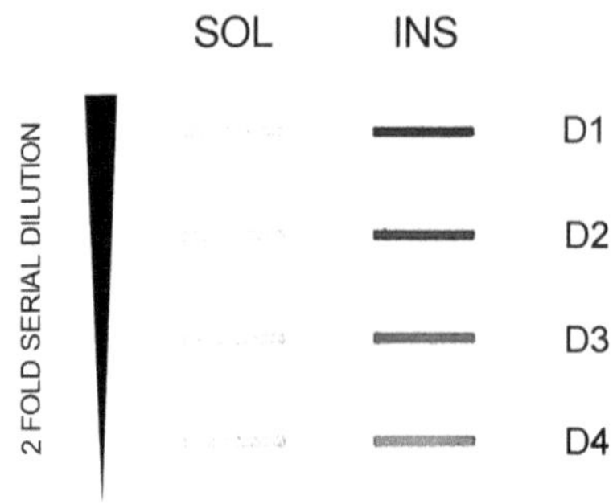

Fig. 1 Filter trap analyses of insoluble protein aggregates. Twofold serial dilutions of the soluble (SOL) and insoluble (INS) protein fractions were filtered through a 0.45 μM PVDF membrane and stained with Coomassie Blue solution. The labels on the *right* indicate the different dilutions (D1–D4)

26. After staining, remove the membrane and place into the destain solution (*see* **Note 10**).
27. Wash the membrane on rocking platform for 10 min, changing solution every 5 min.
28. After destaining, wash the membrane with water and dry completely.
29. Analyze the resulting bands via densitometry (Fig. 1). (We use the ImageJ software for densitometry analyses.)

3.5 Immunoprecipitation

1. Add 2 mL of DMF to 60 mg of dry M-270 epoxy Dynabeads.
2. Vortex the beads for 2 min and keep at 4 °C. The beads may be stored indefinitely at 4 °C.
3. Vortex the stock of beads for 1–2 min and transfer 200 μL of beads into a microcentrifuge tube (*see* **Note 11**).
4. Place tube on magnetic stand to separate beads from DMF for 1 min and discard the supernatant.
5. Wash the beads with 1 mL of sodium phosphate buffer: gently resuspend the beads in the buffer by rotation and then place tube on magnet for 1 min.
6. Discard supernatant and repeat **step 5** for additional two times.
7. Immediately, add 230 μL of sodium phosphate buffer to the dry beads.
8. Add 10 μL of antibody (10 mg/mL stock concentration) to the mixture and vortex for 15 s.
9. Add 120 μL of ammonium sulfate buffer to the mixture.
10. Incubate the mixture at 37 °C for 16–24 h on a nutator.
11. Following incubation, wash the beads four times with 1 mL of PBS as described in **step 5**.

12. Discard supernatant and resuspend the beads in 1 mL of PBS containing 1 % (v/v) Triton-X 100 and wash for 10 min on nutator.
13. Discard supernatant and wash the beads twice with 1 mL PBS for a 1 min.
14. Discard supernatant and perform four washes, 10 min each with 1 mL of 0.1 M citric acid.
15. Discard supernatant and repeat **step 13**.
16. Resuspend the antibody-conjugated beads in 400 μL of PBS.
17. Store beads at 4 °C. The antibody-conjugated beads should be used within 14 days of preparation.
18. Grow cells to OD_{660} between 0.6–0.8 and collect as described in Subheading 3.1.
19. Resuspend the cells in 300 μL of buffer C containing 0.5 % (v/v) protease inhibitors.
20. Lyse the cells and elute as described in Subheading 3.2, **steps 4–10**.
21. Centrifuge lysate at 20,000 × g for 15 min at 4 °C.
22. Collect supernatant and transfer to a fresh microcentrifuge tube.
23. Determine protein concentration.
24. Aliquot 15 mg of protein lysate.
25. Add 50 μL of conjugated magnetic beads to lysate.
26. Mix sample via rotation at 4 °C for 3 h.
27. Place sample on magnetic stand and wash the beads with 1 mL of cold buffer C for a total of five times.
28. After final wash, discard the residual liquid.
29. Resuspend the beads in 40 μL of buffer D.
30. Elute co-purified protein by heating the mixture at 65 °C for 10 min.
31. Place tube on the magnet to separate the beads.
32. Transfer supernatant to a new microcentrifuge tube.
33. Add 2-β-mercaptoethanol to a final concentration of 100 mM.
34. Boil sample at 100 °C for 5 min and analyze via SDS-PAGE and silver staining.

3.6 Vacuolar Staining

1. Grow cells in 5 mL of YPD medium to an OD_{660} between 0.4 and 0.5.
2. Collect the cells by centrifugation at 4,600 × g for 5 min at room temperature.
3. Discard supernatant and resuspend the cells in 2 mL of YPD.

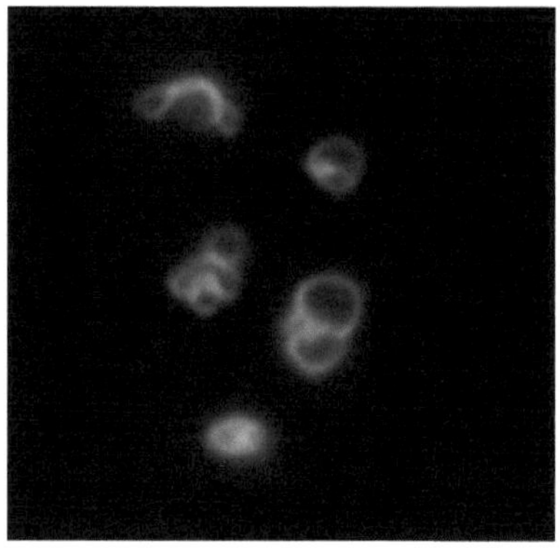

Fig. 2 Vacuolar morphology in yeast. *S. cerevisiae* BY4741 cells were stained with the lipophilic styryl dye, FM 4–64 for 1 h. Images of stained cells were captured using AxioVision 4.8 software on a Zeiss Axio Observer Z1 microscope (63× magnification). The above representation is a magnification of the area of interest depicting variable morphology and number of vacuoles

4. Dissolve the FM dye to a stock concentration of 100 μM in 1× PBS, pH 7.4. Protect from light and place on ice.
5. Add 20 μl of the stock to each sample (final concentration of FM dye is 1 μM).
6. Cover tubes with aluminum foil to protect dye from light.
7. Label vacuoles by further incubating cells with dye at 30 °C, 225 rpm for 1 h (*see* **Note 12**).
8. After incubation with dye, immediately place the cells on ice.
9. Add 270 μL of 37 % (v/v) formaldehyde solution to the sample and further incubate on ice for 15 min.
10. Transfer mixture to microcentrifuge tube and collect the cells by centrifugation 5,600 × *g* for 2 min.
11. Discard supernatant and resuspend the cells in 500 μL of 1× PBS.
12. Collect the cells via centrifugation as in **step 10**.
13. Resuspend the cells in 200 μL of 1× PBS. The cells may be stored at 4 °C for 24 h.
14. Spot 15–20 μL of fixed cells on microscope slide.
15. Place coverslip. Protect sample from light.
16. View and take pictures of the cells on the microscope (Fig. 2).

3.7 DNA Staining

1. Grow cells in 5 mL of YPD media to an OD_{660} between 0.4 and 0.5.
2. Transfer 1 mL of the culture to a microcentrifuge and collect cells via centrifugation at 1,400 × *g* for 5 min.
3. Resuspend the cells in 1 mL of cold 70 % (v/v) ethanol.

4. Incubate at 4 °C for 12 h (*see* **Note 13**).
5. Discard supernatant and resuspend the cells in 1 mL of 50 mM sodium citrate.
6. Collect the cells via centrifugation at 1,400 × *g* for 5 min at 4 °C.
7. Resuspend the cells in 500 μL of 50 mM sodium citrate containing 0.2 mg/mL RNase A.
8. Incubate cells at 37 °C for 20–24 h.
9. Add SYTOX Green dye to a final concentration of 3.7 μM.
10. Incubate at room temperature for 1 h in the dark.
11. Collect the cells via centrifugation in microcentrifuge at 5,600 × *g* for 2 min.
12. Discard supernatant and resuspend the cells in 400 μL of 1× PBS.
13. Collect the cells and resuspend in 400 μL of 1× PBS.
14. The cells may be viewed under the microscope or sorted via flow cytometry.

4 Notes

1. We buffer our YPD medium to a pH 3.5 using HCl. This medium is used throughout all experimentation. To prepare 500 mL of acidic YPD medium, dissolve 5 g of yeast extract and 10 g of Bacto Peptone in 425 mL of water. Add HCl to the medium until the pH reaches 3.5. The solution will have a "grainy" appearance. Make up the volume to 450 mL and autoclave. Make 20 % (w/v) dextrose solution and filter-sterilize. Add 50 mL of the dextrose solution to the autoclaved solution (2 % final concentration) to make the final acidic YPD medium.
2. The constant or crystallizable Fc region of Rabbit IgG binds with high affinity to the "protein A" component of the TAP tag. This antibody can be replaced with other antibodies that bind protein A, another epitope within the TAP tag or a different protein of interest.
3. When collecting cells or starting any treatment, ensure that the OD_{660} is similar across all samples ($OD_{660} \pm 0.05$).
4. To ensure credibility of heat-stressed samples, limit the time taken to collect the sample to 10 min. Cells may be collected at a higher centrifugal force (6,000–8,000 × *g*) for less amount of time (2–3 min) for this time point.
5. Buffer B is more hypotonic with a lower NP-40 concentration and is ideal for preparation and analysis of soluble and insoluble

protein fractions. For conducting normal immunoblots, prepare extracts using buffer A.

6. For efficient elution, ensure that the needle completely pierces the microcentrifuge tube. The out pour of the liquid when removing the needle is indicative of whether or not the tube has been sufficiently pierced. If elution is not complete, repeat the centrifugation in **step 9** for 15–20 s.
7. During the equilibration step, the 30 s filtration time may be excessive depending on the vacuum strength. This time may be adjusted accordingly. More importantly, monitor the filtration to ensure that the wells do not dry out completely. Partial or complete drying out of the membrane will affect binding of proteins to the membrane. Once the buffer in a majority of the wells has filtered through, turn off the vacuum and remove any excess liquid from the remaining wells. This step will also be indicative of the uniformity of the vacuum.
8. During filtration, ensure that there is no buildup of fluid within the tubing and around the flow valve to prevent backflow which may ruin the experiment if not corrected quickly. This can be prevented by increasing the vacuum strength by blocking the atmosphere line in the flow valve. We highly recommend getting acquainted with the apparatus before proceeding with the experiment. We usually see buildup around the flow valve after assembling the apparatus due to excess retention of buffer in the filter paper. In such instances, open the valve to drain the buffer before loading the samples.
9. Sealing off the wells helps maintain uniform vacuum across the apparatus and prevents drying. We find that sometimes the filtration can be length depending on the amount of protein used. In such instances keep adding buffer (few drops from a P1000 pipet) to the sample wells that have filtered through to prevent drying of the well until all samples have filtered through. This should be done in addition to the wash step.
10. To avoid nonuniform destaining of the membrane, do not pour the destain solution directly onto the membrane. Always place membrane onto the solution. Monitor the bands on the membrane to make sure that the higher dilutions (D3, D4) do not destain completely. The incubation time in the destain solution can be adjusted accordingly.
11. Keep all of the samples and buffers which are to be utilized on ice throughout the procedure.
12. Incubating the cells with the FM stain for a longer duration of time (1.5–2 h) may yield in better staining of the vacuoles. However, this may be accompanied by increased background staining. The concentration of the stain in culture may be increased but we do recommend a minimum 1 h incubation

period with the stain. In order to avoid high background staining, additional washes may be performed as described in **steps 11** and **12**. For staining vacuoles through the stress-recovery time course described in Subheading 3.1, incubate the cells with the stain for a minimum of 1 h before the 42 °C for 1 h treatment.

13. Ethanol-fixed cells can be stored for 36–48 h at 4 °C or for longer period (5 days) at −20 °C.

References

1. Aravind L, Koonin EV (2002) Classification of the caspase–hemoglobinase fold: detection of new families and implications for the origin of the eukaryotic separins. Proteins 46: 355–367
2. Wong AH, Yan C, Shi Y (2012) Crystal structure of the yeast metacaspase Yca1. J Biol Chem 287:29251–29259
3. Uren AG, O'Rourke K, Aravind L et al (2000) Identification of paracaspases and metacaspases: two ancient families of caspase-like proteins one of which plays a key role in MALT lymphoma. Mol Cell 6:961–967
4. Tsiatsiani L, Van Breusegem F, Gallois P et al (2011) Metacaspases. Cell Death Differ 18: 1279–1288
5. Madeo F, Herker E, Maldener C et al (2002) A caspase-related protease regulates apoptosis in yeast. Mol Cell 9:911–917
6. Carmona-Gutierrez D, Fröhlich KU, Kroemer G et al (2010) Metacaspases are caspases. Doubt no more. Cell Death Differ 17:377
7. Lee REC, Puente LG, Kaern M et al (2008) A non-death role of the yeast metacaspase: Yca1p alters cell cycle dynamics. PLoS ONE 3:e2956
8. Madeo F, Carmona-Gutierrez D, Ring J et al (2009) Caspase-dependent and caspase-independent cell death pathways in yeast. Biochem Biophys Res Commun 382:227–231
9. Guscetti F, Nath N, Denko N (2005) Functional characterization of human pro-apoptotic molecules in yeast S. cerevisiae. FASEB J 19:464–466
10. Fernando P, Megeney LA (2007) Is caspase-dependent apoptosis only cell differentiation taken to the extreme? FASEB J 21:8–17
11. Shrestha A, Megeney LA (2012) The non-death role of metacaspase proteases. Front Oncol 2:78
12. Mitchell L, Lambert JP, Gerdes M et al (2008) Functional dissection of the NuA4 histone acetyltransferase reveals its role as a genetic hub and that Eaf1 is essential for complex integrity. Mol Cell Biol 28:2244–2256
13. Lee REC, Brunette S, Puente LG et al (2010) Metacaspase Yca1 is required for clearance of insoluble protein aggregates. Proc Natl Acad Sci U S A 107:13348–13353
14. Shrestha A, Brunette S, Puente LG et al (2013) The role of Yca1 in proteostasis. Yca1 regulates the composition of the insoluble proteome. J Proteomics 81:24–30
15. Parsell DA, Kowal AS, Singer MA et al (1994) Protein disaggregation mediated by heat-shock protein Hsp104. Nature 372:475–478
16. Glover JR, Lindquist S (1998) Hsp104, Hsp70, and Hsp40: a novel chaperone system that rescues previously aggregated proteins. Cell 94:73–82
17. Cashikar AG, Duendald M, Lindquist SL (2005) A chaperone pathway in protein disaggregation. Hsp26 alters the nature of protein aggregates to facilitate reactivation by Hsp104. J Biol Chem 280:23869–23875
18. Haslbeck M, Miess A, Stromer T et al (2005) Disassembling protein aggregates in the yeast cytosol. The cooperation of Hsp26 with Ssa1 and Hsp104. J Biol Chem 280:23861–23868
19. Rand JD, Grant CM (2006) The thioredoxin system protects ribosomes against stress induced aggregation. Mol Biol Cell 17:387–401
20. Journo D, Winter G, Abeliovich H (2008) Monitoring autophagy in yeast using FM4-64 fluorescence. Methods Enzymol 451:79–88

Chapter 15

Plant Metacaspase Activation and Activity

Elena A. Minina, Simon Stael, Frank Van Breusegem, and Peter V. Bozhkov

Abstract

Metacaspases are essential for cell death regulation in plants. Further understanding of biochemistry of metacaspases and their molecular function in plant biology requires a set of robust methods for detection of metacaspase activation and quantitative analysis of corresponding proteolytic activity. Here we describe methods for purification of recombinant metacaspases, measurement of enzymatic activity of recombinant and endogenous metacaspases in vitro and in cell lysates, respectively, and finally detection of metacaspase activation in vivo. Additionally, an in vitro metacaspase protein substrate cleavage assay based on the cell-free production of substrate protein followed by proteolysis with recombinant metacaspase is presented. These methods have been originally developed for type II metacaspases from *Arabidopsis* and Norway spruce (*Picea abies*), but they can be used as templates for type I metacaspases, as well as for type II metacaspases from other species.

Key words Cell lysate, Metacaspase, Natural substrate, Programmed cell death, Proteolytic activity, Synthetic substrate

1 Introduction

Most plants have two types of structurally distinct metacaspases [1–3]. Type I metacaspases have a variable sized N-terminal extension upstream of the p20 caspase-like domain and a short (<40 aa-long) linker between the p20 and p10 domains. Such structure is not unique to plant metacaspases and is shared with metacaspases from fungal and protozoan lineages [3]. Type II metacaspases lack a prodomain, while possessing a longer (>100 aa-long) interdomain linker. This type of metacaspases is plant-specific, potentially derived from primary endosymbiosis [3], and therefore cannot be found in other organisms. The number of metacaspase genes in the genomes of different plant species varies considerably, from one or two in some of green algae to several homologues of each type (I and II) in angiosperms [2].

Peter V. Bozhkov and Guy Salvesen (eds.), *Caspases, Paracaspases, and Metacaspases: Methods and Protocols*, Methods in Molecular Biology, vol. 1133, DOI 10.1007/978-1-4939-0357-3_15, © Springer Science+Business Media New York 2014

While fungal and protozoan metacaspases have been implicated in a range of biological processes [2, 4], the physiological role of plant metacaspases has so far been only linked to cell death regulation. In *Arabidopsis*, type I metacaspases AtMC1 and AtMC2 antagonistically control hypersensitive response-associated cell death activated by intracellular immune receptors [5], whereas type II metacaspases are positive regulators of cell death induced by abiotic (AtMC4 and AtMC8) and biotic stress (AtMC4) [6, 7]. mcII-Pa type II metacaspase of Norway spruce is required for vacuolar cell death in the embryo-suspensor, which is essential for correct embryonic patterning [8].

Considering a growing evidence for metacaspases to represent bona fide plant cell-death proteases functionally equivalent to metazoan caspases, and assuming potential involvement of plant metacaspases in other, cell death unrelated, processes (by analogy with fungal and protozoan metacaspases [9], there is a need for a set of robust methods to detect activation and to measure proteolytic activity of recombinant and endogenous plant metacaspases. These methods are also required to advance our understanding of molecular mechanisms regulating metacaspase activation.

Similar to caspases, maturation of metacaspase enzyme involves autocatalytic processing of the zymogen [2]; thus autoprocessing can be used as a marker of activated metacaspase. Detection of processed form of metacaspase in cell lysates or in situ should however be interpreted with caution, since inhibitor binding (e.g., serpin [10]) can antagonize autoprocessing-mediated activation. Active metacaspase is a monomer [11, 12] for type I metacaspases (PV Bozhkov and AP Smertenko, unpublished results, for type II metacaspase mcII-Pa), a feature diverging metacaspases and caspases, because the latter require dimerization as a part of activation mechanism [13]. Another feature contrasting metacaspases and caspases is a requirement of Ca^{2+} for activation. While caspases are Ca^{2+}-independent proteases, a vast majority of metacaspases studied so far require millimolar concentrations of Ca^{2+} to become active in vitro [14–16]. All above-described biochemical characteristics should be taken into consideration when setting up assays for detecting active forms of metacaspases and measuring their activity.

However, the chief property of a protease that must be known for correct measurement of its catalytic activity is substrate cleavage specificity. In contrast to Asp-specific caspases, metacaspases cleave peptide bonds C-terminal to Arg or Lys residues [2, 14]. We have previously pointed to inapplicability of caspase substrates for metacaspase activity assays and urged to use peptidic substrates with Arg or Lys at the P1 position [2] (*see* **Note 1**).

There are two further points that require special attention when measuring metacaspase activity in the total or fractionated plant cell extracts. First, it remains unknown how much the extended substrate specificity (i.e., P2–P4 and P1′–P4′) differs between type I and type II metacaspases, as well as between different metacaspases of the same type. Yet extended specificity of individual plant metacaspases appears to be rather loose, as follows from screening of combinatorial peptide library with recombinant AtMC9 [10], AtMC9 degradome [17] and analysis of four mcII-Pa cleavage sites present in its natural target protein Tudor staphylococcal nuclease (TSN) [18]. This implies that a given substrate could be the target of two or several metacaspases and also provides explanation for a high degree of redundancy among metacaspases. Second, synthetic peptides with Arg(P1) or Lys(P1) could be cleaved by proteases other than metacaspases, e.g., separases, cathepsins, or subtilisins. Genetic knockouts or knockdowns of metacaspase gene(s) are far and away the best tools to assess which fraction of Arg- and/or Lys-specific proteolytic activity is attributed to metacaspases [15]. The bottom line is that no truly specific synthetic substrates (and for the same reasons, inhibitors) are available and that strict caution must be exercised in ascribing observed activity to metacaspases in general, to their specific type (I or II) or to individual metacaspase. Accordingly, quantitative data obtained in the assays measuring proteolysis of Arg(P1)- or Lys(P1)-containing synthetic peptides reflect "metacaspase-*like* activity", unless there is an experimental evidence for the absence of other types of Arg- and/or Lys-specific proteases in the given biological system. In general it is preferably to assess metacaspase activity by using a combination of different methods as described in this chapter.

Two main approaches are currently used to detect metacaspase activation and measure its proteolytic activity: (1) detection of processed metacaspase and/or reduced zymogen forms by immunoblotting and (2) substrate cleavage assays using synthetic peptidic substrates conjugated with fluorescent moiety. New evidences about in vivo substrates of metacaspases are appearing [17, 18]. In case of Norway spruce embryos where type II metacaspase mcII-Pa is known to cleave TSN protein, immunoblotting analysis of full-length TSN and its fragments is an additional tool to evaluate mcII-Pa activity. In the following sections, we describe above-mentioned methods step by step and how they are used in our laboratories.

2 Materials

2.1 Production of Recombinant Metacaspases

1. Sterile conical flasks (50–500 mL).
2. Shaker at 37 and 28 °C.
3. Spectrophotometer to measure OD at 600 nm.
4. Sonicator or French press.
5. Nickel-nitrilotriacetic acid (Ni-NTA) resin (Qiagen, 30210 or similar).
6. Isopropyl β-D-thiogalactoside (IPTG) (Thermo Scientific, R1171 or similar).

7. Luria Broth (LB) medium: tryptone 10 g/L, yeast extract 5 g/L, NaCl 10 g/L. Add 15 g/L of Bacto Agar to solidify the medium, when required. Autoclave for 20 min at 120 °C.
8. Bacterial strain for IPTG-inducible expression (Stratagene, 230245 or similar).
9. Sonication buffer: 100 mM Tris–HCl pH 8.0, 300 mM NaCl, 10 mM ethylenediaminetetraacetic acid (EDTA), 100 μM phenylmethanesulfonyl-fluoride (PMSF), 20 μM pepstatin A, 10 μM leupeptin.
10. Binding buffer: 50 mM NaH_2PO_4 pH 8.0, 300 mM NaCl, 10 mM imidazole, 10 mM β-mercaptoethanol.
11. Washing buffers for native conditions: 50 mM NaH_2PO_4 pH 8.0, 10 mM β-mercaptoethanol, 50–100 mM imidazole.
12. Elution buffer for native conditions: 50 mM NaH_2PO_4 pH 8.0, 10 mM β-mercaptoethanol, 250 mM imidazole.
13. Lysis buffer for denaturing conditions: 100 mM NaH_2PO_4 pH 8.0, 100 mM Tris–HCl, 6 M guanidine–HCl, 100 μM PMSF, 20 μM pepstatin A, 10 μM leupeptin. Adjust pH to 8.0.
14. Washing buffers for denaturing conditions: 100 mM NaH_2PO_4 pH 8.0, 10 mM β-mercaptoethanol, 8 M urea. Prepare a series of buffers with a pH gradient from pH 8.0 to pH 6.0.
15. Elution buffer for denaturing conditions: 100 mM NaH_2PO_4 pH 8.0, 10 mM β-mercaptoethanol, 8 M urea. Adjust pH to 4.5.
16. Bradford reagent (Thermo Scientific, 23200 or similar).
17. Dialysis buffer: 50 mM 4-(2-hydroxyethyl)-1-piperazineethanesulfonic acid (HEPES) pH7.5, 150 mM NaCl, 5 mM dithiothreitol (DTT).
18. Dialysis tubing or cassette.

2.2 Measurement of Recombinant Metacaspase Activity

1. Reaction buffer: 50 mM HEPES (if optimal pH is not known a range of pH from 4.0 to 9.0 with 0.5 increment should be used for optimization of the reaction), 0.1 % (w/v) 3-[(3-cholamido-propyl)dimethylammonio]-1-propanesulfonate (CHAPS), 5 mM DTT, 50 mM $CaCl_2$ (do not add for negative controls).
2. 7-Amino-4-methylcoumarin (AMC)-labeled substrate, for example Boc-Glu(OBzl)-Gly-Arg-AMC-HCl (Bachem I-1545 or similar) or Ac-VRPR-AMC (Bachem I-1965). Make a 10 mM stock solution in DMSO, aliquot, and store at −20 °C.
3. AMC (Bachem Q-1025 or similar). Make a 100 mM stock solution in DMSO, aliquot, and store at −20 °C. Before use dilute additionally with reaction buffer to 24 μM in order to prepare standard solutions.

4. 96-well flat bottom tissue culture plates. Using black opaque plates will decrease the crosstalk and background noise during fluorescence detection.
5. Plate reader to detect fluorescence at excitation/emission 360–380 nm/440–460 nm (Omega Fluostar, BMG or similar).

2.3 Measurement of Metacaspase Activity in Cell Lysates

1. Materials described under **items 1–5** in Subheading 2.2.
2. Table centrifuge at 4 °C.

2.4 Detection of Metacaspase Activation In Vivo

1. Reaction buffer: 50 mM HEPES pH 7.5, 0.1 % (w/v) CHAPS, 5 mM DTT, 100 μM PMSF, 10 μM pepstatin A.
2. 1× Laemmli buffer: 50 mM Tris–HCl pH 6.8, 1 % (w/v) sodium dodecylsulfate (SDS), 4 % (v/v) glycerol, 0.75 M β-mercaptoethanol, 0.005 % bromophenol blue. Prepare a 5× stock, store at room temperature. Add β-mercaptoethanol to an aliquot right before use.
3. 12 % Resolving gel: 12 % Acrylamide–Bis-acrylamide (29:1), 375 mM Tris–HCl pH 8.8, 0.1 % (w/v) SDS.
4. 4 % Stacking gel: 4 % Acrylamide–Bis-acrylamide (29:1), 125 mM Tris–HCl pH 6.8, 0.1 % (w/v) SDS.
5. SDS-PAGE Running buffer: 25 mM Tris base, 24 mM glycine, 0.1 % (w/v) SDS.
6. Transfer buffer: 25 mM Tris base, 24 mM glycine, 10 % ethanol.
7. PBST: 4 mM Na_2HPO_4, 4 mM NaH_2PO_4, 1.5 mM KH_2PO_4, 137 mM NaCl pH 7.2, 0.1 % (v/v) Tween-20.
8. Secondary antibody (Amersham, RPN 4201, RPN 4301, or similar).
9. ECL (Amersham RPN2232 or similar).
10. Coomassie Brilliant Blue Solution: 1/4.5/4.5 (v/v/v) glacial acetic acid–ethanol–water, 0.05 mg/mL Coomassie Brilliant Blue G250.
11. Stripping buffer: 25 mM glycine, 1 % (w/v) SDS.
12. Equipment for SDS-PAGE and protein transfer.
13. PVDF membrane (Bio-Rad 162-0177 or similar).
14. Table centrifuge.

2.5 Metacaspase Protein Substrate Cleavage Assay

1. TNT® Coupled Transcription/Translation System (Promega) (*see* **Note 2**).
2. Plasmid (0.5 μg/μl) containing the coding sequence (CDS) for the metacaspase protein substrate of interest in a suitable expression vector (*see* **Note 2**).

3. Radiolabeled L-[^{35}S]methionine (PerkinElmer EasyTag™ Cat. # NEG709A, or similar.
4. RNasin® Ribonuclease Inhibitor (Promega).
5. Recombinant metacaspase protein stock solution. Prepare dilutions in an 8-well eppi strip at 4× the final reaction concentration in buffer containing 50 % glycerol and 25 mM HEPES (pH 7.5). For recombinant *Arabidopsis thaliana* metacaspase 4 (rAtMC4), we standardly use a dilution series (final concentration) of 4.0, 1.0, 0.5, 0.25, 0.125, and 0.0 μM. Inactive rAtMC4, in which the active site cysteine is mutated to alanine (rAtMC4-C139A), acts as a negative control and is used at 4.0 μM. Store at −20 °C.
6. 4× Reaction buffer: 200 mM HEPES pH 7.5, 600 mM NaCl, 40 % glycerol, 200 mM $CaCl_2$. Add 40 mM DTT to an aliquot right before use.
7. 5× Laemmli buffer + EGTA: 250 mM Tris–HCl pH 6.8, 10 % (w/v) SDS, 20 % (v/v) glycerol, 0.1 % bromophenol blue, 250 mM EGTA. Add 10 % β-mercaptoethanol to an aliquot right before use. This is aliquoted in an 8-well eppi strip.
8. 12 % Resolving gel: 12 % Acrylamide–Bis-acrylamide (29:1), 375 mM Tris–HCl pH 8.8, 0.1 % (w/v) SDS.
9. 4 % Stacking gel: 4 % Acrylamide–Bis-acrylamide (29:1), 125 mM Tris–HCl pH 6.8, 0.1 % (w/v) SDS.
10. SDS-PAGE Running buffer: 25 mM Tris base, 24 mM glycine, 0.1 % (w/v) SDS.
11. Whatman 3MM paper (Whatman).
12. SDS-PAGE gel dryer (Bio-Rad, Model 583 or similar).
13. 8-well eppi strips and multichannel pipette (volume 5–50 μl).
14. 96-well heat block or water bath at 30 and 85 °C.
15. Phosphor Imager™ 445 SI (Amersham Pharmacia Biotech Benelux) and Storage Phosphor Screen (Molecular Dynamics), or similar.

3 Methods

3.1 Production of Recombinant Metacaspases

Recombinant metacaspases can be used for various assays, e.g., analysis of substrate specificity, preparation of antibodies, crystallography, mutagenesis for functional mapping. The most time- and cost-efficient method of production of recombinant metacaspase is expression in bacterial cells. Since metacaspases undergo autoprocessing and subsequent self-inactivation in vitro, the conditions for metacaspase expression and purification must be adjusted to decrease proteolytic activity. For instance, induction of expression at low temperatures gives higher yield of purified zymogen. While prolongation of induction time leads to accumulation of high amount of metacaspase in bacterial cells, this also induces intermolecular cleavage of the protease significantly reducing the yield of enzymatically active protease.

Both N- and C-terminal tags can be used for purification of metacaspases without disturbing their proteolytic activity. Introducing metacaspase-specific cleavage sites should be avoided while cloning.

We have established a protocol for producing recombinant Norway spruce metacaspase mcII-Pa using His-tagged system. This system allows purification under either native or denaturing conditions. Although denaturing conditions give higher amounts of pure zymogen that can be used for raising antibodies, they are not recommended for producing metacaspases for enzymatic activity assays due to the need for protein refolding.

1. To avoid expression of metacaspase before induction, clone metacaspase gene under control of lac promoter and make sure that either the plasmid with the metacaspase gene or the *E. coli* strain contain lacI gene. For example, we commonly use pET11 vectors (Novagen) or pDEST17 vector (Gateway, Invitrogen®) with BL21(DE3) Rosetta (Novogen) or BL21(DE3)pLysE (Invitrogen) cells. It is preferable to use freshly transformed cells to avoid any random mutations in the expressed gene.
2. Grow bacterial culture overnight in a small volume at 37 °C.
3. Next morning dilute the culture till $OD_{600} = 0.1$ and incubate it at the same growth conditions until OD_{600} reaches 0.4–0.6.
4. Let the cell culture cool down on ice for 10–15 min.
5. Add IPTG to the final concentration 0.5 mM and incubate the culture on a shaker in the darkness at 28 °C for 2 h or overnight at 16 °C.
6. Pellet down the cells for 15 min at 4,000 × *g* at 4 °C. The cell pellet can be stored at −20 °C for at least several weeks without affecting the proteolytic activity of metacaspase.

3.2 Purification Under Native Conditions

1. Gently resuspend the cells in a small volume of cold sonication buffer and lysate by sonicating at the amplitude 1 μm for 20 pulses: 20 s on/20 s off. Make sure that the cell suspension is not warmed up during sonication. Pellet the cell debris by centrifugation for 10 min at 17,000 × *g* at 4 °C. Add imidazole, pH 8.0 to the cleared lysate to the final concentration 10 mM and β-mercaptoethanol to the final concentration 10 mM.
2. Alternatively, the cells can be resuspended in binding buffer and lysed using French press.
3. Load the cell lysate on Ni-NTA column and let it pass through the column several times. Incubating for 2 h at 4 °C with Ni-NTA resin can additionally increase the yield.

4. Wash the column with several volumes of binding buffer.
5. Wash the column with washing buffer for native conditions gradually increasing concentration of imidazole from 50 to 100 mM. Efficiency of washing can be estimated by measuring total protein concentration in washing fractions using Bradford reagent.
6. Load one volume of elution buffer and incubate it on the column for 5 min.
7. Collect the elution fraction and repeat elution several times until there is no more protein detectable with Bradford reagent.
8. Dialyze the protein against dialysis buffer (*see* **Note 3**). Fractions can be stored at 4 °C for several days. For longer storage samples can be kept at −20 °C in the presence of 50 % glycerol and 0.02 % NaN_3 or frozen as 30 μL drops in liquid nitrogen and stored at −70 °C. Avoid repeated freezing–thawing cycles, as it affects protein activity.
9. Measure protein concentration using Bradford reagent. For better accuracy it is important to check compatibility of the Bradford kit with high concentration of DTT prior to the test. If needed the samples can be diluted ten times to decrease DTT concentration.

3.3 Purification Under Denaturing Conditions

1. Resuspend the cells in the lysis buffer and incubate for 1 h at room temperature gently shaking.
2. Pellet down cell debris by centrifugation at 17,000 × *g* at room temperature for 15 min.
3. Load the cleared lysate on a Ni-NTA column and let it pass through the column several times. Incubating for 1 h at room temperature with Ni-NTA resin can additionally increase the yield.
4. Wash the column with several volumes of the lysis buffer.
5. Wash the column with washing buffers for denaturing conditions gradually decreasing pH from 8.0 to 6.0. Efficiency of washing can be estimated by monitoring total protein concentration in washing fractions with Bradford reagent.
6. Load one volume of elution buffer and incubate it on the column for 5 min.
7. Collect the elution fraction and repeat elution several times until there is no more protein detectable with Bradford. Elution fractions can be stored at 4 °C for several weeks or at −20 °C for a longer period of time.
8. If required protein can be partially renatured by gradual dialysis against dialysis buffer (*see* **Notes 4** and **5**).

3.4 Measurement of Recombinant Metacaspase Activity

Although different metacaspases have many biochemical features in common (e.g., active as monomers and Ca^{2+}-dependent), the optimal pH range for catalysis varies significantly necessitating optimization of conditions for each individual metacaspase. We have optimized activity assay for recombinant 6× His-tagged mcII-Pa using AMC-labeled fluorogenic substrates FESR-AMC (custom-synthesized; matches mcII-Pa autoprocessing site), EGR-AMC (e.g., from Bachem), or Ac-VRPR-AMC (Bachem I-1965). AMC is a fluorescent compound, which is highly stable in aqueous solutions and changes its maxima of excitation/emission after being released from a peptide upon carboxypeptidase activity [19, 20]. Detection of free AMC fluorescence gives a high precision and sensitivity readout of proteolytic activity in the reaction.

1. Keep all samples on ice before measurement.
2. Dilute substrate with reaction buffer to a final concentration of 1 mM.
3. Prepare a blank sample by mixing 152 μL of reaction buffer with 8 μL of 1 mM substrate.
4. Prepare five standard samples by making serial dilutions of 24 μM AMC stock using reaction buffer to obtain 12, 6, 3, 1.5, and 0.75 μM.
5. Pipet 50 μL of each standard solution and the blank solution into 96-well opaque plate in triplicates.
6. Prepare two dilutions of recombinant metacaspase purified in native conditions (*see* Subheading 3.1) with reaction buffer containing 50 mM $CaCl_2$ and without calcium. Final concentration of the diluted protein should be 80 μg/mL (*see* **Note 6**).
7. Pipet 47.5 μL of the diluted metacaspase into 96-well opaque plate (3 ng of metacaspase per reaction) in triplicates.
8. Add 2.5 μl of 1 mM substrate to each well with diluted metacaspase (final concentration of substrate 50 μM).
9. Set the detector gain to 70 % using the 3-μM AMC standard.
10. Set the temperature in the plate reader chamber in the range 25–28 °C.
11. Shake the plate in the plate reader for 10 s and take reads of fluorescence intensity at excitation/emission 360–380 nm/440–460 nm every min making at least 20 flashes per well. Take reads for at least 15 min.
12. Use the standard curve and linear part of the reaction curve to calculate pmol of AMC released per minute per nanogram of metacaspase.
13. Subtract values obtained for samples without calcium from the values of samples containing calcium (*see* **Note 6**).

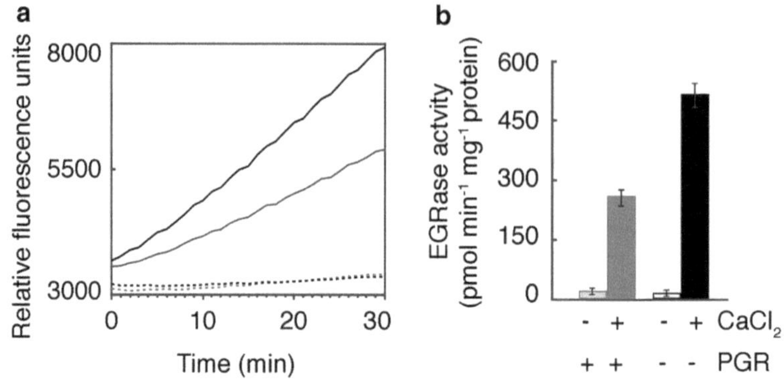

Fig. 1 Cleavage of EGR-AMC in cell lysates prepared from Norway spruce embryogenic culture. Withdrawal of plant growth regulators (PGR), auxin and cytokinin, induces development of early embryos and developmental cell death accompanied by increase in mcII-Pa activity. Cultures were incubated for 5 days either in proliferation medium containing PGR (+PGR) or in embryogenesis inducing medium lacking PGR (−PGR). The cell lysates were prepared with and without 50 mM $CaCl_2$. Proteolytic activity was measured by detecting increase of AMC fluorescence. (**a**) Kinetics of fluorescence increase over time. *Solid line* and *dashed line* curves represent reactions with and without $CaCl_2$, respectively. *Gray* and *black* curves represent reactions containing lysates from cells growing with and without PGR, respectively. (**b**) The kinetics curves shown on panel (**a**) and standard curve were used to convert relative fluorescence units into picomoles of free AMC released in one min per milligram of total protein

3.5 Measurement of Metacaspase Activity in Cell Lysates

This approach is closer to in vivo assay and allows tracking activation or suppression of metacaspase activity in different tissues or at different developmental stages or treatments. One should keep in mind, however, that most organisms contain more than one metacaspase gene and all metacaspases present in the cell lysate will contribute to the detected activity due to low stringency of metacaspase substrate-specificity. The assay can be optimized by using samples where a single metacaspase is expressed or by using multiple metacaspase knockout or knockdown organisms. We have optimized conditions for measuring metacaspase activity in the cell lysates from Norway spruce embryogenic cell cultures (Fig. 1).

1. Grind plant material in liquid nitrogen.
2. Mix 100 mg of the ground material with 100 μL of the reaction buffer without calcium and vortex.
3. Centrifuge the samples at 17,000 × *g* for 15 min at 4 °C.
4. Transfer the supernatant into new Eppendorf tube. If required filter it through four layers of miracloth. Keep samples cold on ice.

5. Dilute 5-μL aliquot of the supernatant ten times with Milli-Q or dH_2O water and use it for estimating total protein concentration using Bradford kit.
6. Make two dilutions of the supernatant with reaction buffer containing 50 mM $CaCl_2$ and without calcium. Final protein concentration in the diluted samples should be 158 μg/mL.
7. Dilute AMC-conjugated substrate with reaction buffer to the final concentration 1 mM.
8. Prepare a blank sample by mixing 152 μL of reaction buffer with 8 μL of 1 mM substrate.
9. Prepare five standard samples by making serial dilutions of 24 μM AMC stock using reaction buffer to obtain 12, 6, 3, 1.5, and 0.75 μM of AMC.
10. Pipet 50 μL of each standard solution and the blank solution into 96-well opaque plate (1,200, 600, 300, 150, 75, and 0 pmol of AMC per well), in triplicates.
11. Pipet 47.5 μL of diluted samples (from **step 6**) into 96-well plate (7.5 μg of total protein per reaction), in triplicates.
12. Add 2.5 μL of 1 mM substrate to each well containing cell lysate samples (final concentration of substrate 50 μM).
13. Set the detector gain to 70 % using the 3-μM AMC standard.
14. Set the temperature in the plate reader in the range 25–28 °C.
15. Shake the plate in the plate reader for 10 s and take reads of fluorescence intensity at excitation/emission 360–380 nm/440–460 nm every min making at least 20 flashes per well. Take reads for at least 15 min.
16. Use the standard curve and linear part of the reaction curve to calculate pmol of AMC released per minute per microgram of total protein.
17. Subtract values obtained for samples without calcium from the values of samples containing calcium, or present the values without calcium as a control (Fig. 1).

3.6 Detection of Metacaspase Activation In Vivo

Most metacaspases require autoprocessing for activation, thus detection of metacaspase auto-cleavage products can be used for assessing its activation in vivo. If metacaspase-specific antibodies are available, autoprocessing can be directly estimated by western blot analysis. Otherwise expression of a tagged metacaspase is required for the assay. Detecting tagged metacaspase instead of using metacaspase-specific antibodies is also preferable if there are multiple metacaspases present in the sample to avoid crosstalk in immunoblotting results.

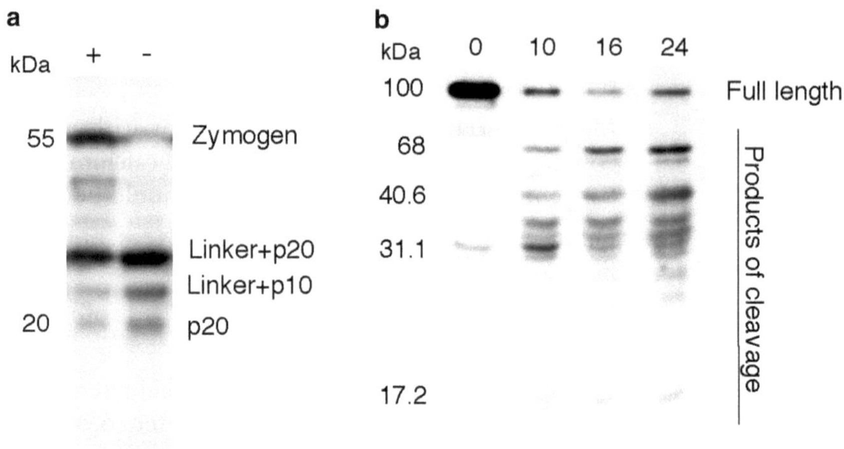

Fig. 2 Autoprocessing of Norway spruce metacaspase mcII-Pa and cleavage of TSN detected by immunoblotting. (**a**) mcII-Pa was activated by transferring embryogenic cells from the medium containing plant growth regulators (PGR; +) to the medium devoid of PGR (–). Zymogen and auto-cleavage products of mcII-Pa were detected by using custom produced mcII-Pa-specific antibody, as described in the protocol. Pattern of the mcII-Pa autoprocessing has been assessed by N-terminal sequencing of products of cleavage [15] (**b**) mcII-Pa was activated by inducing oxidative stress in the 4-day-old cell culture using 10 mM H_2O_2; the samples were taken at 0, 10, 16, and 24 h of treatment. Activity of mcII-Pa was assessed by detecting products of TSN cleavage using custom produced TSN-specific antibody as described in the protocol

Metacaspase activity in vivo can be also assessed by analyzing the efficiency of natural substrate cleavage, e.g., cleavage of TSN by mcII-Pa [18]. TSN is a highly conserved protein and anti-human TSN antibodies are commercially available, but their interaction with TSN from other species should be evaluated in the first place.

We have optimized detection of mcII-Pa activation in vivo during developmental and oxidative stress-induced cell death using custom produced antibodies (Fig. 2).

1. Grind plant material in liquid nitrogen.
2. Mix 100 mg of the ground material with 100 μL of the reaction buffer without calcium and vortex.
3. Add Laemmli buffer and boil the samples for 10 min at 100 °C.
4. Centrifuge the samples at 17,000 × *g* for 10 min.
5. Transfer the supernatant into new Eppendorf tubes.
6. Depending on the tissue, extraction furnishes 1–5 mg/mL of total protein. Load 5–15 μL of each sample on 12 % polyacrylamide gel and run electrophoresis.
7. Transfer proteins on PVDF membrane.
8. Block the membrane for 10 min in 5 % skim milk.

9. Rinse the membrane in PBST buffer and incubate with primary antibody for 1 h at room temperature or overnight at 4 °C.
10. Wash the membrane three times for 5 min in PBST.
11. Incubate with secondary antibody (HRP-conjugated antibody is preferable to get a strong signal).
12. Wash the membrane three times for 10 min in PBST.
13. Detect the signal using ECL substrate.
14. If the membrane was used for detecting metacaspase auto-cleavage products it can be reused to detect cleavage of TSN and vice versa (*see* **Note 7**). Strip the antibodies from the membrane by shaking it in stripping buffer at 37 °C for 15 min. Change the buffer twice.
15. Block the membrane in skim milk as at **step 8** and proceed with staining as usual.
16. For loading control, a housekeeping protein can be detected by western blot analysis using the same samples or the membrane can be stained with Coomassie. After developing the chemiluminescent signal rinse the membrane in PBST and incubate in Coomassie Brilliant Blue solution for 10–15 min.
17. Rinse the membrane in 70 % ethanol to get rid of excessive staining (*see* **Note 8**).
18. Let the membrane dry at room temperature.
19. Scan the membrane. If required loading can be quantified using ImageJ or any similar software (*see* **Note 9**).

3.7 Metacaspase Protein Substrate Cleavage Assay

Metacaspases, as all proteases, exert their function through the cleavage of protein substrates. The following assay is a rapid and easy way to prove the cleavage of potential metacaspase protein substrates by recombinant metacaspase in vitro. Due to the minimal requirements of input material (basically, a suitable plasmid for cell-free protein expression containing the CDS for the protein substrate of interest), a relative high-throughput can be achieved. This is particularly interesting to screen many potential substrate proteins coming from proteome-wide degradome studies, such as from COFRADIC technology [17, 21], or to screen point-mutated variants of a single substrate protein. The readout is based on autoradiography of radiolabeled L-[^{35}S]methionine containing proteins, so best practices for work with radioactive isotopes should be upheld. However, the readout can be replaced by western blotting when using appropriate epitope tags. Recombinant *Arabidopsis thaliana* metacaspase 4 (rAtMC4) is used as an example here (Fig. 3), but the assay should be widely applicable to other metacaspases with similar reaction conditions. Because of the rAtMC4 concentration range the efficacy of cleavage can be assessed between protein substrates in a relatively quantitative manner.

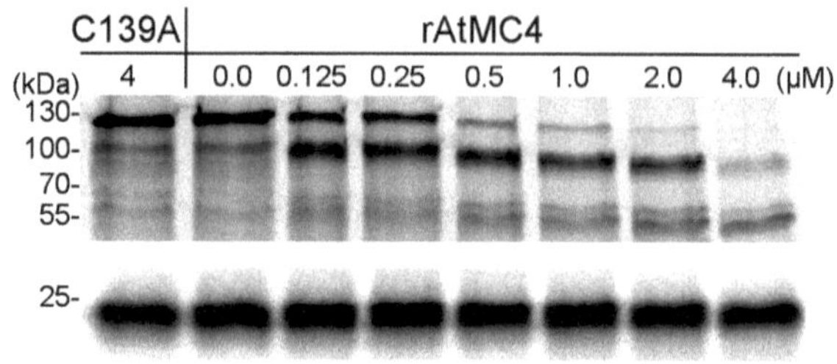

Fig. 3 In vitro metacaspase protein substrate cleavage assay. A potential substrate protein is cleaved by an increasing concentration of recombinant *Arabidopsis thaliana* metacaspase 4 (rAtMC4) to discrete cleavage products (*top panel*). Glutathione S-transferase (GST) does not get cleaved by metacaspase (*bottom panel*). Concentrations are indicated in the figure. Inactive rAtMC4, in which the active site cysteine is mutated to alanine (C139A), acts as a negative control

1. Produce the protein substrate in the cell free TNT® Coupled Transcription/Translation System (TNT® mix) according the manufacturer's instructions (Promega). A choice can be made here to use radiolabeled L-[^{35}S]methionine for subsequent readout by autoradiography or unlabeled methionine and epitope tags for western blotting.
2. Prepare the metacaspase reaction master mix by mixing 80 μL of 4× reaction buffer with 120 μL of dH_2O. Divide 25 μL per well over an 8-well eppi strip.
3. Add 5 μL of the TNT® mix to the side of each well. Take care: the drop of TNT® mix should not touch the metacaspase reaction mix.
4. The reaction is started by the simultaneous addition of 10 μL of the rAtMC4 dilution series to the 8-well eppi strip with the help of a multichannel pipette and by pipetting the metacaspase reaction and TNT® mix up and down. The reaction is carried out at 30 °C for 30 min.
5. The reaction is stopped by the simultaneous addition of 10 μL of the 5× Laemmli buffer + EGTA (*see* **Note 10**) to each well. The samples are heated for 5 min at 85 °C and can be stored at −20 °C or immediately used in the subsequent steps.
6. Load 15 μL of each sample on a 12 % SDS-PAGE gel and run the electrophoresis.
7. After electrophoresis, rinse the gel shortly in dH_2O and dry for 1 h on a Whatman paper in a gel dryer when using autoradiography as a readout, or proceed to western blot as detailed in Subheading 3.4 with the appropriate antibody against the used epitope tag.
8. For autoradiography, expose the dried gel for 48 h to a phosphor screen and make the readout on a phosphor imager with appropriate settings (*see* **Note 11**).

4 Notes

1. P1 refers to the residue situated N-terminus to the scissile bond cleaved by the protease extending towards the N-terminus (positions P2, P3, etc.) of the peptide, whereas prime positions (P1′, P2′, etc.) are situated C-terminus to the cleaved bond [22].
2. We have only used the TNT® Rabbit Reticulocyte Lysate so far, but the TNT® Wheat Germ Lysate should be equivalent. We routinely use the pDEST (Gateway®, Invitrogen) vectors and pUNI51 vectors (SSP consortium). Any vector with the minimal requirements of a T3 or T7 promoter and terminator should in theory be suitable.
3. At this concentration DTT effectively prevents oxidation of metacaspase cysteins and also prevents formation of intra- and intermolecular disulfide bonds, thus inhibiting aggregation of the protein. Metacaspase structure can be additionally stabilized in the presence of 50 mM $CaCl_2$, but it might also lead to increased autoprocessing.
4. To prevent aggregation of the protein it is preferable to perform dialysis at 4 °C gradually decreasing the concentration of urea in the dialysis buffer and increasing its volume.
5. Metacaspases are highly active proteases, thus already 3 ng of purified metacaspase is enough to produce detectable amount of cleaved fluorescent product in activity assay using AMC-tagged substrates. Still the exact amount of metacaspase required for the assay might vary depending on purity of the preparation.
6. Even minute contamination with calcium will significantly influence results of the experiment. An inactive mutant metacaspase with substitution of Cys and/or His in the active center to Ala or Gly can be used as an alternative negative control in this assay. It is also known that autoprocessing of metacaspases is essential for their proteolytic activity; thus uncleavable mutant of metacaspase can be used as another negative control in the activity assay.
7. This stripping method works for PVDF membranes and usually allows to efficiently strip only secondary antibodies. It is therefore preferable to use for second round of staining primary antibody produced in a different organism than antibody used at the first staining.
8. Avoid uneven destaining, which might impact loading quantification. If necessary the membrane can be re-stained. Let it dry in vertical position.
9. Depending on tissue or treatment, expression pattern in the sample can vary significantly. Take this into consideration while selecting bands for loading quantification to compare between samples.

10. EGTA is added to the Laemmli buffer to sequester the Ca^{2+} and improve the SDS-PAGE.
11. Don't cover the gel with plastic foil as this will obstruct the signal and keep the gel as straight and close to the screen as possible in order to enhance the sharpness of the image (for example by the addition of additional papers behind the gel).

Acknowledgements

S.S. is indebted to the Special Research Fund of Ghent University for a postdoctoral fellowship. F.V.B acknowledges support from grants of the Ghent University Multidisciplinary Research Partnership "Ghent BioEconomy" 27 (project 01MRB510W), the Belgian Science Policy Office (project IAP7/29), and the Research Foundation Flanders (FWO-Vlaanderen; project G.0038.09). E.A.M. and P.V.B. acknowledge support from grants of the Swedish Research Council (VR), Knut and Alice Wallenberg Foundation, the Swedish Foundation for Strategic Research (SSF), Pehrssons Fund and Olle Engkvist Byggmästare Foundation.

References

1. Uren AG, O'Rourke K, Aravind L et al (2000) Identification of paracaspases and metacaspases: two ancient families of caspase-like proteins, one of which plays a key role in MALT lymphoma. Mol Cell 6:961–967
2. Tsiatsiani L, Van Breusegem F, Gallois P et al (2011) Metacaspases. Cell Death Differ 18:1279–1288
3. Choi CJ, Berges JA (2013) New types of metacaspases in phytoplankton reveal diverse origins of cell death proteases. Cell Death Dis 4:e490
4. Bozhkov PV, Smertenko AP, Zhivotovsky B (2010) Aspasing out metacaspases and caspases: proteases of many trades. Sci Signal 3:pe48
5. Coll NS, Vercammen D, Smidler A et al (2010) Arabidopsis type I metacaspases control cell death. Science 330:1393–1397
6. He R, Drury GE, Rotari VI et al (2008) Metacaspase-8 modulates programmed cell death induced by ultraviolet light and H2O2 in Arabidopsis. J Biol Chem 283:774–783
7. Watanabe N, Lam E (2011) Arabidopsis metacaspase 2d is a positive mediator of cell death induced during biotic an abiotic stresses. Plant J 66:969–982
8. Suarez MF, Filonova LH, Smertenko A et al (2004) Metacaspase-dependent programmed cell death is essential for plant embryogenesis. Curr Biol 14:R339–R340
9. Shrestha A, Megeney LA (2012) The non-death role of metacaspase proteases. Front Oncol 2:78
10. Vercammen D, Belenghi B, van de Cotte B et al (2006) Serpin1 of Arabidopsis thaliana is a suicide inhibitor for metacaspase 9. J Mol Biol 364:625–636
11. McLuskey K, Rudolf J, Proto WR et al (2012) Crystal structure of a Trypanosoma brucei metacaspase. Proc Natl Acad Sci U S A 109:7469–7474
12. Wong AH-H, Yan C, Shi Y (2012) Crystal structure of the yeast metacaspase Yca1. J Biol Chem 287:29251–29259
13. Pop C, Salvesen GS (2009) Human caspases: activation, specificity, and regulation. J Biol Chem 284:21777–21781
14. Vercammen D, van de Cotte B, De Jaeger G et al (2004) Type II metacaspases Atmc4 and Atmc9 of Arabidopsis thaliana cleave substrates after arginine and lysine. J Biol Chem 279:45329–45336
15. Bozhkov PV, Suarez MF, Filonova LH et al (2005) Cysteine protease mcII-Pa executes programmed cell death during plant embryogenesis. Proc Natl Acad Sci U S A 102: 14463–14468
16. Watanabe N, Lam E (2011) Calcium-dependent activation and autolysis of

Arabidopsis metacaspase 2d. J Biol Chem 286: 10027–10040

17. Tsiatsiani L, Timmerman E, De Bock P-J et al (2013) The Arabidopsis metacaspase9 degradome. Plant Cell 25(8):2831–2847
18. Sundström JF, Vaculova A, Smertenko AP et al (2009) Tudor staphylococcal nuclease is an evolutionarily conserved component of the programmed cell death degradome. Nat Cell Biol 11:1347–1354
19. Zimmerman M, Yurewicz E, Patel G (1976) A new fluorogenic substrate for chymotrypsin. Anal Biochem 70:258–262
20. Zimmerman M, Ashe B, Yurewicz E, Patel G (1977) Sensitive assays for trypsin, elastase, and chymotrypsin using new fluorogenic substrates. Anal Biochem 78:45–71
21. Staes A, Impens F, Van Damme P, Ruttens B et al (2011) Selecting protein N-terminal peptides by combined fractional diagonal chromatography. Nat Protoc 6:1130–1141
22. Schechter I, Berger M (1967) On the size of the active site in proteases. Biochem Biophys Res Commun 27:157–162

Chapter 16

Preparation of *Arabidopsis thaliana* Seedling Proteomes for Identifying Metacaspase Substrates by N-terminal COFRADIC

Liana Tsiatsiani, Simon Stael, Petra Van Damme, Frank Van Breusegem, and Kris Gevaert

Abstract

Proteome-wide discovery of in vivo metacaspase substrates can be obtained by positional proteomics approaches such as N-terminal COFRADIC, for example by comparing the N-terminal proteomes (or N-terminomes) of wild-type plants to transgenic plants not expressing a given metacaspase. In this chapter we describe a protocol for the preparation of plant tissue proteomes, including differential isotopic labelling allowing for a comparison of in vivo N-terminomes that serves as the starting point for N-terminal COFRADIC studies.

Key words Metacaspases, Positional proteomics, N-terminal COFRADIC, Protease substrates, Neo-N-termini, Tissue samples, Degradomics

1 Introduction

Identification of protease substrates and characterization of protease substrate specificities in utmost detail rely nowadays mostly on mass spectrometry driven proteomics (recently reviewed in [1]). When sampling whole proteomes, protease substrates are identified either based on their altered mobility during gel electrophoresis (e.g., the PROTOMAP technology introduced by the Cravatt lab [2]) or by exploiting the chemical reactivity of the alpha-amino groups that, amongst others, are introduced when proteases cleave their substrates and are referred to as neo-N-termini. The latter technologies involve the enzymatic or chemical labelling of these reactive groups with affinity tags (e.g., the subtiligase approach [3]), scavenging all other peptides on solid supports (e.g., the TAILS approach [4]) or depleting these non N-terminal peptides using consecutive chromatography steps. This last approach was introduced in 2003 and termed N-terminal COmbined FRActional

Peter V. Bozhkov and Guy Salvesen (eds.), *Caspases, Paracaspases, and Metacaspases: Methods and Protocols*, Methods in Molecular Biology, vol. 1133, DOI 10.1007/978-1-4939-0357-3_16,

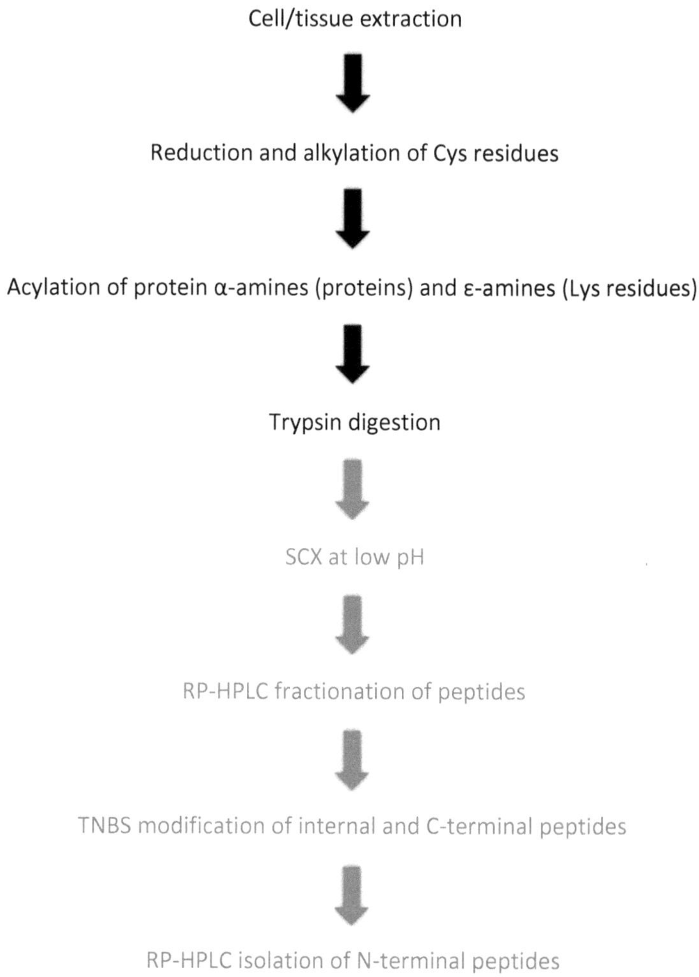

Fig. 1 Schematic workflow of the N-terminal COFRADIC protocol. The steps not outlined in this chapter are shown in *grey*

DIagonal Chromatography (COFRADIC, [5]). More recently, the Overall lab and our lab have published positional proteomics approaches that enable the enrichment of protein C-terminal peptides which also serve as proxies for protease substrates, and thus yield complementary information on proteolytic events [6, 7].

The general N-terminal COFRADIC procedure is schematically depicted in Fig. 1. Briefly, prior to digestion of the sampled proteins with a specific protease such as trypsin, all primary amino groups in proteins—N-terminal α-amino groups and lysine ε-amino groups—are chemically blocked, for instance by butyrylation (e.g., [8]). Note that by using isotopic variants of butyric acid (here), this essential step in the overall COFRADIC procedure allows for a direct comparison of two samples. Because of the way that lysine is chemically modified, trypsin will now only cleave at arginine residues and essentially render two types of peptides; protein N-terminal peptides, including neo-N-terminal peptides, carrying an acetylated α-amino group (in vivo) or a butyrylated α-amino group (in vitro), and non N-terminal peptides (i.e., internal peptides and C-terminal peptides) carrying a primary α-amino group. Strong cation exchange chromatography (SCX), when performed at acidic pH (pH < 3) enriches for N-terminal peptides as well as for C-terminal peptides devoid of basic amino acids [9]. Inevitably, some internal peptides are co-enriched by this SCX step, but they are later removed together with the C-terminal peptides by the actual COFRADIC step. Now, peptides are separated by reverse-phase chromatography (RP-HPLC) in a distinct number of fractions. Peptides in these fractions are treated with 2,4,6-trinitrobenzene sulfonic acid (TNBS) which reacts highly with primary amino groups and attaches a very hydrophobic trinitrophenyl group onto internal peptides and C-terminal peptides. Following a series of identical RP-HPLC separations, in each initial peptide fraction, the N-terminal peptides are separated from the modified, more hydrophobic internal and C-terminal peptides and in this way isolated, enriched and ready for subsequent analysis by mass spectrometry.

Many of the protease degradomics studies done so far with the N-terminal COFRADIC method used mammalian cell cultures that are readily metabolically labelled (e.g., using isotopic variants of essential amino acids (SILAC) [10]) and thus allow for a direct comparison of two or more N-terminal proteomes (e.g., [11]). Although plant proteins can also be metabolically labelled for subsequent proteome studies (e.g., using nitrogen-15 enriched nitrogen salts [12]), post-identification analysis of data can be cumbersome and needs specific data analysis software tools (e.g., [13]).

In this chapter, we describe a protocol for preparing plant proteomes—here described for Arabidopsis seedling proteomes, but generally applicable to other plant tissues as well—including post-metabolic labelling events that allow for a direct comparison of the N-terminomes of two different plant tissue proteomes. When comparing proteomes of wild-type plants with plants not expressing a given metacaspase (or any other protease), using the whole procedure, N-terminal peptides indicative for substrate processing by this metacaspase in wild-type plants, can be identified [14].

2 Materials

2.1 Proteome Extraction from Arabidopsis thaliana Seedlings

1. Proteome extraction buffer: 1 % (w/v) 3-((3-cholamidopropyl)dimethylammonio)-1-propanesulfonic acid (CHAPS), 0.5 % (w/v) deoxycholate, 0.1 % (w/v) SDS, 5 mM ethylenediaminetetraacetic acid (EDTA), and 10 % glycerol in phosphate-buffered saline (PBS) (pH 7.5). Add the suggested amount of a mixture of protease inhibitors according to the manufacturer's instructions (e.g., Complete Protease Inhibitor Cocktail Tablets from Roche Applied Science).
2. Guanidinium hydrochloride.

2.2 Preparation of Proteomes for Differential N-Terminal COFRADIC Analysis

1. Tris(2-carboxyethyl)phosphine (TCEP) stock solution: 150 mM TCEP (*see* **Note 1**).
2. Iodoacetamide stock solution: 300 mM iodoacetamide.
3. NAP™-10 columns (GE Healthcare Life Sciences) or similar.
4. Light and heavy labelling solutions: freshly prepare 10 mM of *N*-hydroxysuccinimide esters of either $^{12}C_4$-butyric acid (light) or $^{13}C_4$-butyric acid (heavy) in 50 % of acetonitrile (*see* **Note 2**).
5. 1 M glycine.
6. Hydroxylamine.
7. 10 mM ammonium bicarbonate (pH 8). Prepare fresh.
8. Trypsin (e.g., sequencing-grade modified trypsin from Promega).

3 Methods

3.1 Proteome Extraction from Arabidopsis thaliana Seedlings

1. Collect seedling tissue from the two plant lines studied (e.g., wild-type plants versus transgenic plants that do not express a particular metacaspase), freeze in liquid nitrogen and grind into a fine powder with mortar and pestle (*see* **Note 3**).
2. Resuspend 0.5 g of frozen ground tissue in 1 mL of pre-chilled (4 °C) proteome extraction buffer and allow to thaw (*see* **Note 4**). Transfer to a fresh tube.
3. Centrifuge the sample for 10 min at 16,000 × *g* at 4 °C, carefully aspirate the supernatant transfer to a fresh tube and discard the pellet.
4. Repeat the previous step to remove any remaining debris.
5. Measure the protein concentration and add solid guanidinium hydrochloride to the cleared lysate to a final concentration of 4 M (*see* **Note 5**).

3.2 Preparation of Proteomes for Differential N-Terminal COFRADIC Analysis

1. Add TCEP stock solution and iodoacetamide stock solution to the protein mixtures to obtain final concentrations of 15 and 30 mM, respectively.
2. Incubate for 30 min at 37 °C in the dark.
3. Desalt the protein mixtures over, e.g., NAP™-10 columns in 1.5 mL of 1.4 M of guanidinium hydrochloride in 50 mM sodium phosphate buffer at pH 7.5.
4. Add 50 μL of either light (e.g., to the seedling proteome preparation of wild-type plants) or heavy (e.g., to the seedling proteome preparation of metacaspase knockout plant) labelling solution to the desalted protein mixtures.
5. Incubate for 1 h at 30 °C.

6. Repeat **steps 4** and **5** (*see* **Note 6**).
7. Add 1.25 μL of 1 M glycine to quench the non-reacted *N*-hydroxysuccinimide esters.
8. Incubate for 10 min at room temperature.
9. Add 10 μL of hydroxylamine (*see* **Note 7**).
10. Incubate for 15 min at room temperature.
11. Reduce the volume of the solution to 1 mL by centrifugal vacuum drying.
12. Desalt the protein mixtures over, e.g., NAP™-10 columns in 1.5 mL of 10 mM freshly prepared ammonium bicarbonate (pH 8).
13. Measure the protein concentration of both samples and mix equal amounts of samples.
14. Incubate this protein mixture for 10 min at 95 °C and transfer to ice for 10 min incubation.
15. Add trypsin to an enzyme/substrate ratio of 1/50 (w/w).
16. Incubate overnight at 37 °C.
17. Centrifuge the peptide mixture for 10 min at 16,000 × *g*, collect the supernatant, vacuum dry and store at −20 °C until further use (*see* **Note 8**).

4 Notes

1. TCEP is available as TCEP·HCl, and dissolving this product releases HCl which causes a drop in pH. Therefore, the pH of the TCEP stock solution must be raised to pH 7.5. Typically, 500 μl of TCEP stock solution requires adding 30 μl of 5 M NaOH to obtain a pH of 7.5.
2. These labelling solutions allow for incorporation of differential isotopes in the proteomes to be analyzed. The protocol for synthesizing *N*-hydroxysuccinimide esters of butyric acid variants has been published [15], though note that other *N*-hydroxysuccinimide esters are commercially available (e.g., those of acetate and trideutero-acetate).
3. Note that seeds of *Arabidopsis thaliana* Columbia (Col-0) were overnight gas sterilized with HCl and NaOCl, then sowed on half strength Murashige and Skoog (MS) media plates containing 0.8 % agar and 1 % sucrose. Two plates were sown per plant line. The plates were kept at 4 °C in the dark for seed stratification and after 3 days, were transferred to 21 °C with a 16 h light–8 h dark photoperiod, light intensity of 80–100 μmol/m^2/s and 70 % humidity. The duration of seedling growth at 21 °C varies and depends on the expression

level of the selected metacaspase. Subsequently, seedlings were harvested and frozen in liquid nitrogen for protein extraction.

4. This amount of tissue yields on average about 5 mg of protein material.
5. The recommended amount of protein material per sample is 4 mg.
6. The total amount of labelling reagents added suffices to label the equivalent of up to 4 mg of protein material.
7. A typical side-reaction of labelling protein primary amino groups with *N*-hydroxysuccinimide esters is the acetylation of hydroxyl groups in proteins (serines, threonine, and tyrosine). Hydroxylamine is efficient in reverting this O-acetylation.
8. At this stage, samples can be subjected to the chromatography part of the N-terminal COFRADIC procedure as detailed in ref. 14.

Acknowledgments

L.T. acknowledges support from the VIB International PhD Program and the Netherlands Proteomics Centre, a program embedded in The Netherlands Genomics Initiative. P.V.D. is a Postdoctoral Fellow of the Research Foundation Flanders (FWO-Vlaanderen) and S.S. is indebted to the Special Research Fund of Ghent University for a postdoctoral fellowship. F.V.B acknowledges support from grants of the Ghent University Multidisciplinary Research Partnership "Ghent BioEconomy" 27 (project no. 01MRB510W) and of the Belgian Science Policy Office (project IAP7/29). F.V.B. and K.G. acknowledge support from the Research Foundation Flanders (FWO-Vlaanderen), research project G.0038.09.

References

1. Plasman K et al (2013) Contemporary positional proteomics strategies to study protein processing. Curr Opin Chem Biol 17: 66–72
2. Dix MM et al (2008) Global mapping of the topography and magnitude of proteolytic events in apoptosis. Cell 134:679–691
3. Mahrus S et al (2008) Global sequencing of proteolytic cleavage sites in apoptosis by specific labeling of protein N termini. Cell 134:866–876
4. Kleifeld O et al (2010) Isotopic labeling of terminal amines in complex samples identifies protein N-termini and protease cleavage products. Nat Biotechnol 28:281–288
5. Gevaert K et al (2003) Exploring proteomes and analyzing protein processing by mass spectrometric identification of sorted N-terminal peptides. Nat Biotechnol 21:566–569
6. Schilling O et al (2010) Proteome-wide analysis of protein carboxy termini: C terminomics. Nat Methods 7:508–511
7. Van Damme P et al (2010) Complementary positional proteomics for screening substrates of endo- and exoproteases. Nat Methods 7:512–515
8. de Poot SA et al (2011) Human and mouse granzyme M display divergent and species-specific substrate specificities. Biochem J 437: 431–442

9. Staes A et al (2008) Improved recovery of proteome-informative, protein N-terminal peptides by combined fractional diagonal chromatography (COFRADIC). Proteomics 8:1362–1370
10. Ong SE et al (2002) Stable isotope labeling by amino acids in cell culture, SILAC, as a simple and accurate approach to expression proteomics. Mol Cell Proteomics 1:376–386
11. Plasman K et al (2011) Probing the efficiency of proteolytic events by positional proteomics. Mol Cell Proteomics 10(M110):003301
12. Skirycz A et al (2011) A reciprocal 15 N-labeling proteomic analysis of expanding Arabidopsis leaves subjected to osmotic stress indicates importance of mitochondria in preserving plastid functions. J Proteome Res 10:1018–1029
13. MacCoss MJ et al (2003) A correlation algorithm for the automated quantitative analysis of shotgun proteomics data. Anal Chem 75:6912–6921
14. Tsiatsiani L et al (2013) The Arabidopsis metacaspase9 degradome. The Plant Cell 25(8): 2831–2847
15. Staes A et al (2011) Selecting protein N-terminal peptides by combined fractional diagonal chromatography. Nat Protoc 6:1130–1141

INDEX

Peter V. Bozhkov and Guy Salvesen (eds.), *Caspases, Paracaspases, and Metacaspases: Methods and Protocols*, Methods in Molecular Biology, vol. 1133, DOI 10.1007/978-1-4939-0357-3, © Springer Science+Business Media New York 2014

MIX
Papier aus verantwortungsvollen Quellen
Paper from responsible sources
FSC® C105338

If you have any concerns about our products,
you can contact us on
ProductSafety@springernature.com

In case Publisher is established outside the EU,
the EU authorized representative is:
Springer Nature Customer Service Center GmbH
Europaplatz 3, 69115 Heidelberg, Germany

Printed by Libri Plureos GmbH
in Hamburg, Germany